PERGAMON INT
of Science, Technology,

The 1000-volume original paperback library in aid of education,
industrial training and the enjoyment of leisure

Publisher: Robert Maxwell, M.C.

MATRIX METHODS APPLIED TO ENGINEERING RIGID BODY MECHANICS

THE PERGAMON TEXTBOOK
INSPECTION COPY SERVICE

An inspection copy of any book published in the Pergamon International Library will gladly be sent to academic staff without obligation for their consideration for course adoption or recommendation. Copies may be retained for a period of 60 days from receipt and returned if not suitable. When a particular title is adopted or recommended for adoption for class use and the recommendation results in a sale of 12 or more copies, the inspection copy may be retained with our compliments. The Publishers will be pleased to receive suggestions for revised editions and new titles to be published in this important International Library.

Other Titles of Interest

BENSON & WHITEHOUSE
Internal Combustion Engines (in 2 volumes)

DIXON
Fluid Mechanics, Thermodynamics of Turbomachinery, 3rd Edition

DUNN & REAY
Heat Pipes, 2nd Edition

HAYWOOD
Analysis of Engineering Cycles, 3rd Edition

HEARN
Mechanics of Materials (in 2 volumes)

HOPKINS & SEWELL
Mechanics of Solids

LAI et al
Introduction to Continuum Mechanics, SI Edition

LIVESLEY
Matrix Methods of Structural Analysis, 2nd Edition

NEMAT-NASSER
Mechanics Today, Volumes 1-5

NEMAT-NASSER
Variational Methods in the Mechanics of Solids

REAY & MACMICHAEL
Heat Pumps

Related Pergamon Journals

(Free Specimen Copies Available on Request)

COMPUTERS AND STRUCTURES
INTERNATIONAL JOURNAL OF ENGINEERING SCIENCES
INTERNATIONAL JOURNAL OF MECHANICAL SCIENCES
MECHANISM AND MACHINE THEORY

MATRIX METHODS APPLIED TO ENGINEERING RIGID BODY MECHANICS

T. CROUCH
B.Sc.Mech.Eng., M.I.Mech.E., C.Eng.
*Lecturer in the Department of Mechanical Engineering
Coventry (Lanchester) Polytechnic*

PERGAMON PRESS
OXFORD · NEW YORK · TORONTO · SYDNEY · PARIS · FRANKFURT

U.K.	Pergamon Press Ltd., Headington Hill Hall, Oxford OX3 0BW, England
U.S.A.	Pergamon Press Inc., Maxwell House, Fairview Park, Elmsford, New York 10523, U.S.A.
CANADA	Pergamon of Canada, Suite 104, 150 Consumers Road, Willowdale, Ontario M2J 1P9, Canada
AUSTRALIA	Pergamon Press (Aust.) Pty. Ltd., P.O. Box 544, Potts Point, N.S.W. 2011, Australia
FRANCE	Pergamon Press SARL, 24 rue des Ecoles, 75240 Paris, Cedex 05, France
FEDERAL REPUBLIC OF GERMANY	Pergamon Press GmbH, 6242 Kronberg-Taunus, Hammerweg 6, Federal Republic of Germany

Copyright © 1981 T. Crouch

All Rights Reserved. No part of this publication may be reproduced, stored in a retrieval system or transmitted in any form or by any means: electronic, electrostatic, magnetic tape, mechanical, photocopying, recording or otherwise, without permission in writing from the publishers.

First edition 1981

British Library Cataloguing in Publication Data
Crouch, T
Matrix methods applied to engineering rigid body mechanics. - (Pergamon international library).
1. Mechanics, Applied   2. Vector analysis
3. Matrices
I. Title
531'.01'51563   TA350   80-41186
ISBN 0 08 024245 6 (Hardcover)
ISBN 0 08 024246 4 (Flexicover)

In order to make this volume available as economically and as rapidly as possible the authors' typescripts have been reproduced in their original forms. This method has its typographical limitations but it is hoped that they in no way distract the reader.

Printed in Great Britain by A. Wheaton & Co. Ltd., Exeter

Preface

The purpose of this book is to present the solution of a range of rigid body mechanics problems using a matrix formulation of vector algebra. The treatment has other notable features. It employs a coherent letter and number suffix notation and also exploits the relationship between the orthogonal transformation matrix and angular velocity. Particular emphasis is placed upon the positioning of appropriate frames of reference and specifying their relative position.

In writing this text it has been assumed that the reader will have a knowledge of mathematics and mechanics normally associated with the first year of an Engineering Degree course.

The plan of the book is simple. There are four chapters, Chapter 1 Kinematics, Chapter 2 Dynamics, Chapter 3 Solution of Kinematics Problems and Chapter 4 Solution of Dynamics Problems. Chapters 1 and 2 give a succinct statement of the essential theory formulated in terms of matrix algebra, while Chapters 3 and 4 give a selection of solved problems and problems for solution. The reader is therefore advised to study the problems to which reference is made at various points in the text as they occur. A proper approach to the solution of dynamics problems demands that kinematic considerations have priority. It is suggested, therefore, that the reader studies Chapters 1 and 3 before proceeding the Chapters 2 and 4. Answers to the problems for solution are provided, with some indication of the salient features of their solution in most cases.

Coventry 1980   T. Crouch

Contents

Principal Symbols and Notation		xiii
Chapter 1 Kinematics		1
1.1	The Position Vector	1
1.2	The Relative Position Vector	2
1.3	Transformation of Vectors	3
1.4	The Rotation Matrix for Simple Rotations	8
1.5	Consecutive Rotations	10
1.6	Successive Transformation of Vectors	12
1.7	The Velocity and Acceleration of a Point	14
1.8	Small Rotations	18
1.9	Angular Velocity and the Derivative of the Rotation Matrix	20
1.10	The Relative Velocity of Points Fixed in a Rigid Body	23
1.11	The Central and Instantaneous Axes	26
1.12	The Relative Acceleration of Points Fixed in a Rigid Body	29
1.13	The Relative Acceleration of Coincident Points Which Have Relative Motion	33
1.14	Differentiation of the Angular Velocity Vector, a Special Case	34
1.15	Relative Angular Velocity	35
1.16	Relative Angular Acceleration	37
Chapter 2 Dynamics		38
2.1	Newton's Laws of Motion	38
2.2	The Measurement of Force	39
2.3	Work, Potential and Kinetic Energy	40
2.4	The Activity of a Force and its Relationship to the Rate of Change of Kinetic Energy	43
2.5	Impulse and Momentum	43

2.6	Centre of Mass	44
2.7	Force Moment, Moment of Momentum and Moment of Rate of Change of Momentum	45
2.8	The Linear Momentum of a Rigid Body	47
2.9	The Moment of Momentum of a Rigid Body About its Centre of Mass	47
2.10	The Relationship Between Moments of Inertia Measured in Different Frames	50
2.11	The Rate of Change of Angular Momentum of a Rigid Body About its Centre of Mass	52
2.12	The Moment of Momentum of a Rigid Body About Any Point Q and the Rate of Change of Moment of Momentum About That Point	53
2.13	The Relationship Between the Moment of the External Forces and Couples on a Rigid Body About Any Point Q and the Rate of Change of Moment of Momentum About That Point	54
2.14	The Kinetic Energy of a Rigid Body	55
2.15	The Rate of Change of Kinetic Energy of a Rigid Body	56
2.16	The Special Case of the Motion of a Solid of Revolution	57
2.17	Rotation About a Fixed Axis	58
2.18	Principal Axes and Principal Moments of Inertia of a Rigid Body With a Plane of Symmetry	58
2.19	Principal Axes and Principal Moments of Inertia For Any Rigid Body	61

Chapter 3 Solution of Kinematics Problems   65

3.1	Solved Problems	65

Problem

3.1	Velocity and Acceleration of a Point Moving in a Circular Path	65
3.2	Summation of Finite Angles of Rotation	69
3.3	Velocity and Acceleration of a Point Moving on a Straight Rotating Path	71
3.4	Angular Velocity Determination From Linear Velocity Data	74
3.5	Velocity Determination For a Mechanism	77
3.6	Plane Motion of a Disc	80
3.7	Velocity and Acceleration of a Point on a Disc in Conic Rolling	85
3.8	Velocity and Acceleration Determination For a Mechanism	92
3.9	Velocity and Acceleration Determination For a Linkage	97
3.10	Velocity and Accleration of a Point Moving in Earth Fixed Axes	101
3.11	Velocity and Acceleration of a Point in a 'Plane' Epicyclic Gear Train	103
3.12	Angular Velocity and Acceleration of a Precessing Rotor	106
3.13	Angular Velocity and Angular Acceleration of Ball in a Ball Thrust Race	109

3.14	Angular Velocity and Angular Acceleration of a Roller in a Taper-Roller Thrust Race	113
3.15	Angular Acceleration of a Nutating and Precessing Rotor	115
3.16	Rubbing Velocities Associated With a Rolling Rotor	118
3.17	Discs in Rolling Contact	120
3.18	Rolling Disc With Bevel Gear Drive	123
3.19	Bevel Wheel Epicyclic Gear Train	126
3.20	Hooke's Joint	128
3.21	Rolling Disc With Constant Velocity Joint	131
3.22	Euler's Theorem on the Motion of Rigid Bodies	133
3.2	Problems For Solution	141

Problem

3.23	Velocity and Acceleration of a Point Moving on a Rotating Circular Path	141
3.24	A Vector Which is Unchanged by a Transformation	
3.25	Differentiation of the Rotation Matrix For Simple Rotations	142
3.26	Relative Motion of Aircraft 1	143
3.27	Relative Motion of Aircraft 2	
3.28	Derivation of a Transformation Matrix 1	144
3.29	Derivation of a Transformation Matrix 2	144
3.30	Aircraft Tracking	145
3.31	Derivation of Angular Velocity From Linear Velocity Data	146
3.32	Location of Central Axis 1	147
3.33	Location of Central Axis 2	
3.34	Euler Angles	149
3.35	Bryant Angles	150
3.36	Acceleration of a Point on a Rotating Disc 1	
3.37	Acceleration of a Point on a Rotating Disc 2	151
3.38	Acceleration of a Point on a Rotating Disc 3	
3.39	Acceleration of a Point on a Rotating Disc 4	
3.40	Rolling Wheel on Moving Surface	152
3.41	Wheels and Axle	153
3.42	Rotating and Extending Antenna	
3.43	Articulated Trailer	154
3.44	Precession of Rotating Cylinder	156
3.45	Thrust Bearing	
3.46	Automotive Differential	157
3.47	A Bevel Wheel Gear Train	158
3.48	Rotor on Rotating Pivoted Axle	159
3.49	Rotating Epicyclic	160
3.50	Velocity and Acceleration Determination For a Mechanism	
3.51	Velocities and Acceleration of a Rolling Disc	161

Chapter 4 Solution of Dynamics Problems 164

   4.1 Solved Problems 164

Problem

   4.1 The Conditions Which a Moving Reference Frame

	Must Satisfy to Allow it to be Treated as an Inertial Frame	164
4.2	Gravitational Potential	165
4.3	Strain Potential 1	166
4.4	Strain Potential 2	168
4.5	Work Done by a Non-conservative Force 1	171
4.6	Work Done by a Non-conservative Force 2	174
4.7	Gravitational and Elastic Potential	179
4.8	Del V Referred to a Rotating Frame	181
4.9	Work Done and Change of Kinetic Energy When Force is a Given Function of Time	183
4.10	Potential and Kinetic Energy of a Spring and Mass System	186
4.11	The Motion of a Particle on a Helical Path	188
4.12	The Motion of a Particle Falling Freely Near the Surface of the Earth	191
4.13	The Motion of a Rotating Spring and Mass System	192
4.14	Forced Pendulum Motion	194
4.15	The Foucault Pendulum	198
4.16	Location of Centre of Mass 1	201
4.17	Location of Centre of Mass 2	204
4.18	Force Moment	205
4.19	Static Equilibrium 1	206
4.20	Static Equilibrium 2	208
4.21	Static Equilibrium 3	212
4.22	The Wrench	214
4.23	The Perpendicular and Parallel Axis Theorems	217
4.24	Inertia Matrix For a Thin Disc	218
4.25	Inertia Matrix For a Three Bladed Airscrew	219
4.26	Determination of Inertia Matrix, Angular Momentum and Rate of Change of Angular Momentum 1	221
4.27	Determination of Inertia Matrix, Angular Momentum and Rate of Change of Angular Momentum 2	225
4.28	Motion of a Rigid Body About its Centre of Mass	229
4.29	Stability of the Free Motion of a Rigid Body	232
4.30	Free Motion of a Rigid Body Having Axial Symmetry	234
4.31	Motion of a Constrained Body 1	240
4.32	Motion of a Constrained Body 2	243
4.33	Forces on a Ball in a Ball Thrust Race	245
4.34	Motion of a High Speed Rotor Mounted in Gimbals	248
4.35	Relationships Between Kinetic Energy and Momentum For a Constrained Rotor	251
4.36	Forces and Moments Due to Constraints on a Rotor	255
4.37	Motion of a Constrained Rod 1	257
4.38	Motion of a Constrained Rod 2	260
4.39	Motion of a Constrained Rod 3	263
4.40	Forces in a Mechanism	266
4.41	Inertia Determination For a Hemisphere	270
4.42	Inertia Determination For a Thin Rod	275
4.43	Inertia Determination For a Rod System	277

4.2 Problems For Solution   282

Problem

4.44	Activity of a Force	282
4.45	Work Done by a Conservative Force	
4.46	Work and Potential in a Conservative System	
4.47	Del V Referred to a Rotating Frame	283
4.48	A Property of the Couple	
4.49	A Component of Force Moment	
4.50	Resultants of a Given Force System	285
4.51	Equilibrium of a Simple Structure	286
4.52	Equilibrium of a Mechanism	
4.53	The Wrench	
4.54	Inertia Determination For a Rectangular Parallelepiped	288
4.55	Inertia Determination For a Right Circular Cone	
4.56	Couple Due to a Two Bladed Airscrew	
4.57	Couple Due to Aircraft Control Surfaces	
4.58	Couple Due to Wobble of Circular Saw Blade	289
4.59	Measurement of Angular Velocity Relative to an Inertial Reference	290
4.60	Bearing Forces Due to Rotor Precession	
4.61	Bearing Forces Due to Rotor Misalignment	292
4.62	Attitude of Pivoted Rectangular Parallelepiped	
4.63	Frequency of Vibrations of Thin Pivoted Rod	
4.64	Frequency of Vibrations of Constrained Rod	294
4.65	Contact Force Due to the Precession of a Rotor 1	
4.66	Contact Force Due to the Precession of a Rotor 2	295
4.67	Forces on a Rolling Cone	296
4.68	Contact Force Due to the Precession of a Rotor 3	297
4.69	Steady Motion of a Top	
4.70	Motion of a Precessing Rotor 1	298
4.71	Motion of a Precessing Rotor 2	299
4.72	Motion of a Pendulously Mounted Rotor 1	300
4.73	Motion of a Spring Controlled Gimbal Mounted Rotor	301
4.74	Motion of a Pendulously Mounted Rotor 2	
4.75	Bearing Forces Due to Rotor Motion 1	302
4.76	Bearing Forces Due to Rotor Motion 2	303
4.77	A Torsional Vibration Absorber	
4.78	A Rudimentary Gyro-compass	305
4.79	Steady Motion of a Disc	306
4.80	Initial Motion of a Thin Rod	
4.81	Vibration of a Spinning Projectile	308
4.82	Motion of a Constrained Disc	
4.83	Conservation of Energy and Momentum in Free Motion	309
4.84	Inertia Determination For a Body Having Plane Symmetry	
4.85	Inertia Determination For a Body Without Plane Symmetry 1	310
4.86	Inertia Determination For a Body Without Plane Symmetry 2	311

Answers to Problems For Solution Chapter 3   313

Answers to Problems For Solution Chapter 4   321

Bibliography   339

Principal Symbols and Notation

The following lists give only the principal use of the symbols for scalar quantities. A given symbol might be used to denote a variety of physical quantities. The interpretation to be given to a symbol will be clear from the context in which it is employed.

Kinematics

1 Scalars

$a, b, c, d, u, v, w, r, s, t$	Length, components of vectors
$\alpha, \beta, \gamma, \theta, \phi, \psi$	Angles
ω, Ω	Components of angular velocity
$\dot{\omega}, \dot{\Omega}$	Components of angular acceleration
ℓ	Direction cosine

2 Vectors

With the exception of the lower case Greek letter omega, upper case letters written inside braces are used to designate vector quantities as follows:

$\{R\}$	Position and relative position
$\{V\}$	Linear velocity
$\{A\}$	Linear acceleration
$\{B\}$	Any vector
$\{\omega\}$	Angular velocity
$\{\dot{\omega}\}$	Angular acceleration

These general symbols for vector quantities are qualified in two ways by appropriate suffixes. Thus

$$\{R_A\}_1 \quad \text{or} \quad \{R_{AO_1}\}_1$$

specifies the position of the point A measured in frame 1, where O_1, often omitted, is the origin of frame 1, while

$$\{R_{BA}\}_4$$

specifies the position of the point B relative to the point A measured in frame 4. It so happens the the relative position vector is independent of the frame in which its is measured, but the number suffix is retained for reasons explained in the text.

Similarly,

$$\{V_A\}_1 \quad \text{or} \quad \{V_{AO_1}\}_1 \quad \text{and} \quad \{V_{PQ}\}_3$$

specify the velocity of the point A relative to O_1 measured in frame 1 and the velocity of the point P relative to the point Q measured in frame 3 respectively.

Also

$$\{A_A\}_1 \quad \text{or} \quad \{A_{AO_1}\} \quad \text{and} \quad \{A_{DC}\}_1$$

specifiy the acceleration of the point A relative to the point O_1 measured in frame 1 and the acceleration of the point D relative to the point C measured in frame 1 respectively.

Numbers are also used as suffixes inside the braces to qualify position, velocity and acceleration. Thus

$$\{R_4\}_1 \quad , \quad \{V_4\}_1 \quad \text{and} \quad \{A_4\}_1$$

specify the position, velocity and acceleration respectively of the centre of mass of body 4 measured in frame 1.

The angular velocity vector is qualified by number suffixes. Thus

$$\{\omega_3\}_2 \quad \text{or} \quad \{\omega_{32}\}$$

specify the angular velocity of body 3 measured with respect to body 2 or the angular velocity of body 3 relative to body 2. A similar notation is used for angular acceleration. The angular velocity and acceleration vectors can be further qualified by lower case superscript letters inside the braces. Thus

$$\{\omega_2^n\}_1 \quad \text{and} \quad \{\omega_2^p\}_1$$

specify, respectively, the components of the angular velocity vector normal to and parallel to to some line joining points(specified in a particular context) fixed in body 2. Similarly,

$$\{A_{BA}^n\}_1 \quad \text{and} \quad \{A_{BA}^p\}_1$$

specify, respectively, the components of the linear acceleration of B relative to A normal to and parallel to the line joining B and A.

The usual modulus notation is employed to indicate the magnitude of a vector. Thus

$|R_A|_1$, $|V_{BA}|_1$ and $|\omega_2|_1$ or $|\omega_{21}|$

are magnitudes of the corresponding vectors. In the case of the relative position vector, which is independent of the frame used for its measurement the number suffix is omitted. Thus the magnitude of

$$\{R_{BA}\}_1 = \{R_{BA}\}_4 = \{R_{BA}\}_n$$

is written

$$|R_{BA}|.$$

A vector can be described by resolving it along the axes of a particular reference frame, when it is said to be referred to that frame. The frame to which a vector is referred is written outside the braces after the first number suffix and separated from it by a solidus or oblique stroke. Thus

$$\{V_{BA}\}_{1/3} = \begin{bmatrix} u_3 \\ v_3 \\ w_3 \end{bmatrix}$$

is the column matrix which describes the velocity of B relative to A, measured in frame 1, in frame 3.

3 The transformation or rotation matrix

The transformation matrix is a 3×3 orthogonal matrix of direction cosines written

$$[\ell].$$

It is used to change the frame to which a vector is *referred*. If, for example, a vector $\{B\}_n$ is referred to frame 1, then the transformation matrix which changes the reference frame to frame 2 is

$$[\ell_1]_2.$$

Thus

$$\{B\}_{n/2} = [\ell_1]_2 \{B\}_{n/1}.$$

The transformation matrix can be regarded as the matrix which specifies a rotation, or sequence of rotations, which a frame undergoes to align it with another. If, for example, frame 1 is to be aligned with frame 2, then the rotation matrix would be written

$$[\ell_2]_1.$$

If this alignment is achieved by a sequence of simple rotations about a single axis of appropriately positioned intermediate frames 3 and 4, then this operation would be specified by the product of rotation matrices

$$[\ell_2]_1 = [\ell_3]_1 [\ell_4]_3 [\ell_2]_4.$$

Dynamics

1 Scalars

A, B, C, D, E, F, I, J	Terms in the inertia matrix
k	Spring rate, constant
m	Mass
T	Kinetic energy
V	Potential energy
g	Magnitude of gravitational acceleration
W	Work

The general symbols can be qualified by appropriate suffixes.

I can take the suffixes xx, xy, xz etc. to denote the axes involved. A, B, C etc. can take number suffixes to denote the reference frame.

m can take a suffix P to indicate that it refers to a particle, or a number suffix to indicate the body to which it refers.

T can take a suffix P to indicate that it refers to a particle, or a number suffix to indicate the body to which it refers. It can be further qualified to indicate that the energy is evaluated at some particular position. Thus, for example

$$T_{4\,rot}\Big|_{\alpha}$$

is the rotational kinetic energy in body 4 when in some position defined by the angle α. V can be qualified in a similar manner.

W can take suffix statements such as $A \rightarrow B \rightarrow C$ to specify the path traced out by the point of application of the force involved.

2 Vectors

Upper case letters written inside braces are used to designate vector quantities as follows:

$\{F\}$	Force
$\{G\}$	Linear momentum
$\{H\}$	Angular momentum (Angular momentum)
$\{L\}$	Couple moment
$\{M\}$	Force moment
$\{W\}$	Weight
$\{\nabla\}$	Vector operator del

The general symbols for vector quantities are qualified in two ways by appropriate suffixes and also by superscripts.

In the case of the force vector, number suffixes inside the braces are used to specify a contact force between two bodies. As, for example,

$$\{F_{23}\}$$

which is the force *on* body 2 *due to* body 3. Similarly,

$$\{L_{23}\}$$

is the couple on body 2 due to body 3. Also

$$\{F_2\}$$

is the external force on body 2. It might be, for example,

$$\{F_2\} = \{F_{23}\} + \{F_{24}\} + \{F_{25}\} + \ldots$$

where 3, 4 and 5 are bodies which exert a force on body 2. A similar notation can be used in respect of couples. Components of $\{F\}$ and $\{L\}$ can be singled out by writing an appropriate superscript inside the braces, as for example,

$$\{F^x\} \text{ and } \{L^y\}$$

or

$$\{F^n\} \text{ and } \{L^p\}$$

where the superscripts n and p refer to components parallel to some reference direction.

A number suffix is used outside the braces to specify the frame to which the vector is referred. Thus

$$\{F_{34}\}_3$$

is the column matrix which describes the force on body 3 due to body 4 which is referred to frame 3. The $\{L\}$ can be similarly subscripted.

In the case of linear momentum a number suffix inside the braces specifies the body concerned and the first number suffix outside the braces specifies the frame in which the momentum is measured. This frame will invariably be an inertial reference frame which, in this text, is always designated 1. It is always included by way of emphasis. The second number suffix outside the braces, written after a solidus, specifies the frame to which the vector is referred. Thus

$$\{G_3\}_{1/4}$$

is the column matrix which describes the linear momentum of body 3, measured with respect to frame 1, the vector being referred to frame 4.

In the case of angular momentum of a body about its centre of mass, a number suffix inside the braces specifies the body concerned and the number suffixes outside the braces have the same significance as in the case of linear momentum. Thus

$$\{H_3\}_{1/4}$$

is the column matrix which describes the angular momentum of body 3 about its centre of mass, measured with respect to frame 1, the vector being referred to frame 4. If the angular momentum about a point other than the centre of mass is to be specified, point Q say, then this is written

$$\{H_{3Q}\}_{1/4} .$$

In the case of the moment vector, a single letter suffix is used to specify the point about which moments are taken and a single number suffix outside the braces specifies the frame to which both force and position vectors are referred. Thus

$$\{M_A\}_3$$

is the column matrix which describes the moment of a force, or system of forces and couples about A, the vector being referred to frame 3.

In the case of the weight vector, a number suffix specifies the body to which it refers and a single number suffix outside the braces specifies the frame to which the vector is referred. Hence

$$\{W_4\}_2$$

is the column matrix which describes the weight of body 4, the vector being referred to frame 2.

3 The inertia matrix

The inertia matrix is a 3×3 symmetric matrix written

$$[I]$$

Number suffixes are used in the same way as for vectors. Thus

$$[I_3]_{3/3}$$

describes the inertia of body 3, measured with respect to frame 3 and referred to frame 3. Unless expressly stated otherwise, the centre of of mass of body 3 will be at the origin of frame 3. Similarly,

$$[I_3]_{4/5}$$

describes the inertia of body 3, measured with respect to frame 4 and referred to frame 5.

Chapter 1

Kinematics

1.1. The Position Vector

The position of a point depends upon the datum used for its measurement. Consider three bodies of a system of bodies designated 1, 2, 3 etc. Let a system of co-ordinate axes be fixed in a convenient position in each of the bodies as shown in Fig. 1.1. A point P_3 in body 3 can have its position measured relative to each set of axes or frame of reference. The vector $\overrightarrow{O_1P_3}$ is the position of P_3 measured in frame 1, the vector $\overrightarrow{O_2P_3}$ is the position of P_3 measured in frame 2 and the vector $\overrightarrow{O_3P_3}$ is the position of P_3 measured in in frame 3.

Let $\{R\}$ be used to represent the position vector, and in particular represent

$\overrightarrow{O_1P_3}$ by $\{R_{P_3O_1}\}_1$ or simply $\{R_{P_3}\}_1$,

$\overrightarrow{O_2P_3}$ by $\{R_{P_3O_2}\}_2$ or simply $\{R_{P_3}\}_2$,

$\overrightarrow{O_3P_3}$ by $\{R_{P_3O_3}\}_3$ or simply $\{R_{P_3}\}_3$

and so on. The suffix outside the braces is used to indicate the frame in which the position of P_3 has been measured.

The position vector can be specified by components along any one set of co-ordinate axes or reference frame, when it is said to be *referred* to that set of axes or reference frame. The reference frame is indicated by a second suffix, so that

$$\{R_{P_3}\}_{1/1} = \begin{bmatrix} a_1 \\ b_1 \\ c_1 \end{bmatrix}$$

represents the column matrix which specifies the position of P_3

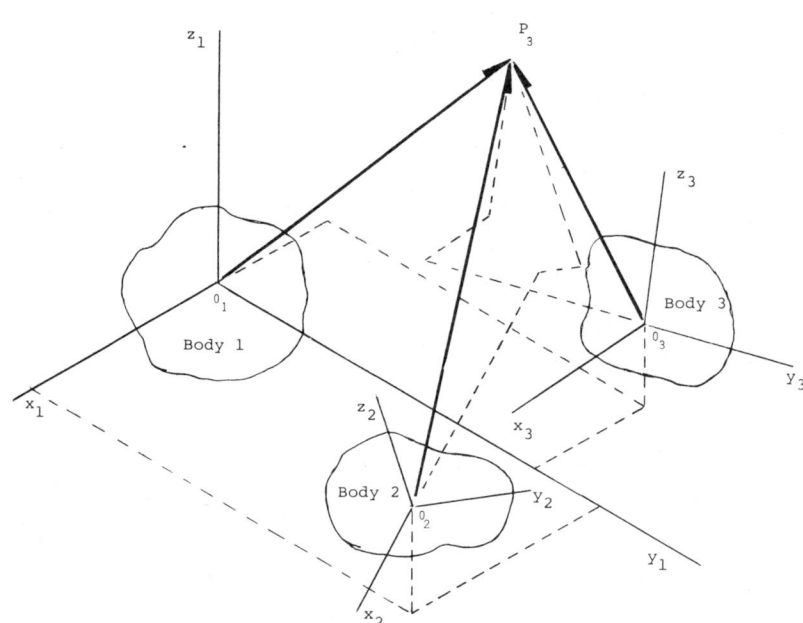

Fig. 1.1

measured in frame 1 and referred to frame 1, where a_1, b_1 and c_1 are the components of the vector along x_1, y_1 and z_1 respectively. Similarly,

$$\{R_{P_3}\}_{2/3} = \begin{bmatrix} a_3 \\ b_3 \\ c_3 \end{bmatrix}$$

represents the column matrix which specifies the position of P_3 measures in frame 2 and referred to frame 3.

While the particular case of the position vector has been considered, it will be clear that, in general, a vector canonly be completely specified by a column matrix when the frame used for its measurement and the frame used for reference are quoted. Some vectors are, however, independent of any reference body and in such cases a single suffix outside the braces can be used to indicate the frame to which the components of the vector are referred.

1.2. The Relative Position Vector

Let A and B be two points fixed in body 3. Then by reference to Fig. 1.2, which is drawn two dimensionally for ease of illustration,

Kinematics

$$\{R_B\}_3 = \{R_A\}_3 + \{R_{BA}\}_3 \tag{1.1}$$

and

$$\{R_B\}_2 = \{R_A\}_2 + \{R_{BA}\}_2 . \tag{1.2}$$

Also

$$\{R_{O2}\}_3 + \{R_B\}_2 = \{R_B\}_3 \tag{1.3}$$

and

$$\{R_{O2}\}_3 + \{R_A\}_2 = \{R_A\}_3 . \tag{1.4}$$

Subtraction of Eq. 1.4 from Eq. 1.3 gives

$$\{R_B\}_2 - \{R_A\}_2 = \{R_B\}_3 - \{R_A\}_3$$

and therefore

$$\{R_{BA}\}_2 = \{R_{BA}\}_3 . \tag{1.5}$$

It is thus clear that the relative position vector does not depend upon the reference body used for its measurement, but the matrix which specifies the vector will depend upon the frame to which it is referred. Thus, for example, while

$$\{R_{BA}\}_{1/5} = \{R_{BA}\}_{3/5} = \{R_{BA}\}_{5/5} \text{ etc.},$$

$$\{R_{BA}\}_{1/3} \neq \{R_{BA}\}_{1/2} \neq \{R_{BA}\}_{2/1} \text{ etc.}$$

Strictly therefore, in the case of the relative position vector, the first suffix which denotes the reference body used for its measurement is not necessary, but it is wise to retain it because when time derivatives are considered it will be found that, for example

$$\frac{d}{dt}\{R_{BA}\}_{2/2} \neq \frac{d}{dt}\{R_{BA}\}_{1/2} .$$

Refer to Problem 3.1 and Problem 3.23.

1.3. Transformation of Vectors

Let $\{B\}_n$ be any vector where n is the reference body used for its measurement. If the frame to which it is referred is designated 1, then the column matrix representing the vector would be written

$$\{B\}_{n/1} = \begin{bmatrix} u_1 \\ v_1 \\ w_1 \end{bmatrix}$$

and if the vector was referred to frame 2 then its column matrix representation would be written

$$\{B\}_{n/2} = \begin{bmatrix} u_2 \\ v_2 \\ w_2 \end{bmatrix}$$

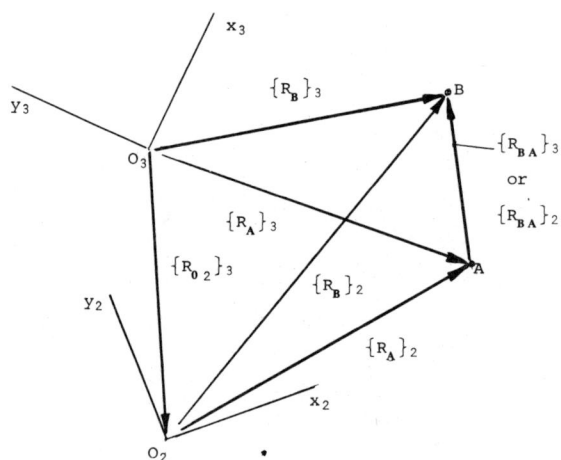

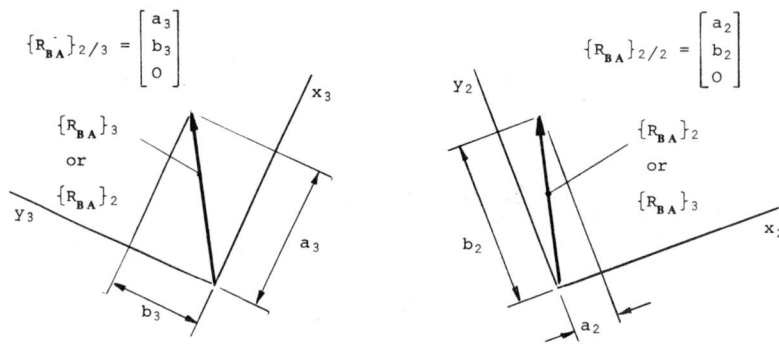

Fig. 1.2

as shown in Fig. 1.3. It is necessary, given $\{B\}_{n/1}$ to find $\{B\}_{n/2}$ and vice versa. Consider the u component of $\{B\}_{n/1}$ as a vector in frame 2 as shown in Fig. 1.4a. Let

$$\ell_{x_2 x_1}, \quad \ell_{y_2 x_1} \quad \text{and} \quad \ell_{z_2 x_1}$$

be the direction cosines of u_1 with respect to the x_2, y_2 and z_2 axes respectively. Then the components of u_1 along x_2, y_2 and z_2 are respectively

$$u_1 \ell_{x_2 x_1}, \quad u_1 \ell_{y_2 x_1} \quad \text{and} \quad u_1 \ell_{z_2 x_1}.$$

Similarly, by reference to Fig. 1.4b, the components of v_1 along x_2, y_2 and z_2 are respectively

Kinematics

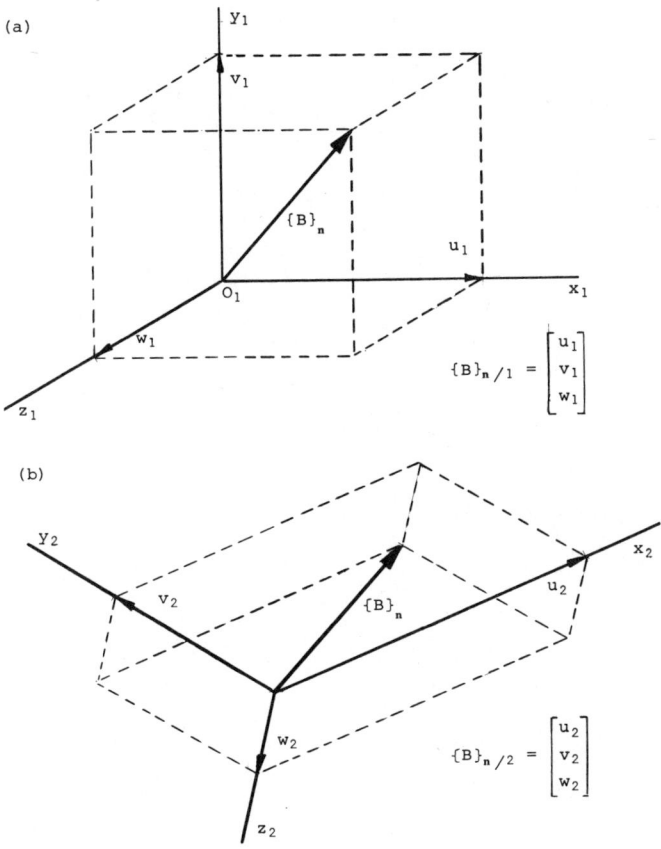

Fig. 1.3

$v_1 \ell_{x_2 y_1}$, $v_1 \ell_{y_2 y_1}$ and $v_1 \ell_{z_2 y_1}$.

Also, the components of w_1 along x_2, y_2 and z_2 are respectively

$w_1 \ell_{x_2 z_1}$, $w_1 \ell_{y_2 z_1}$ and $w_1 \ell_{z_2 z_1}$.

Adding corresponding components gives

$$u_2 = u_1 \ell_{x_2 x_1} + v_1 \ell_{x_2 y_1} + w_1 \ell_{x_2 z_1}$$
$$v_2 = u_1 \ell_{y_2 x_1} + v_1 \ell_{y_2 y_1} + w_1 \ell_{y_2 z_1}$$
$$w_2 = u_1 \ell_{z_2 x_1} + v_1 \ell_{z_2 y_1} + w_1 \ell_{z_2 z_1}.$$

These equations can be written in the form

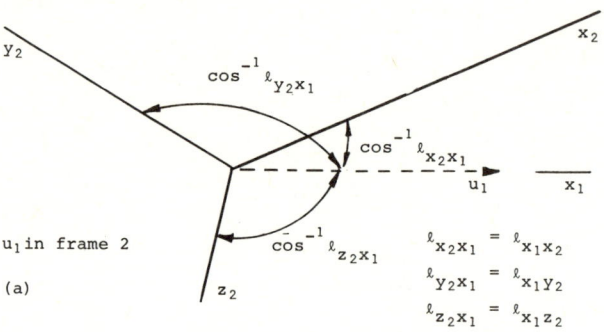

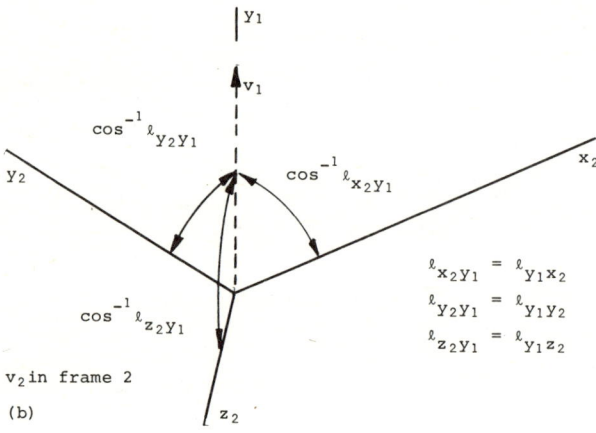

Fig. 1.4

$$\begin{bmatrix} u_2 \\ v_2 \\ w_2 \end{bmatrix} = \begin{bmatrix} \ell_{x_2 x_1} & \ell_{x_2 y_1} & \ell_{x_2 z_1} \\ \ell_{y_2 x_1} & \ell_{y_2 y_1} & \ell_{y_2 z_1} \\ \ell_{z_2 x_1} & \ell_{z_2 y_1} & \ell_{z_2 z_1} \end{bmatrix} \begin{bmatrix} u_1 \\ v_1 \\ w_1 \end{bmatrix} \quad (1.6)$$

or

$$\{B\}_{n/2} = [\ell_1]_2 \{B\}_{n/1} \quad (1.7)$$

where

$$[\ell_1]_2 = \begin{bmatrix} \ell_{x_2 x_1} & \ell_{x_2 y_1} & \ell_{x_2 z_1} \\ \ell_{y_2 x_1} & \ell_{y_2 y_1} & \ell_{y_2 z_1} \\ \ell_{z_2 x_1} & \ell_{z_2 y_1} & \ell_{z_2 z_1} \end{bmatrix} \quad (1.8)$$

is the transformation matrix which transforms the components of a a vector from frame 1 to frame 2. Remember, the 'direction' of the of the transformation is *from 1 to 2*

$$\begin{bmatrix} \ell \\ 1 \end{bmatrix}_2 .$$

Similarly, it can be shown that

$$\begin{bmatrix} u_1 \\ v_1 \\ w_1 \end{bmatrix} = \begin{bmatrix} \ell_{x_1 x_2} & \ell_{x_1 y_2} & \ell_{x_1 z_2} \\ \ell_{y_1 x_2} & \ell_{y_1 y_2} & \ell_{y_1 z_2} \\ \ell_{z_1 x_2} & \ell_{z_1 y_2} & \ell_{z_1 z_2} \end{bmatrix} \begin{bmatrix} u_2 \\ v_2 \\ w_2 \end{bmatrix} \qquad (1.9)$$

or

$$\{B\}_{n/1} = [\ell_2]_1 \{B\}_{n/2} \qquad (1.10)$$

where

$$[\ell_2]_1 = \begin{bmatrix} \ell_{x_1 x_2} & \ell_{x_1 y_2} & \ell_{x_1 z_2} \\ \ell_{y_1 x_2} & \ell_{y_1 y_2} & \ell_{y_1 z_2} \\ \ell_{z_1 x_2} & \ell_{z_1 y_2} & \ell_{z_1 z_2} \end{bmatrix} . \qquad (1.11)$$

Reference to Fig. 1.4 and Eqs. 1.8 and 1.11 shows that

1st. column of $[\ell_1]_2$ — $\begin{bmatrix} \ell_{x_2 x_1} \\ \ell_{y_2 x_1} \\ \ell_{z_2 x_1} \end{bmatrix} \begin{matrix} = \\ = \\ = \end{matrix} \begin{bmatrix} \ell_{x_1 x_2} \\ \ell_{x_1 y_2} \\ \ell_{x_1 z_2} \end{bmatrix}$ — 1st. row of $[\ell_2]_1$,

Direction cosines of x_1 relative to frame 2

2nd. column of $[\ell_1]_2$ — $\begin{bmatrix} \ell_{x_2 y_1} \\ \ell_{y_2 y_1} \\ \ell_{z_2 y_1} \end{bmatrix} \begin{matrix} = \\ = \\ = \end{matrix} \begin{bmatrix} \ell_{y_1 x_2} \\ \ell_{y_1 y_2} \\ \ell_{y_1 z_2} \end{bmatrix}$ — 2nd. row of $[\ell_2]_1$

Direction cosines of y_1 relative to frame 2

and

3rd. column of $[\ell_1]_2$ — $\begin{bmatrix} \ell_{x_2 z_1} \\ \ell_{y_2 z_1} \\ \ell_{z_2 z_1} \end{bmatrix} \begin{matrix} = \\ = \\ = \end{matrix} \begin{bmatrix} \ell_{z_1 x_2} \\ \ell_{z_1 y_2} \\ \ell_{z_1 z_2} \end{bmatrix}$ — 3rd. row of $[\ell_2]_1$.

Direction cosines of z_1 relative to frame 2

Thus

$$[\ell_1]_2^T = [\ell_2]_1 \qquad (1.12)$$

where the superscript T indicates the transpose. Also

$$[\ell_2]_1^T = [\ell_1]_2 . \qquad (1.13)$$

It should be particularly noted that transforming a vector changes

8 Matrix Methods in Engineering Mechanics

only the frame to which the vector is referred and has no influence on the frame in which the measurement is made.

1.4. The Rotation Matrix for Simple Rotations

The transformation matrix $[\ell_2]_1$ is the matrix which transforms a vector from frame 2 to frame 1. It can be usefully regarded as having another role. Consider the case of two frames, 1 and 2 which have their x axes aligned and arranged so that a positive rotation of frame 1 through α about the x_1 axis aligns y_1 with y_2 and z_1 with z_2 as shown in Fig. 1.5a.

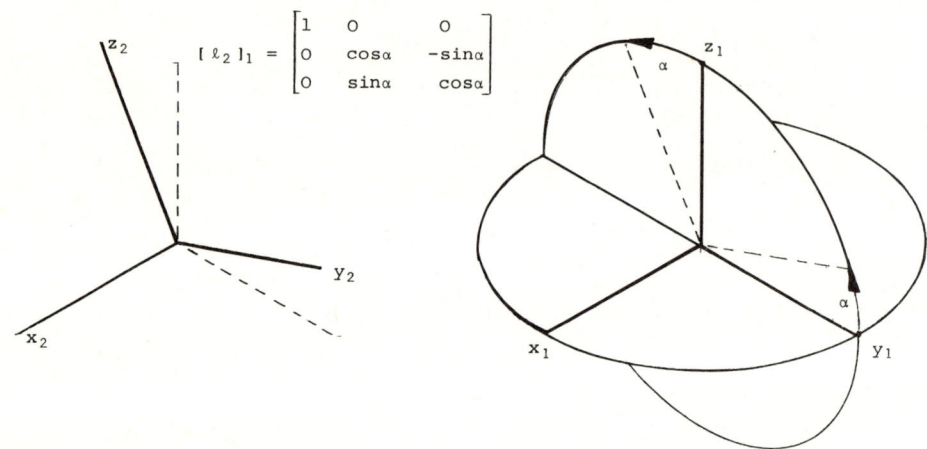

$$[\ell_2]_1 = \begin{bmatrix} 1 & 0 & 0 \\ 0 & \cos\alpha & -\sin\alpha \\ 0 & \sin\alpha & \cos\alpha \end{bmatrix}$$

Frame 1 rotates α about x_1 to align with frame 2

$1 \xrightarrow{\alpha \text{ about } x_1} 2$

Fig. 1.5a

By reference to Eq. 1.11 the terms in $[\ell_2]_1$ for this particular case are

$\ell_{x_1x_2} = \cos 0 = 1, \quad \ell_{x_1y_2} = \cos\pi/2 = 0, \quad \ell_{x_1z_2} = \cos\pi/2 = 0,$

$\ell_{y_1x_2} = \cos\pi/2 = 0, \quad \ell_{y_1y_2} = \cos\alpha, \quad \ell_{y_1z_2} = \cos(\pi/2 + \alpha) = -\sin\alpha$

$\ell_{z_1x_2} = \cos\pi/2 = 0, \quad \ell_{z_1y_2} = \cos(\pi/2 - \alpha) = \sin\alpha, \quad \ell_{z_1z_2} = \cos\alpha$

and therefore

$$[\ell_2]_1 = \begin{bmatrix} 1 & 0 & 0 \\ 0 & \cos\alpha & -\sin\alpha \\ 0 & \sin\alpha & \cos\alpha \end{bmatrix}. \qquad (1.14)$$

Thus, $[\ell_2]_1$ can be taken to represent the positive rotation α about the x_1 axis which frame 1 must undergo to align it with frame 2, or alternatively, the rotation of frame 2 about the x_2 axis in moving from alignment with frame 1 to its given position. The direction of rotation indicated by the rotation matrix $[\ell_2]_1$ is thus *from* 1 *to* 2

$$[\ell_2^{\curvearrowright}]_1 \ .$$

Compare this with its transformation interpretation.

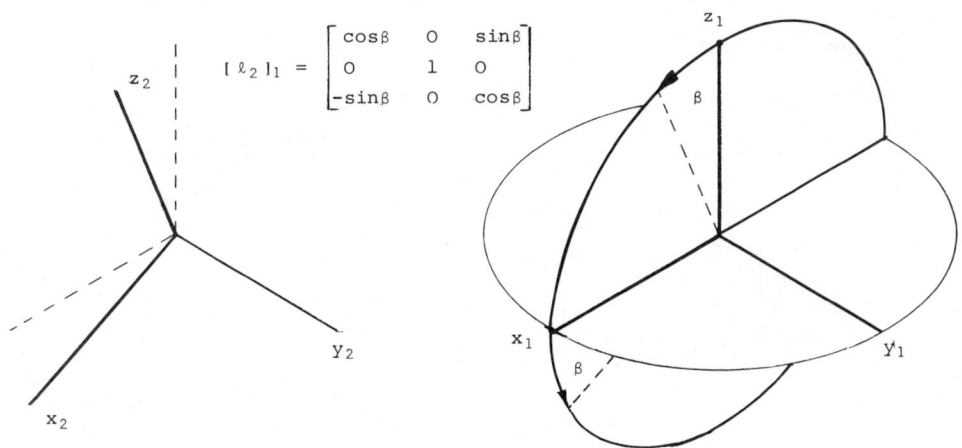

Frame 1 rotates β about y_1 to align with frame 2

1 $\xrightarrow{\beta \text{ about } y_1}$ 2

Fig. 1.5b

Figure 1.5b shows the case in which frames 1 and 2 have their y axes aligned and arranged so that frame 1 aligns with frame 2 when frame 1 undergoes the positive rotation β about the y_1 axis. In this case

$$[\ell_2]_1 = \begin{bmatrix} \cos\beta & 0 & \sin\beta \\ 0 & 1 & 0 \\ -\sin\beta & 0 & \cos\beta \end{bmatrix} \ . \tag{1.15}$$

Similarly, in Fig. 1.5c a rotation of frame 1 through γ about the z_1 axis aligns frame 1 with frame 2 giving

$$[\ell_2]_1 = \begin{bmatrix} \cos\gamma & -\sin\gamma & 0 \\ \sin\gamma & \cos\gamma & 0 \\ 0 & 0 & 1 \end{bmatrix} \ . \tag{1.16}$$

Equations 1.14, 1.15 and 1.16 are important results which are used repeatedly in the solution of problems and they must therefore be committed to memory.

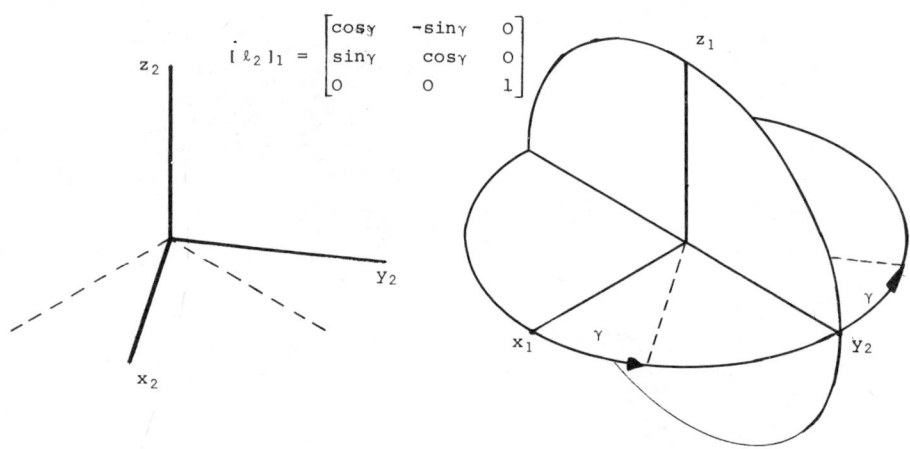

Frame 1 rotates γ about z_1 to align with frame 2

1 γ about z_1 2

Fig. 1.5c

1.5. Consecutive Rotations

If the angular position of frame 4 relative to frame 1 is defined by the rotation matrix

$$[\ell_4]_1 = \begin{bmatrix} \ell_{x_1 x_4} & \ell_{x_1 y_4} & \ell_{x_1 z_4} \\ \ell_{y_1 x_4} & \ell_{y_1 y_4} & \ell_{y_1 z_4} \\ \ell_{z_1 x_4} & \ell_{z_1 y_4} & \ell_{z_1 z_4} \end{bmatrix}, \qquad (1.17)$$

then it is always possible to replace the matrix by a product of three matrices such as, for example,

$$[\ell_4]_1 = \begin{bmatrix} c\gamma & -s\gamma & 0 \\ s\gamma & c\gamma & 0 \\ 0 & 0 & 1 \end{bmatrix} \begin{bmatrix} 1 & 0 & 0 \\ 0 & c\alpha & -s\alpha \\ 0 & s\alpha & c\alpha \end{bmatrix} \begin{bmatrix} c\beta & 0 & s\beta \\ 0 & 1 & 0 \\ -s\beta & 0 & c\beta \end{bmatrix},$$

each of which represents a simple rotation of frame 1. Note that here c has been written for cos and s for sin to effect an economy of space and effort.

Figure 1.6 shows frames 1, 2, 3 and 4. The origins of frames 2 and 3 are coincident with the origin of frame 1. Frame 4 is also shown copied with its origin coincident with that of frame 1. Frame 2 is positioned such that the z_2 axis is coincident with the z_1 axis and the x_2 axis is perpendicular to the y_4 axis. Thus the rotation of frame 2 about the x_2 axis makes it possible to align the y_4 axis with the y_4 axis. Frame 3 is positioned such that the x_3 axis is

Kinematics

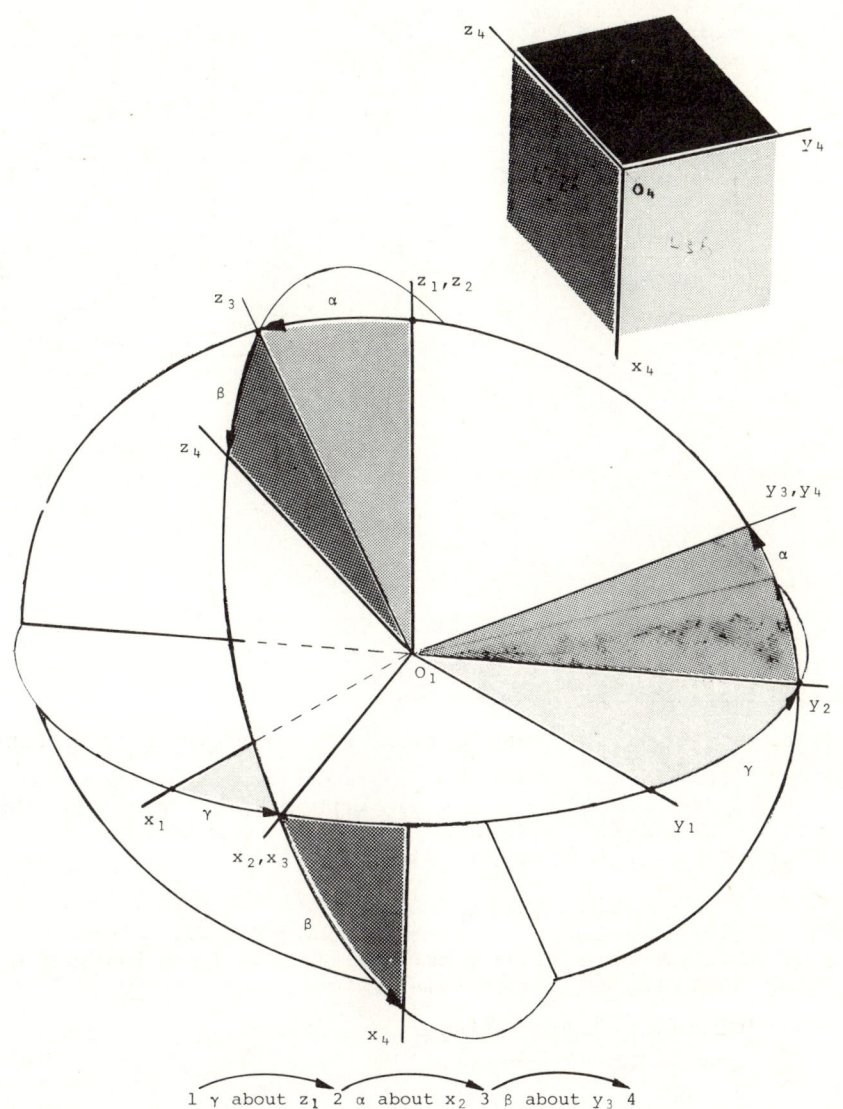

1 γ about z_1 → 2 α about x_2 → 3 β about y_3 → 4

Fig. 1.6

coincident with the x_2 axis and the y_3 axis is coincident with the copied y_4 axis. The z_3 and x_3 axes are thus in the z_4x_4 plane of the copied frame 4. Thus frame 1 can be aligned with frame 4 by the following simple rotations

1 γ about z_1 → 2 α about x_2 → 3 β about y_3 → 4

giving

$$[\ell_4]_1 = [\ell_2]_1[\ell_3]_2[\ell_4]_3$$

$$= \begin{bmatrix} c\gamma & -s\gamma & 0 \\ s\gamma & c\gamma & 0 \\ 0 & 0 & 1 \end{bmatrix} \begin{bmatrix} 1 & 0 & 0 \\ 0 & c\alpha & -s\alpha \\ 0 & s\alpha & c\alpha \end{bmatrix} \begin{bmatrix} c\beta & 0 & s\beta \\ 0 & 1 & 0 \\ -s\beta & 0 & c\beta \end{bmatrix}$$

$$= \begin{bmatrix} c\gamma c\beta - s\gamma s\alpha s\beta & -s\gamma c\alpha & c\gamma s\beta + s\gamma s\alpha c\beta \\ s\gamma c\beta + c\gamma s\alpha s\beta & c\gamma c\alpha & s\gamma s\beta - c\gamma s\alpha c\beta \\ -c\alpha s\beta & s\alpha & c\alpha c\beta \end{bmatrix} . \quad (1.18)$$

Hence, by Eqs. 1.17 and 1.18

$$\sin\alpha = \ell_{z_1 y_4}, \quad (1.19)$$

$$-\tan\gamma = \ell_{x_1 y_4}/\ell_{y_1 y_4} \quad (1.20)$$

and

$$-\tan\beta = \ell_{z_1 x_4}/\ell_{z_1 z_4} . \quad (1.21)$$

Refer to Problem 3.2 and Problems 3.24 and 3.25.

1.6. Successive Transformation of Vectors

It is frequently necessary to transform vectors from one frame to another in cases where the alignment between frames cannot readily be achieved by a simple rotation about an axis of one of the frames. In such cases it is possible to choose convenient intermediate frames which can be readily aligned by simple rotations.

Consider the system of Fig. 1.7 in which body 2 rotates about the z_1 axis fixed in body 1 and body 3 turns on body 2 about the y_2 axis fixed in body 2 and also rolls on body 1. Frame 1 can be aligned with frame 3 by the following simple consecutive rotations

$$1 \;\;\gamma\; \text{about}\; z_1 \;\; 2 \;\; -\beta\; \text{about}\; y_2 \;\; 3$$

and therefore

$$[\ell_3]_1 = [\ell_2]_1[\ell_3]_2 = \begin{bmatrix} c\gamma & -s\gamma & 0 \\ s\gamma & c\gamma & 0 \\ 0 & 0 & 1 \end{bmatrix} \begin{bmatrix} c\beta & 0 & -s\beta \\ 0 & 1 & 0 \\ s\beta & 0 & c\beta \end{bmatrix}.$$

Thus, for P a point in body 3

$$\{R_{PO_2}\}_{1/1} = \{R_{O_3 O_2}\}_{1/1} + \{R_{PO_3}\}_{1/1}$$

$$= [\ell_2]_1\{R_{O_3 O_2}\}_{2/2} + [\ell_2]_1[\ell_3]_1\{R_{PO_3}\}_{3/3}$$

Kinematics

$$= \begin{bmatrix} c\gamma & -s\gamma & 0 \\ s\gamma & c\gamma & 0 \\ 0 & 0 & 1 \end{bmatrix} \begin{bmatrix} 0 \\ a \\ 0 \end{bmatrix}$$

$$+ \begin{bmatrix} c\gamma & -s\gamma & 0 \\ s\gamma & c\gamma & 0 \\ 0 & 0 & 1 \end{bmatrix} \begin{bmatrix} c\beta & 0 & -s\beta \\ 0 & 1 & 0 \\ s\beta & 0 & c\beta \end{bmatrix} \begin{bmatrix} b \\ c \\ d \end{bmatrix} \quad (1.22)$$

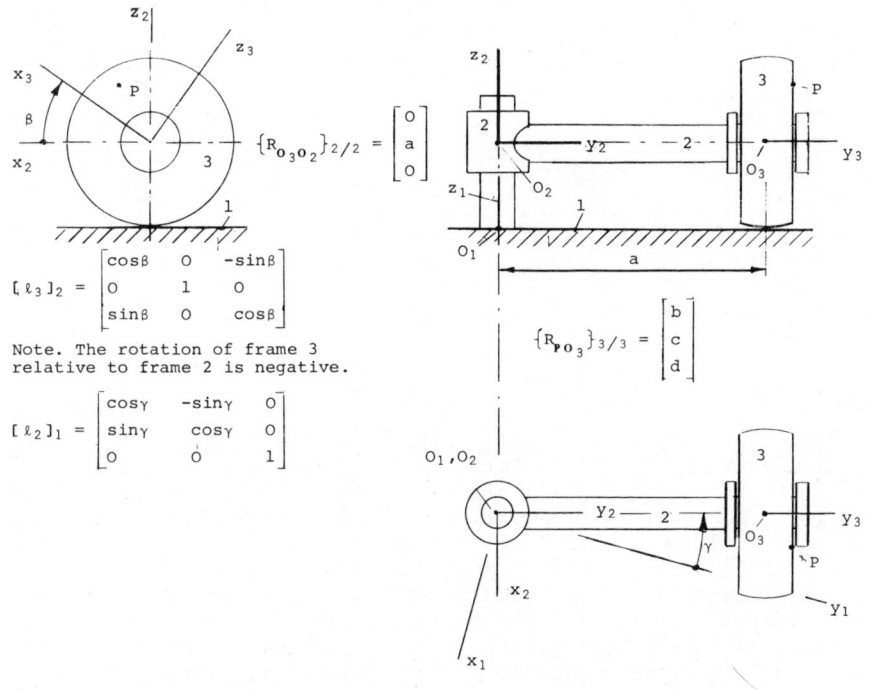

Fig. 1.7

If γ is known in terms of time (and if body 3 rolls without slip on body 1, β is also known) then the final expression for

$$\{R_{PO_2}\}_{1/1}$$

is readily differentiated with respect to time, even when b, c and d are known functions of time, to obtain a general expression for the velocity of P. It is thus possible to obtain a general expression for velocity and acceleration of points in complicated systems when less systematic methods would require considerably more ingenuity and have less chance of producing the correct result.

Suppose it is necessary, in a more general case, to find $\{B\}_{n/1}$ from

$\{B\}_{n/5}$. Then

$$\{B\}_{n/4} = [\ell_5]_4 \{B\}_{n/5} \, ,$$

$$\{B\}_{n/3} = [\ell_4]_3 \{B\}_{n/4} = [\ell_4]_3 [\ell_5]_4 \{B\}_{n/5} \, ,$$

$$\{B\}_{n/2} = [\ell_3]_2 \{B\}_{n/3} = [\ell_3]_2 [\ell_4]_3 [\ell_5]_4 \{B\}_{n/5}$$

and

$$\{B\}_{n/1} = [\ell_2]_1 \{B\}_{n/2} = [\ell_2]_1 [\ell_3]_2 [\ell_4]_3 [\ell_5]_4 \{B\}_{n/5}$$

$$= [\ell_5]_1 \{B\}_{n/5} \, . \qquad (1.23)$$

There is, of course, no need for the frames to be numbered consecutively. Thus $\{B\}_{n/1}$ might equally well be given by

$$\{B\}_{n/1} = [\ell_8]_1 [\ell_2]_8 [\ell_3]_2 [\ell_7]_3 [\ell_5]_7 \{B\}_{n/5} \qquad (1.24)$$

where

$$[\ell_5]_1 = [\ell_2]_1 [\ell_3]_2 [\ell_4]_3 [\ell_5]_4 \qquad (1.25)$$

or

$$[\ell_5]_1 = [\ell_8]_1 [\ell_2]_8 [\ell_3]_2 [\ell_7]_3 [\ell_5]_7 \, . \qquad (1.26)$$

The transformation thus proceeds from right to left, 5 to 4 to 3 to 2 to 1, or 5 to 7 to 3 to 2 to 8 to 1.

In the alternative rotation role, the matrix product is read from left to right. Frame 1 can be aligned with frame 5 by the following successive rotations
   rotate frame 1 to align with frame 2,
   rotate frame 2 to align with frame 3,
   rotate frame 3 to align with frame 4
and
   rotate frame 4 to align with frame 5.

1.7. The Velocity and Acceleration of a Point

Figure 1.8 shows a point P at time t when it is coincident with a a point A fixed in frame 1 and also at time t + Δt when it is coincident with point B fixed in frame 1. The change in position of P which occurs in time Δt is represented by the vector

$$\{\Delta R_p\}_1$$

when the change is measured with respect to or measured in frame 1. Thus

$$\{\Delta R_p\}_1 = \{R_p\}_1 \Big|_{t+\Delta t} - \{R_p\}_1 \Big|_t$$

$$= \{R_B\}_1 - \{R_A\}_1 = \{R_{BA}\}_1 \qquad (1.27)$$

Kinematics

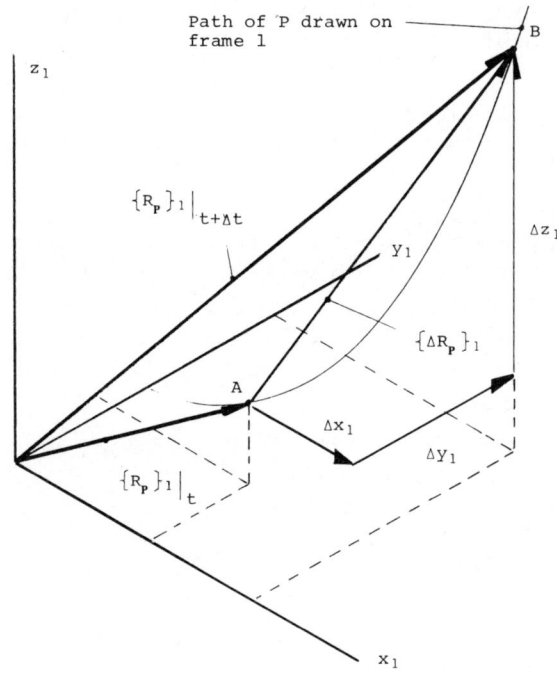

Fig. 1.8

and the velocity is given by

$$\{V_p\}_1 = \underset{\Delta t \to 0}{\text{Limit}} \frac{\{\Delta R_p\}_1}{\Delta t} \ . \tag{1.28}$$

If $\{\Delta R_p\}_1$ is specified by

$$\{\Delta R_p\}_{1/1} = \begin{bmatrix} \Delta x_1 \\ \Delta y_1 \\ \Delta z_1 \end{bmatrix}$$

then

$$\{V_p\}_{1/1} = \underset{\Delta t \to 0}{\text{Limit}} \frac{\{\Delta R_p\}_{1/1}}{t} = \underset{\Delta t \to 0}{\text{Limit}} \begin{bmatrix} \Delta x_1/\Delta t \\ \Delta y_1/\Delta t \\ \Delta z_1/\Delta t \end{bmatrix} = \begin{bmatrix} \dot{x}_1 \\ \dot{y}_1 \\ \dot{z}_1 \end{bmatrix}$$

$$= \{\dot{R}_p\}_{1/1} = \frac{d}{dt}\{R_p\}_{1/1} \tag{1.29}$$

The acceleration of P is given by

$$\{A_P\}_{1/1} = \frac{d}{dt}\{V_P\}_{1/1} = \{\dot{V}_P\}_{1/1} = \{\ddot{R}_P\}_{1/1}$$

$$= \underset{\Delta t \to 0}{\text{Limit}} \frac{\{\Delta V_P\}_{1/1}}{\Delta t} = \underset{\Delta t \to 0}{\text{Limit}} \begin{bmatrix} \Delta \dot{x}_1/\Delta t \\ \Delta \dot{y}_1/\Delta t \\ \Delta \dot{z}_1/\Delta t \end{bmatrix} = \begin{bmatrix} \ddot{x}_1 \\ \ddot{y}_1 \\ \ddot{z}_1 \end{bmatrix} \quad (1.30)$$

If the change in position of P is measured with respect to another frame, say frame 2, which is moving relative to frame 1, then by considering the situation shown in Fig. 1.9, drawn two dimensionally for ease of illustration,

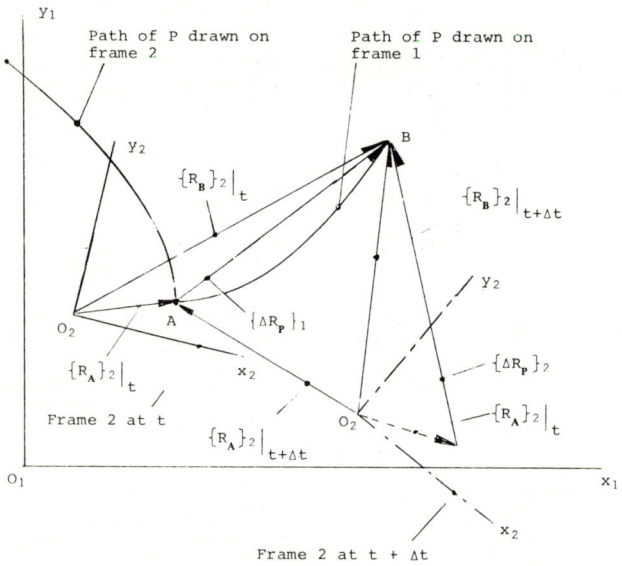

Fig. 1.9.

$$\{\Delta R_P\}_2 = \{R_P\}_2\big|_{t+\Delta t} - \{R_P\}_2\big|_t$$

$$= \{R_B\}_2\big|_{t+\Delta t} - \{R_A\}_2\big|_t , \quad (1.31)$$

$$\{\Delta R_P\}_1 = \{R_B\}_2\big|_t - \{R_A\}_2\big|_t$$

$$= \{R_B\}_2\big|_{t+\Delta t} - \{R_A\}_2\big|_{t+\Delta t}$$

Kinematics

and since
$$\{R_A\}_2\Big|_t \neq \{R_A\}_2\Big|_{t+\Delta t},$$
$$\{\Delta R_P\}_1 \neq \{\Delta R_P\}_2 .$$

Therefore, in general
$$\{V_P\}_1 \neq \{V_P\}_2 .$$

The vector
$$\{\Delta R_P\}_1$$
can be referred to frame 2 so that
$$\{\Delta R_P\}_{1/2} = [\ell_1]_2 \{\Delta R_P\}_{1/1} = \begin{bmatrix} \Delta x_2 \\ \Delta y_2 \\ \Delta z_2 \end{bmatrix} \quad (1.32)$$

Hence
$$\{V_P\}_{1/2}$$
can be defined as
$$\{V_P\}_{1/2} = \underset{\Delta t \to 0}{\text{Limit}} \frac{\{\Delta R_P\}_{1/2}}{\Delta t} = \underset{\Delta t \to 0}{\text{Limit}} \begin{bmatrix} \Delta x_2/\Delta t \\ \Delta y_2/\Delta t \\ \Delta z_2/\Delta t \end{bmatrix}$$

$$= [\ell_1]_2 \underset{\Delta t \to 0}{\text{Limit}} \frac{\{\Delta R_P\}_{1/1}}{\Delta t} = [\ell_1]_2 \{V_P\}_{1/1} \quad (1.33)$$

Now
$$\{V_P\}_{1/2}$$
can also be written
$$\{\dot{R}_P\}_{1/2} = [\ell_1]_2 \{\dot{R}_P\}_{1/1} = [\ell_1]_2 \frac{d}{dt}\{R_P\}_{1/1} = [\ell_1]_2 \{V_P\}_{1/1},$$

but it is important to note that while
$$\{\dot{R}_P\}_{1/1} = \frac{d}{dt}\{R_P\}_{1/1} , \quad \{\dot{R}_P\}_{1/2} \neq \frac{d}{dt}\{R_P\}_{1/2} .$$

In general, a vector specified by a column matrix has its time derivative determined by differentiating each element of the matrix with respect to time only if it is referred to the frame used for its measurement. Thus
$$\frac{d}{dt}\{B\}_{n/n} = \{\dot{B}\}_{n/n} \quad (1.34)$$

where the dot is taken as an instruction to differentiate each term in the $\{B\}_{n/n}$ matrix only when the suffixes are the same. The vector $\{\dot{B}\}_{1/2}$ is not obtained by differentiating each term of $\{B\}_{1/2}$ with respect to time since $\{\dot{B}\}_{1/2}$ is defined by

$$\{\dot{B}\}_{1/2} = [\ell_1]_2 \{\dot{B}\}_{1/1} .$$

The relationship between

$$\{\dot{B}\}_{1/2} \text{ and } \frac{d}{dt}\{B\}_{1/2}$$

is readily determined by differentiating the relationship

$$\{B\}_{1/2} = [\ell_1]_2 \{B\}_{1/1} .$$

Thus

$$\frac{d}{dt}\{B\}_{1/2} = \left[\frac{d}{dt}[\ell_1]_2\right]\{B\}_{1/1} + [\ell_1]_2 \{\dot{B}\}_{1/1}$$

$$\left[\frac{d}{dt}[\ell_1]_2\right]\{B\}_{1/1} + \{\dot{B}\}_{1/2} \qquad (1.35)$$

and therefore only if the first term on the right hand side of Eq. 1.35 is a null matrix will

$$\{\dot{B}\}_{1/2}$$

be equal to

$$\frac{d}{dt}\{B\}_{1/2} .$$

Notably this is true for certain descriptions of the angular velocity vector by virtue of the relationship between the derivative of the rotation matrix and angular velocity.(See section 1.9).

Refer to Problem 3.3 and Problems 3.26 to 3.29.

1.8. Small Rotations

A body, body 4, moves relative to body 1 such that frame 4 moves from alignment with frame 2, which is fixed relative to frame 1, to alignment with frame 3 which is also fixed relative to frame 1. It is always possible to move frame 4 from alignment with frame 2 to alignment with frame 3 by a maximum of three simple rotations such as

2 α_1 about x_2 i β_1 about y_i j γ_1 about z_j 3

where frames i and j are frames intermediate between frames 2 and 3.

Kinematics

Thus

$$[\ell_3]_2 = [\ell_i]_2 [\ell_j]_i [\ell_3]_j$$

and if the angles α_1, β_1 and γ_1 are small, then

$$[\ell_3]_2 = \begin{bmatrix} 1 & 0 & 0 \\ 0 & 1 & -\alpha_1 \\ 0 & \alpha_1 & 1 \end{bmatrix} \begin{bmatrix} 1 & 0 & \beta_1 \\ 0 & 1 & 0 \\ -\beta_1 & 0 & 1 \end{bmatrix} \begin{bmatrix} 1 & -\gamma_1 & 0 \\ \gamma_1 & 1 & 0 \\ 0 & 0 & 1 \end{bmatrix}$$

$$= \begin{bmatrix} 1 & -\gamma_1 & \beta_1 \\ \gamma_1 - \alpha_1\beta_1 & 1 - \alpha_1\beta_1\gamma_1 & -\alpha_1 \\ \alpha_1\gamma_1 - \beta_1 & \alpha_1 + \gamma_1\beta_1 & 1 \end{bmatrix}.$$

When second and higher order products are neglected

$$[\ell_3]_2 = \begin{bmatrix} 1 & -\gamma_1 & \beta_1 \\ \gamma_1 & 1 & -\alpha_1 \\ -\beta_1 & \alpha_1 & 1 \end{bmatrix}.$$

An alternative combination of rotations could be

2 α_2 about x_2 i γ_2 about z_i j β_2 about y_j 3

in which case

$$[\ell_3]_2 = \begin{bmatrix} 1 & 0 & 0 \\ 0 & 1 & -\alpha_2 \\ 0 & \alpha_2 & 1 \end{bmatrix} \begin{bmatrix} 1 & -\gamma_2 & 0 \\ \gamma_2 & 1 & 0 \\ 0 & 0 & 1 \end{bmatrix} \begin{bmatrix} 1 & 0 & \beta_2 \\ 0 & 1 & 0 \\ -\beta_2 & 0 & 1 \end{bmatrix}$$

$$= \begin{bmatrix} 1 & -\gamma_2 & \beta_2 \\ \gamma_2 + \alpha_2\beta_2 & 1 - \alpha_2\beta_2\gamma_2 & -\alpha_2 \\ \alpha_2\gamma_2 - \beta_2 & \alpha_2 - \gamma_2\beta_2 & 1 \end{bmatrix}$$

and when second and higher order products are neglected

$$[\ell_3]_2 = \begin{bmatrix} 1 & -\gamma_2 & \beta_2 \\ \gamma_2 & 1 & -\alpha_2 \\ -\beta_2 & \alpha_2 & 1 \end{bmatrix}.$$

Examination of the other four possible combinations of rotations would show that the form of $[\ell_3]_2$ is the same as the two previous cases and therefore independent of the order of 'addition' of the small rotations. It is therefore possible to write

$$[\ell_3]_2 = \begin{bmatrix} 1 & -\gamma & \beta \\ \gamma & 1 & -\alpha \\ -\beta & \alpha & 1 \end{bmatrix}.$$

This vector like property can be used to treat small rotations like vectors. Thus

$$\{\Delta\theta_4\}_2$$

can be regarded as a vector of magnitude

$$|\Delta\theta_4|_2 = \sqrt{(\alpha^2 + \beta^2 + \gamma^2)}$$

which is the angle through which body 4 rotates when frame 4 moves from alignment with frame 2 to alignment with frame 3, the angle being measured with respect to frame 2. The column matrix representation of this vector is thus

$$\{\Delta\theta_4\}_{2/2} = \begin{bmatrix} \alpha \\ \beta \\ \gamma \end{bmatrix} \qquad (1.36)$$

but since frame 2 is fixed relative to frame 1, $\{\Delta\theta_4\}_2$ is equally well $\{\Delta\theta_4\}_1$ and therefore

$$\{\Delta\theta_4\}_{1/2} = \begin{bmatrix} \alpha \\ \beta \\ \gamma \end{bmatrix}. \qquad (1.37)$$

1.9. Angular Velocity and the Derivative of the Rotation Matrix

Consider the motion of body 4 relative to a reference body 1. Let A and B be points fixed in body 4. Then

$$\{R_{BA}\}_{1/1} = \{R_{BA}\}_{4/1} = [\ell_4]_1 \{R_{BA}\}_{4/4} .$$

Let frames 2 and 3, fixed relative to frame 1, be arranged such that frame 4 (fixed in body 4) moves from alignment with frame 2 at time t to alignment with frame 3 at time $t + \Delta t$. At time t

$$[\ell_4]_1 = [\ell_2]_1 ,$$

and

$$\{R_{BA}\}_{4/4} = \{R_{BA}\}_{4/2}$$

$$\{R_{BA}\}_{1/1} = [\ell_2]_1 \{R_{BA}\}_{4/4} .$$

Similarly, at time $t + \Delta t$

$$\{R_{BA}\}_{1/1} = [\ell_3]_1 \{R_{BA}\}_{4/4} .$$

Now

$$\{\Delta R_{BA}\}_{1/1} = \{R_{BA}\}_{1/1}\Big|_{t+\Delta t} - \{R_{BA}\}_{1/1}\Big|_{t}$$

$$= \left[[\ell_3]_1 - [\ell_2]_1\right]\{R_{BA}\}_{4/4} .$$

Kinematics

The change in angular position of frame 4 which occurs in time Δt is

$$[\Delta \ell_4]_1 = [\ell_4]_1 \Big|_{t+\Delta t} - [\ell_4]_1 \Big|_t = [\ell_3]_1 - [\ell_2]_1$$

and since

$$[\ell_3]_1 = [\ell_2]_1 [\ell_3]_2$$

$$[\Delta \ell_4]_1 = [\ell_2]_1 \Big[[\ell_3]_2 - [1]\Big].$$

Now

$$[\ell_3]_2 - [1] = \begin{bmatrix} 0 & -\gamma & \beta \\ \gamma & 0 & -\alpha \\ -\beta & \alpha & 0 \end{bmatrix}$$

which is a skew-symmetric matrix which can be formed from the column matrix of Eq. 1.37. It is thus possible to define a skew-symmetric form of the angular rotation matrix as

$$[\Delta \theta_4]_{1/2} = \begin{bmatrix} 0 & -\gamma & \beta \\ \gamma & 0 & -\alpha \\ -\beta & \alpha & 0 \end{bmatrix} \qquad (1.38)$$

and therefore

$$[\Delta \ell_4]_1 = [\ell_2]_1 [\Delta \theta_4]_{1/2}.$$

Hence, by reference to the footnotes on pages 22 and 23, † which show how the transformation of the skew-symmetric form of a column matrix is effected,

$$[\Delta \theta_4]_{1/2} = [\ell_1]_2 [\Delta \theta_4]_{1/1} [\ell_1]_2^T$$

so that

$$[\Delta \ell_4]_1 = [\ell_2]_1 [\ell_1]_2 [\Delta \theta_4]_{1/1} [\ell_1]_2^T = [\Delta \theta_4]_{1/1} [\ell_2]_1. \qquad (1.39)$$

The angular velocity of body 4 relative to body 1 is defined by

$$\{\omega_4\}_1 = \underset{\Delta t \to 0}{\text{Limit}} \frac{\{\Delta \theta_4\}_1}{\Delta t} \qquad (1.40)$$

and therefore

$$[\dot{\ell}_4]_1 = \left[\underset{\Delta t \to 0}{\text{Limit}} \frac{[\Delta \theta_4]_{1/1}}{\Delta t}\right] [\ell_2]_1 = [\omega_4]_{1/1} [\ell_2]_1$$

$$= [\omega_4]_{1/1} [\ell_4]_1. \qquad (1.41)$$

Thus, in general

$$\frac{d}{dt}[\ell_m]_n = [\dot{\ell}_m]_n = [\omega_m]_{n/n}[\ell_m]_n . \qquad (1.42)$$

Now
$$\{\Delta R_{BA}\}_{1/1} = [\Delta \ell_4]_1 \{R_{BA}\}_{4/4} = [\Delta \theta_4]_{1/1}[\ell_2]_1 \{R_{BA}\}_{4/4}$$
so that
$$\{V_{BA}\}_{1/1} = \underset{\Delta t \to 0}{\text{Limit}} \frac{[\Delta \theta_4]_{1/1} [\ell_2]_1 \{R_{BA}\}_{4/4}}{\Delta t}$$
$$= [\omega_4]_{1/1}[\ell_2]_1 \{R_{BA}\}_{4/2} = [\omega_4]_{1/1}\{R_{BA}\}_{1/1}. \qquad (1.43)$$

If
$$\{\omega_4\}_{1/1} = \begin{bmatrix} \omega_x \\ \omega_y \\ \omega_z \end{bmatrix} \quad \text{and} \quad \{R_{BA}\}_{1/1} = \begin{bmatrix} x \\ y \\ z \end{bmatrix} ,$$
then
$$\{V_{BA}\}_{1/1} = \begin{bmatrix} 0 & -\omega_z & \omega_y \\ \omega_z & 0 & -\omega_x \\ -\omega_y & \omega_x & 0 \end{bmatrix} \begin{bmatrix} x \\ y \\ z \end{bmatrix} = \begin{bmatrix} z\omega_y - y\omega_z \\ x\omega_z - z\omega_x \\ y\omega_x - x\omega_y \end{bmatrix} .$$

Vectors corresponding to the above statements are shown in Fig. 1.10.

---

† The relationship between the transpose and inverse of the transformation matrix

Now $\{B\}_{n/1} = [\ell_2]_1 \{B\}_{n/2}$ and since $|\{B\}_{n/1}| = |\{B\}_{n/2}|$,

$$\{B\}^T_{n/1}\{B\}_{n/1} = \{B\}^T_{n/2}\{B\}_{n/2} .$$

Also
$$\{B\}_{n/1} = [\ell_2]_1 \{B\}_{n/2} \quad \text{and} \quad \{B\}^T_{n/1} = \{B\}^T_{n/2}[\ell_2]^T_1 .$$

Hence
$$\{B\}^T_{n/2}[\ell_2]^T_1[\ell_2]_1\{B\}_{n/2} = \{B\}^T_{n/2}\{B\}_{n/2}$$

which requires that
$$[\ell_2]^T_1[\ell_2]_1 = [1] \quad \text{or} \quad [\ell_2]^T_1 = [\ell_2]^{-1}_1 .$$

1.10. The Relative Velocity of Points Fixed in a Rigid Body

Consider two points A and B fixed in body 2 which is moving relative to body 1. The velocity of B relative to A, measured in frame 1 and referred to frame 1 is given by

$$\{V_{BA}\}_{1/1} = [\omega_2]_{1/1}\{R_{BA}\}_{1/1} \tag{1.44}$$

as shown in section 1.9 and the relative disposition of the vectors corresponding to this statement is illustrated in Fig. 1.11. Note particularly that the relative velocity vector

$$\{V_{BA}\}_1$$

is perpendicular to the plane containing the vectors $\{\omega_2\}_1$ and $\{R_{BA}\}$. If

$$|V_{BA}|_1 = v, \quad |\omega_2|_1 = \omega \quad \text{and} \quad |R_{BA}| = r$$

then

$$v = (\omega \sin\theta) r \quad \text{or} \quad v = \omega(r \sin\theta)$$

as can be seen from Fig. 1.12.

The component of angular velocity along AB does not contribute towards

$$\{V_{BA}\}_1$$

and cannot therefore be found from $\{V_{BA}\}_1$. If $\{\omega_2\}_1$ is to be found,

---

† The transformation of the skew-symmetric form of a matrix

Consider the vector product of the vectors described by $\{B\}_{n/1}$ and $\{C\}_{n/1}$

$$\{A\}_{n/1} = [B]_{n/1}\{C\}_{n/1}.$$

Therefore

$$[\ell_1]_2\{A\}_{n/1} = [\ell_1]_2[B]_{n/1}[\ell_1]_2^T[\ell_1]_2\{C\}_{n/1}$$

or

$$\{A\}_{n/2} = [\ell_1]_2[B]_{n/1}[\ell_1]_2^T\{C\}_{n/2}$$

which requires that

$$[B]_{n/2} = [\ell_1]_2[B]_{n/1}[\ell_1]_2^T.$$

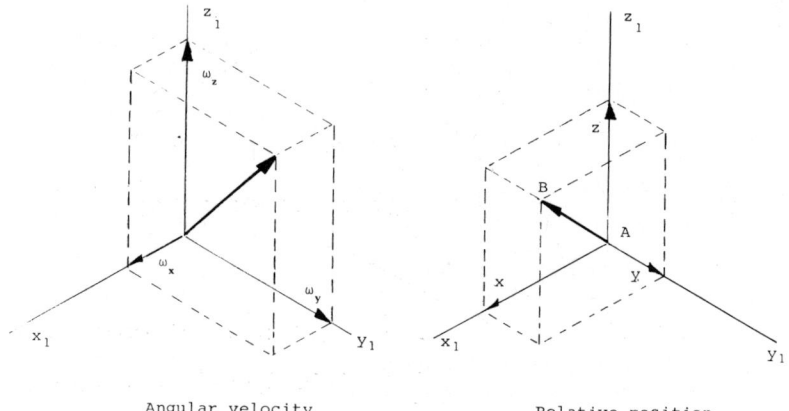

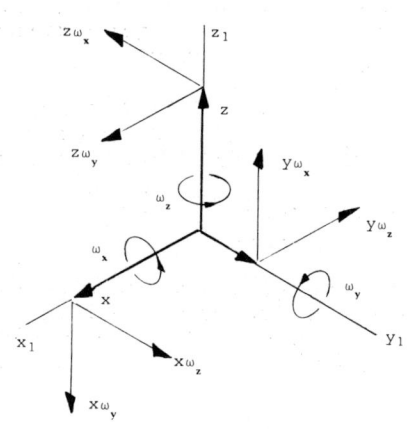

Relative velocity

Fig. 1.10.

and not simply its component perpendicular to AB, then more information about the motion must be provided, for example the component of angular velocity along AB.

Consider the problem of finding $\{\omega_2\}_{1/1}$ given $\{V_{BA}\}_{1/1}$ and $\{R_{BA}\}_{1/1}$. Write

$$\{V_{BA}\}_{1/1} = \{V\}, \quad \{\omega_2\}_{1/1} = \{\omega\} \text{ and } \quad \{R_{BA}\}_{1/1} = \{R\}$$

for economy of space and effort. Now

$$\{V\} = [\omega]\{R\} \tag{1.45}$$

and if this equation is premultiplied by $\lfloor R \rfloor$, then

Kinematics

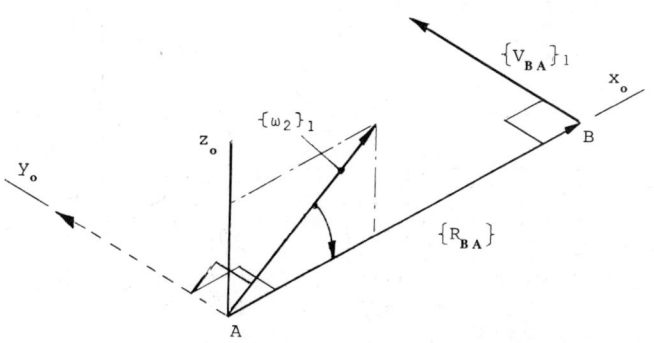

Fig. 1.11.

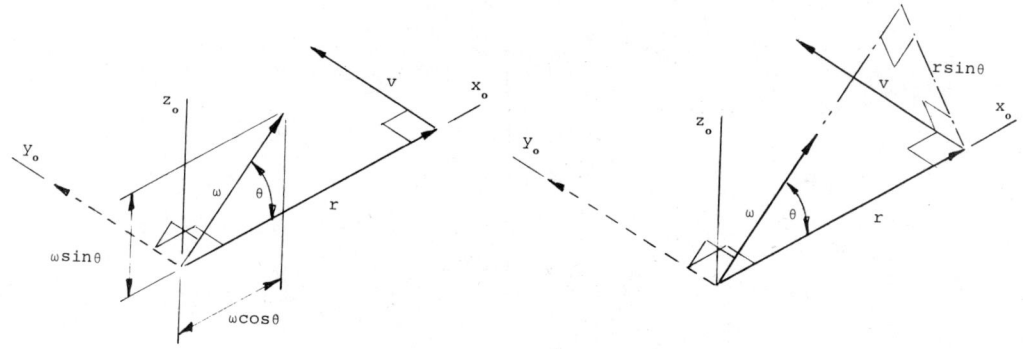

Fig. 1.12.

$$[R]\{V\} = [R][\omega]\{R\}.$$

The right hand side of this equation can be expanded using the relationship

$$[A][B]\{C\} = \{B\}(\{A\}^T\{C\}) - \{C\}(\{A\}^T\{B\})$$

to give

$$[R]\{V\} = \{\omega\}(\{R\}^T\{R\}) - \{R\}(\{R\}^T\{\omega\})$$

and therefore

$$\{\omega\} = \frac{[R]\{V\}}{|R|^2} + \frac{(\{R\}^T\{\omega\})}{|R|^2}\{R\}$$

$$= \{\omega^n\} + \{\omega^p\} \tag{1.46}$$

where $\{\omega^n\}$ is the component of $\{\omega\}$ normal to the plane containing $\{R\}$ and $\{V\}$ which is determinate, while $\{\omega^p\}$ is that component of parallel to $\{R\}$. $\{\omega^p\}$ depends upon $\{\omega\}$, which is to be determined, and is therefore indeterminate from a knowledge of $\{R\}$ and $\{V\}$ alone. The vector $\{\omega^n\}$ is illustrated in Fig. 1.13.

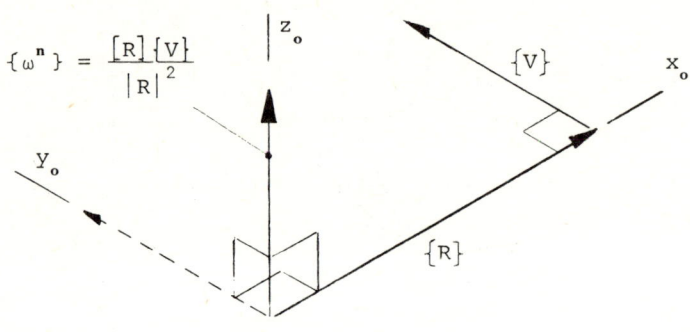

Fig. 1.13.

The reader should appreciate that it is not possible to find $\{\omega\}$ from Eq. 1.45 as follows

$$\{V\} = -[R]\{\omega\}$$

$$-[R]^{-1}\{V\} = \{\omega\}$$

since $[R]$ is singular, its determinant being zero.

If both sides of Eq. 1.45 are premultiplied by $[\omega]$, then it is easy to show that

$$\{R\} = \frac{[\omega]\{V\}}{|\omega|^2} - \frac{(\{\omega\}^T\{R\})}{|\omega|^2}\{\omega\} \tag{1.47}$$

and therefore the component of $\{R\}$ normal to the plane containing $\{\omega\}$ and $\{V\}$ is defined, but not that parallel to $\{\omega\}$.

Refer to Problems 3.4 and 3.5 and Problems 3.30 and 3.31.

1.11. The Central and Instantaneous Axes

Let P be a point in or attached to body 2 which is such that its velocity, measured with respect to body 1, is parallel to the angular velocity vector $\{\omega_2\}_1$. Such a point is said to be on the central axis for the motion of body 2 relative to body 1.

Kinematics

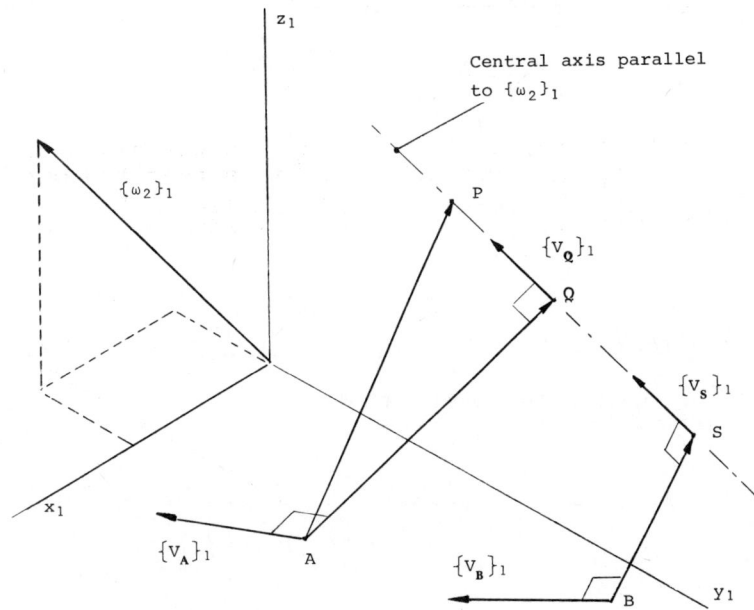

Fig. 1.14.

If A is a point fixed in body 2 which has a known velocity, then

$$\{V_P\}_{1/1} = \{V_A\}_{1/1} + \{V_{PA}\}_{1/1}$$

or

$$\{V_P\}_{1/1} = \{V_A\}_{1/1} + [\omega_2]_{1/1}\{R_{PA}\}_{1/1}$$

which, for the immediate purpose can be abbreviated to

$$\{V_P\} = \{V_A\} + [\omega]\{R_{PA}\}$$

for economy of effort. Since $\{\omega\}$ and $\{V_P\}$ are parallel

$$[\omega]\{V_P\} = \{0\} = [\omega]\{V_A\} + [\omega][\omega]\{R_{PA}\}$$

and therefore, using the expansion for a vector triple product given in Section 1.10,

$$\{0\} = [\omega]\{V_A\} + (\{\omega\}^T\{R_{PA}\})\{\omega\} - (\{\omega\}^T\{\omega\})\{R_{PA}\}$$

giving

$$\{R_{PA}\} = \frac{[\omega]\{V_A\}}{|\omega|^2} + \frac{(\{\omega\}^T\{R_{PA}\})}{|\omega|^2}\{\omega\} \ .$$

Thus, the position of P relative to A can be considered to be made up of two parts as shown in Fig. 1.14. Introduce a point Q on the central axis such that

$$\{R_{QA}\}_{1/1} = \frac{[\omega_2]_{1/1}\{V_A\}_{1/1}}{|\omega_2|_1^2} \cdot \qquad (1.48)$$

QA is perpendicular to both $\{V_A\}_1$ and the central axis, and the vector $\{R_{QA}\}$ is determinate. The vector

$$\{R_{PQ}\}_{1/1} = \frac{(\{\omega_2\}_{1/1}^T\{R_{PA}\}_{1/1})}{|\omega_2|_1^2}\{\omega_2\}_{1/1} \qquad (1.49)$$

is parallel to the angular velocity vector and indeterminate.

The velocity of Q, and therefore that of P, or any point on the central axis, can be found from

$$\{V_Q\}_{1/1} = \{V_A\}_{1/1} + \{V_{QA}\}_{1/1} = \{V_A\}_{1/1} + [\omega_2]_{1/1}\{R_{QA}\}_{1/1}$$

$$= \{V_A\}_{1/1} + \frac{[\omega_2]_{1/1}[\omega_2]_{1/1}\{V_A\}_{1/1}}{|\omega_2|_1^2} \cdot \qquad (1.50)$$

Using the expansion for the vector triple product given in Section 1.10 to expand the second term on the right hand side of this equation

$$\{V_Q\} = \{V_A\} + \frac{(\{\omega\}^T\{V_A\})}{|\omega|^2}\{\omega\} - \frac{(\{\omega\}^T\{\omega\})}{|\omega|^2}\{V_A\}$$

and since the last term on the right hand side of this equation is simply $\{V_A\}$

$$\{V_Q\}_{1/1} = \frac{(\{\omega_2\}_{1/1}^T\{V_A\}_{1/1})}{|\omega_2|_1^2}\{\omega_2\}_{1/1} \qquad (1.51)$$

For a similar point S relative to some point B fixed in body 2

$$\{V_S\}_{1/1} = \frac{(\{\omega_2\}_{1/1}^T\{V_B\}_{1/1})}{|\omega_2|_1^2}\{\omega_2\}_{1/1} \qquad (1.52)$$

and since

$$\{V_S\}_1 = \{V_Q\}_1 ,$$

$$\{\omega_2\}_{1/1}\{V_A\}_{1/1} = \{\omega_2\}_{1/1}\{V_B\}_{1/1} = \text{constant.} \qquad (1.53)$$

Hence the quantity

$$\frac{\{\omega_2\}_{1/1}\{V_A\}_{1/1}}{|\omega_2|_1^2}$$

is also a constant for the motion. It has the dimensions length per unit angle and can thus be though of as representing the distance through which the body would advance along the central axis per unit angle of rotation if the instantaneous motion persisted.

If the instantaneous motion of the central axis is zero, then the central axis is called the instantaneous axis.

If two bodies having relative motion have points in them, or attached to them, which have no relative motion at all times, then such points are on the permanent axis for the relative motion of the bodies.

A determination of the position of the central axis, instantaneous axis for the relative motion of two bodies allows the direction of the vector representing their relative angular velocity to be determined.

Refer to Problem 3.6 and Problems 3.32 and 3.33.

1.12. The Relative Acceleration of Points Fixed in a Rigid Body

For points A and B fixed in a rigid body, body 2, moving relative to body 1

$$\{V_{BA}\}_{1/1} = [\omega_2]_{1/1}[\ell_2]_1\{R_{BA}\}_{2/2} \quad . \tag{1.54}$$

Differentiating to obtain the acceleration of B relative to A

$$\{A_{BA}\}_{1/1} = [\dot{\omega}_2]_{1/1}[\ell_2]_1\{R_{BA}\}_{2/2} +$$

$$[\omega_2]_{1/1}[\omega_2]_{1/1}[\ell_2]_1\{R_{BA}\}_{2/2} \tag{1.55}$$

since $\{R_{BA}\}_{2/2}$ is constant. Hence

$$\{A_{BA}\}_{1/1} = \left[[\dot{\omega}_2]_{1/1} + [\omega_2]_{1/1}^2\right]\{R_{BA}\}_{1/1} \tag{1.56}$$

and since

$$[\omega_2]_{1/1}\{R_{BA}\}_{1/1} = \{V_{BA}\}_{1/1}$$

the acceleration can also be written

$$\{A_{BA}\}_{1/1} = [\dot{\omega}_2]_{1/1}\{R_{BA}\}_{1/1} + [\omega_2]_{1/1}\{V_{BA}\}_{1/1} \tag{1.57}$$

The vector corresponding to

$$[\dot{\omega}_2]_{1/1}\{R_{BA}\}_{1/1}$$

is perpendicular to the plane containing $\{\dot{\omega}_2\}_1$ and $\{R_{BA}\}$. The vector corresponding to

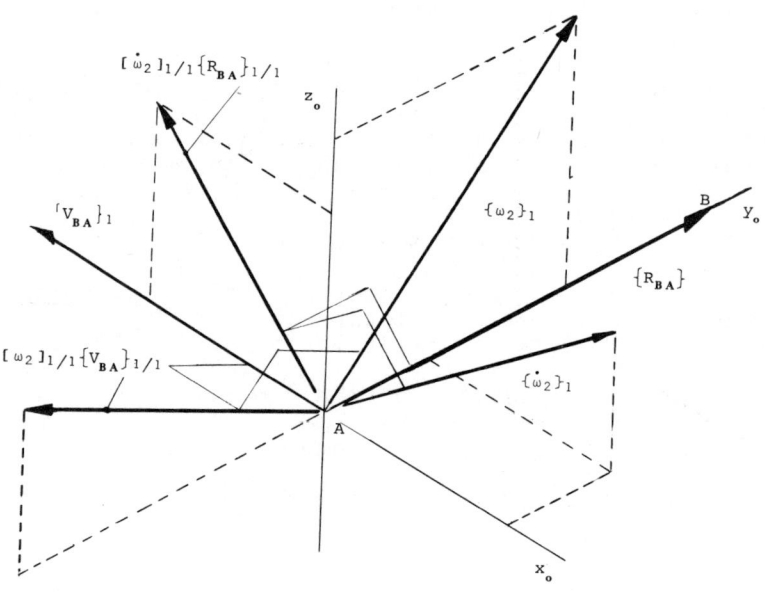

Fig. 1.15.

$$[\dot{\omega}_2]_{1/1}\{V_{BA}\}_{1/1}$$

is perpendicular to the plane containing $\{\omega_2\}_1$ and $\{R_{BA}\}$. These component accelerations are shown in Fig. 1.15.

Alternativley, if the equation

$$\{V_{BA}\}_{1/1} = [\ell_2]_1\{V_{BA}\}_{1/2}$$

is differentiated with respect to time, then

$$\{A_{BA}\}_{1/1} = [\omega_2]_{1/1}[\ell_2]_1\{V_{BA}\}_{1/2} + [\ell_2]_1\frac{d}{dt}\{V_{BA}\}_{1/2}$$

and

$$\{A_{BA}\}_{1/2} = [\omega_2]_{1/2}\{V_{BA}\}_{1/2} + \frac{d}{dt}\{V_{BA}\}_{1/2} \quad . \tag{1.58}$$

Also, the acceleration of B relative to A can conveniently be resolved inot two comonents, one parallel to BA, the other normal to BA. Thus

$$\{A_{BA}\}_{1/1} = \{A^p_{BA}\}_{1/1} + \{A^n_{BA}\}_{1/1} \tag{1.59}$$

as shown in Fig.1.16. The magnitude of $\{A_{BA}\}_1$ parallel to $\{R_{BA}\}$ is

Kinematics

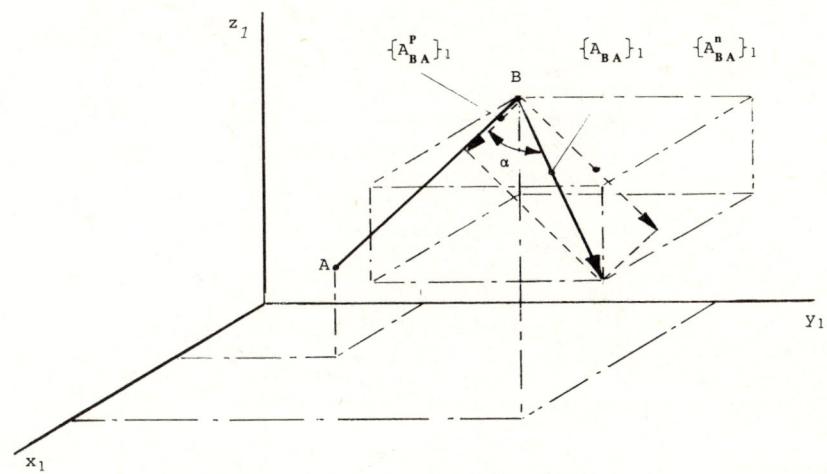

Fig. 1.16.

$$|A_{BA}^p|_1 = |A_{BA}|_1 \cos\alpha = \frac{|R_{BA}||A_{BA}|_1}{|R_{BA}|}\cos\alpha$$

$$= \frac{\{R_{BA}\}_{1/1}^T \{A_{BA}\}_{1/1}}{|R_{BA}|}$$

and therefore

$$\{A_{BA}^p\}_{1/1} = |A_{BA}^p|_1 \frac{\{R_{BA}\}_{1/1}}{|R_{BA}|} = \frac{\{R_{BA}\}_{1/1}^T \{A_{BA}\}_{1/1} \{R_{BA}\}_{1/1}}{|R_{BA}|^2}$$

$$= \frac{\{R_{BA}\}_{1/1}^T}{|R_{BA}|^2} \left\{ \left[[\dot{\omega}_2]_{1/1} + [\omega_2]_{1/1}^2 \right] \{R_{BA}\}_{1/1} \right\} \{R_{BA}\}_{1/1}.$$

As $\{R_{BA}\}$ is perpendicular to the vector corresponding to

$$[\dot{\omega}_2]_{1/1}\{R_{BA}\}_{1/1}, \quad \{R_{BA}\}_{1/1}^T \left\{ [\dot{\omega}_2]_{1/1}\{R_{BA}\}_{1/1} \right\}$$

is zero. Hence

$$\{A_{BA}^p\}_{1/1} = \frac{(\{R_{BA}\}_{1/1}^T [\omega_2]_{1/1} [\omega_2]_{1/1} \{R_{BA}\}_{1/1}) \{R_{BA}\}_{1/1}}{|R_{BA}|^2}$$

$$= \frac{(-\{R_{BA}\}^T_{1/1}[\omega_2]^T_{1/1}\{V_{BA}\}_{1/1})\{R_{BA}\}_{1/1}}{|R_{BA}|^2}$$

$$= \frac{(-\{[\omega_2]_{1/1}\{R_{BA}\}_{1/1}\}^T\{V_{BA}\}_{1/1})\{R_{BA}\}_{1/1}}{|R_{BA}|^2}$$

giving

$$\{A^P_{BA}\}_{1/1} = -\frac{|V_{BA}|^2_1\{R_{BA}\}_{1/1}}{|R_{BA}|^2} \qquad (1.60)$$

as shown in Fig. 1.17.

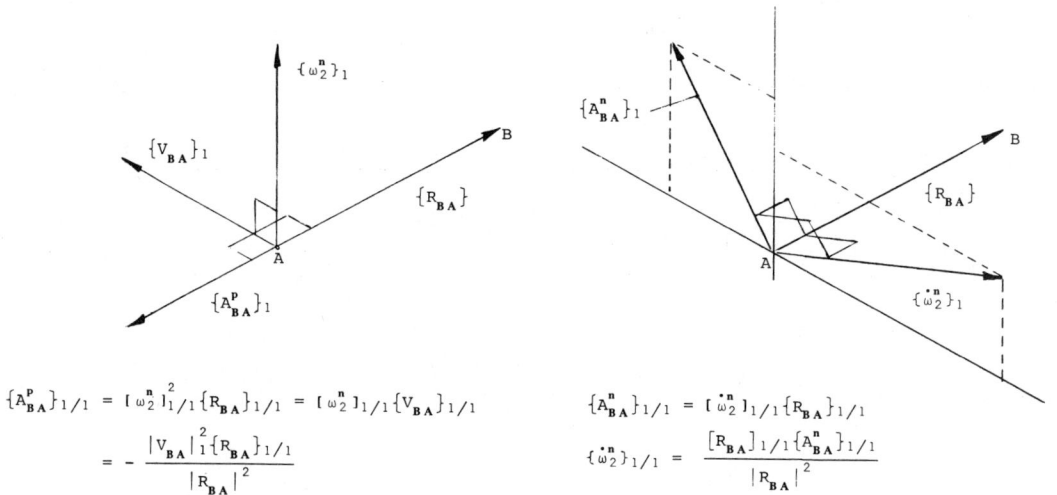

Fig. 1.17.

Also, if

$$\{A^n_{BA}\}_{1/1} = \begin{bmatrix} r_2 \\ s_2 \\ t_2 \end{bmatrix} \quad \text{and} \quad \{R_{BA}\}_{1/1} = \begin{bmatrix} x_2 \\ y_2 \\ z_2 \end{bmatrix},$$

then since the vectors are at right angles to each other

$$\{R_{BA}\}^T_{1/1}\{A^n_{BA}\}_{1/1} = 0 \quad \text{or} \quad x_2 r_2 + y_2 s_2 + z_2 t_2 = 0. \qquad (1.61)$$

The vectors $\{\omega_2\}_1$ and $\{\dot{\omega}_2\}_1$ can be resolved parallel and normal to $\{R_{BA}\}$ so that

Kinematics

$$\{\omega_2\}_1 = \{\omega_2^n\}_1 + \{\omega_2^p\}_1 \tag{1.62}$$

and

$$\{\dot{\omega}_2\}_1 = \{\dot{\omega}_2^n\}_1 + \{\dot{\omega}_2^p\}_1 . \tag{1.63}$$

Thus, in cases where the constraints on the motion are such that the parallel components are zero,

$$\{A_{BA}^p\}_{1/1} = [\omega_2^n]_{1/1}\{V_{BA}\}_{1/1} = [\omega_2^n]^2_{1/1}\{R_{BA}\}_{1/1} \tag{1.64}$$

and

$$\{A_{BA}^n\}_{1/1} = [\dot{\omega}_2^n]_{1/1}\{R_{BA}\}_{1/1} \tag{1.65}$$

or

$$\{\dot{\omega}_2^n\}_{1/1} = \frac{|R_{BA}|_{1/1}\{A_{BA}^n\}_{1/1}}{|R_{BA}|^2} \tag{1.66}$$

as shown in Fig.1.17

1.13. The Relative Acceleration of Coincident Points Which Have Relative Motion

Consider the motion of two bodies, bodies 2 and 3, which are moving relative to a reference body 1, and constrained such that

$$\{\omega_2\}_1 = \{\omega_3\}_1$$

Let A_2 and A_3 be convenient *coincident* points on bodies 2 and 3 respectively which have relative motion. Then

$$\{R_{A_3A_3}\}_{1/1} = \{R_{A_3A_2}\}_{2/1} = [\ell_2]_1\{R_{A_3A_2}\}_{2/2} = \begin{bmatrix} 0 \\ 0 \\ 0 \end{bmatrix} \tag{1.67}$$

since the points are coincident. Differentiating Eq.1.67 with respect to time to obtain the relative velocity of these two coincident points gives

$$\{V_{A_3A_2}\}_{1/1} = [\omega_2]_{1/1}[\ell_2]_1\{R_{A_3A_2}\}_{2/2} + [\ell_2]_1\{\dot{R}_{A_3A_2}\}_{2/2}$$

$$= [\ell_2]_1\{V_{A_3A_2}\}_{2/2} = \{V_{A_3A_2}\}_{2/1} \tag{1.68}$$

since

$$\{R_{A_3A_2}\}_{2/2}$$

is a null matrix. Differentiating Eq. 1.68 with respect to time to obtain the relative acceleration gives

$$\{A_{A_3A_2}\}_{1/1} = [\omega_2]_{1/1}[\ell_2]_1\{V_{A_3A_2}\}_{2/2}$$
$$+ [\omega_2]_{1/1}[\ell_2]_1\{V_{A_3A_2}\}_{2/2} + [\ell_2]_1\{A_{A_3A_2}\}_{2/2}$$

and therefore

$$\{A_{A_3A_2}\}_{1/1} = 2[\omega_2]_{1/1}\{V_{A_3A_2}\}_{2/1} + \{A_{A_3A_2}\}_{2/1}. \qquad (1.69)$$

1.14. Differentiation of the Angular Velocity Vector, a Special Case

Consider the angular velocity vector represented by the matrix

$$\{\omega_2\}_{1/2}$$

which specifies the angular velocity of body 2 measured with respect to body 1 by referring it to frame 2 which is fixed in body 2. Now

$$\{\omega_2\}_{1/2} = [\ell_1]_2\{\omega_2\}_{1/1}$$

and therefore

$$\frac{d}{dt}\{\omega_2\}_{1/2} = [\omega_1]_{2/2}[\ell_1]_2\{\omega_2\}_{1/1} + [\ell_1]_2\{\dot{\omega}_2\}_{1/1}. \qquad (1.70)$$

Since

$$\{\omega_1\}_{2/2} = -\{\omega_2\}_{1/2}$$

Eq. 1.70 can be written

$$\frac{d}{dt}\{\omega_2\}_{1/2} = -[\omega_2]_{1/2}\{\omega_2\}_{1/2} + \{\dot{\omega}_2\}_{1/2}. \qquad (1.71)$$

The vector product

$$[\omega_2]_{1/2}\{\omega_2\}_{1/2}$$

is a null matrix, so that Eq. 1.71 reduces to

$$\{\dot{\omega}_2\}_{1/2} = \frac{d}{dt}\{\omega_2\}_{1/2}. \qquad (1.72)$$

Thus

$$\{\dot{\omega}_2\}_{1/2}$$

is unique in that it is determined by differentiating each element of

$$\{\omega_2\}_{1/2}$$

with respect to time. Note particularly that

$$\{\omega_2\}_{1/n} \neq \frac{d}{dt}\{\omega_2\}_{1/n}$$

unless frame n is fixed relative to frame 2.

Kinematics

1.15. Relative Angular Velocity

Consider, for example, the situation in which bodies 2, 3 and 4 have motion relative to body 1 and relative to each other. Then (Fig. 1.18a)

$$\{\omega_4\}_1 = \{\omega_2\}_1 + \{\omega_3\}_2 + \{\omega_4\}_3 \quad (1.73)$$

$$1 \longrightarrow 4 \quad 1 \longrightarrow 2 \quad 2 \longrightarrow 3 \quad 3 \longrightarrow 4$$

or (Fig.1.18b)

$$\{\omega_3\}_2 = \{\omega_4\}_2 + \{\omega_3\}_4 \quad (1.74)$$

$$2 \longrightarrow 3 \quad 2 \longrightarrow 4 \quad 4 \longrightarrow 3$$

or (Fig.1.18c)

$$\{\omega_3\}_1 = \{\omega_4\}_1 + \{\omega_2\}_4 + \{\omega_3\}_2 \quad (1.75)$$

$$1 \longrightarrow 3 \quad 1 \longrightarrow 4 \quad 4 \longrightarrow 2 \quad 2 \longrightarrow 3$$

and so on.

To write Eqs. 1.73, 1.74 and 1.75 in column matrix form requires that the vectors be referred to the same frame. If, for example, the vectors of Eq. 1.74 are referred to frame 3, then

$$\{\omega_3\}_{2/3} = \{\omega_4\}_{2/3} + \{\omega_3\}_{4/3} \quad (1.76)$$

or

$$\{\omega_3\}_{2/3} = \{\omega_4\}_{2/3} + [\ell_4]_3 \{\omega_3\}_{4/4} . \quad (1.77)$$

As might be expected, equations like Eq. 1.77 can be derived by differentiating the relationship between the appropriate rotation matrices. Take the example of deriving the column matrix form of Eq. 1.73. Now

$$[\ell_4]_1 = [\ell_2]_1 [\ell_3]_2 [\ell_4]_3$$

and on differentiation this becomes

$$[\omega_4]_{1/1} [\ell_4]_1 = [\omega_2]_{1/1} [\ell_4]_1 + [\ell_2]_1 [\omega_3]_{2/2} [\ell_4]_2$$

$$+ [\ell_3]_1 [\omega_4]_{3/3} [\ell_4]_3 .$$

Postmultiplying this equation by $[\ell_1]_4$ gives

$$[\omega_4]_{1/1} = [\omega_2]_{1/1} + [\ell_2]_1 [\omega_3]_{2/2} [\ell_4]_2 [\ell_1]_4$$

$$+ [\ell_3]_1 [\omega_4]_{3/3} [\ell_4]_3 [\ell_1]_4$$

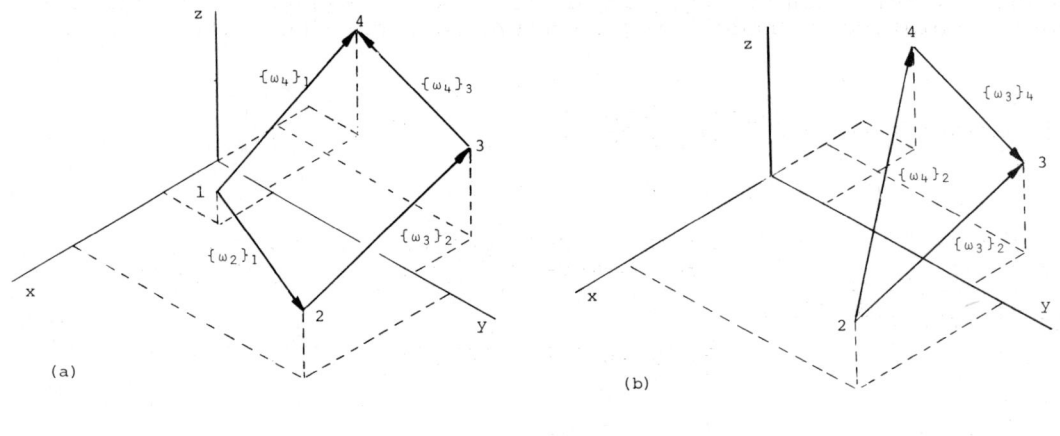

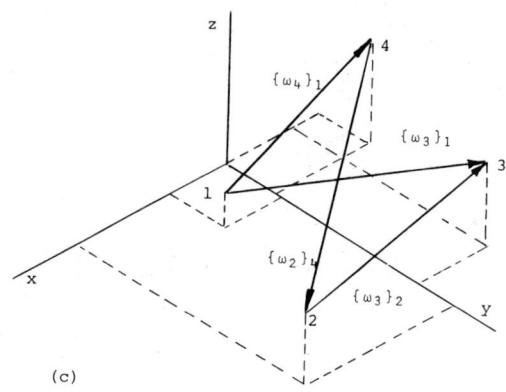

Fig. 1.18.

$$= [\omega_2]_{1/1} + [\ell_2]_1[\omega_3]_{2/2}[\ell_2]_1^T + [\ell_3]_1[\omega_4]_{3/3}[\ell_3]_1^T$$

$$= [\omega_2]_{1/1} + [\omega_3]_{2/1} + [\omega_4]_{3/1}$$

which, written in the column form rather than the skew symmetric form, becomes

$$\{\omega_4\}_{1/1} = \{\omega_2\}_{1/1} + \{\omega_3\}_{2/1} + \{\omega_4\}_{3/1}.$$

Similarly, the column vector form of Eq. 1.75 can be derived from

$$[\ell_3]_1 = [\ell_4]_1[\ell_2]_4[\ell_3]_2.$$

Kinematics

1.16. Relative Angular Acceleration

Consider, for example, the situation in which bodies 2 and 3 have motion relative to body 1 and relative to each other. Then

$$\{\omega_3\}_{1/1} = \{\omega_2\}_{1/1} + [\ell_2]_1\{\omega_3\}_{2/2} \tag{1.78}$$

Differentiating Eq. 1.78 gives

$$\{\dot{\omega}_3\}_{1/1} = \{\dot{\omega}_2\}_{1/1} + [\omega_2]_{1/1}[\ell_2]_1\{\omega_3\}_{2/2} + [\ell_2]_1\{\dot{\omega}_3\}_{2/2} \tag{1.79}$$

The middle term on the right hand side of this equation can be reduced using Eq. 1.78. Thus

$$[\omega_2]_{1/1}[\ell_2]_1\{\omega_3\}_{2/2} = [\omega_2]_{1/1}\{\{\omega_3\}_{1/1} - \{\omega_2\}_{1/1}\}$$

$$[\omega_2]_{1/1}\{\omega_3\}_{1/1}$$

since

$$[\omega_2]_{1/1}\{\omega_2\}_{1/1}$$

is a null matrix. Equation 1.79 can thus be written

$$\{\dot{\omega}_3\}_{1/1} = \{\dot{\omega}_2\}_{1/1} + [\omega_2]_{1/1}\{\omega_3\}_{1/1} + \{\dot{\omega}_3\}_{2/2}. \tag{1.80}$$

Refer to Problems 3.7 to 3.22 and Problems 3.34 to 3.51.

Chapter 2

Dynamics

2.1. Newton's Laws of Motion

Dynamics is concerned with the relationships between force, mass, energy and motion. For Engineering applications, except those dealing with nuclear and fast moving electron phenomena, the Newtonian model of mass, space, time and force is adequate.

Newton (1642 - 1727) in his "*Philosophiae Naturalis Principia Mathematica*" of 1687 enunciated three laws or axioms relating force and motion which can be stated as follows:

1 A particle will continue in a state of rest, or of uniform motion in a straight line, unless it is compelled to change that state by forces impressed upon it.

2 A change of motion with respect to time is proportional to the motive force impressed.

3 For every force acting on a particle, there is a corresponding force exerted by the particle. These forces are equal in magnitude, but opposite in direction.

The first law implies the existence of an inertial frame of reference. Consider the following hypothetical experiment. Erect a set of co-ordinate axes in deep space remote from any other matter and project a particle successively along each axis. If the axes are not accelerating and not rotating, then the force free motion will persist along the axis which it was projected. Such a set of axes is said to be inertial. No set of axes is truly inertial, but a set of axis fixed in the 'fixed' stars are very nearly inertial and must be used, for example, in space ballistics. For most Engineering applications forces can be predicted assuming that a reference frame fixed in the earth is inertial. In this text the inertial reference is always designated 1.

The "motion" of the second law is measured by the momentum of the particle, which is the product of its mass and inertial velocity. Thus, by Newton's second law

Dynamics

$$\{F_p\} \propto d\{m_p\{V_p\}_1\}/dt \propto d\{G_p\}_1/dt \tag{2.1}$$

where $\{F_p\}$ is the external force or impressed force on the particle of mass m_p and $\{V_p\}_1$ is the velocity of the particle measured with respect to an inertial reference frame. For a particle of constant mass, Eq. 2.1 becomes

$$\{F_p\} = km_p\{A_p\}_1 = kd\{G_p\}_1/dt \tag{2.2}$$

which, for consistent units, reduces to

$$\{F_p\} = m_p\{A_p\}_1 = d\{G_p\}_1/dt . \tag{2.3}$$

When the arbitrary units of mass and acceleration chosen are the kilogramme (kg) and the metre/second^2 (m s^{-2}), then the corresponding unit of force is the newton (N). Thus Eq. 2.3 is used to define force and requires that the force vector

$$\{F_p\}$$

be that vector which is identical to the mass-acceleration vector

$$m_p\{A_p\}_1 = d\{G_p\}_1/dt.$$

$\{F_p\}$ is the vector which describes the external force and requires no suffix outside the braces to specify the frame used for its measurement since this is implicit in the accleration which it produces. When Eq. 2.3 is expressed in its column matrix form, the frame to which the force vector is referred is specified by a single suffix outside the braces. Thus, when the vectors are referred, for example, to frame 3, the column matrix form of Eq. 2.3 would be written

$$\{F_p\}_3 = m_p\{A_p\}_{1/3} = d\{G_p\}_{1/3}/dt. \tag{2.4}$$

2.2. The Measurement of Force

While Eq. 2.3 defines force, it is not convenient to use it directly to calibrate force measuring devices or force transducers. The most accurate method is that of using the fact that the force of attraction of the earth on a mass m is mg, where g is the local gravitational accleration. The force mg can be applied to the transducer directly, or with less accuracy, through a system of levers.

Force meters take a wide variety of forms. At one end of the scale there is the dial test indicator type of device which can be used, for example, in the static calibration of materials testing machines, and at the other end of the scale is the piezo-electric force transducer with its sophisticated charge amplifier, capable of measuring force over a frequency range of zero hertz to several kilohertz and of measuring small changes of force in the presence of large mean forces.

2.3 Work, Potential and Kinetic Energy

Let a particle move from a point

$$\{R_P\}_{1/1} = \begin{bmatrix} x_A \\ y_A \\ z_A \end{bmatrix} \quad \text{to the point} \quad \{R_P\}_{1/1} = \begin{bmatrix} x_B \\ y_B \\ z_B \end{bmatrix}$$

under the action of a force

$$\{F_P\}_1 = \begin{bmatrix} F_x \\ F_y \\ F_z \end{bmatrix}.$$

The work done is given by

$$W_{A \to B} = \int_A^B \{F_P\}_1^T d\{R_P\}_{1/1} = \int_A^B (F_x\,dx + F_y\,dy + F_z\,dz) \tag{2.5}$$

where

$$d\{R_P\}_{1/1} = \begin{bmatrix} dx \\ dy \\ dz \end{bmatrix}.$$

If

$$\{F_P\}_1^T d\{R_P\}_{1/1} = -dV,$$

that is it is an exact differential, then

$$W_{A \to B} = \int_A^B -dV = -(V_B - V_A) \tag{2.6}$$

where $V = f(x, y, z)$.

Hence

$$dV = \frac{\partial V}{\partial x}\,dx + \frac{\partial V}{\partial y}\,dy + \frac{\partial V}{\partial z}\,dz$$

$$= -(F_x\,dx + F_y\,dy + F_z\,dz)$$

giving

$$F_x = -\frac{\partial V}{\partial x}, \quad F_y = -\frac{\partial V}{\partial y} \quad \text{and} \quad F_z = -\frac{\partial V}{\partial z} \tag{2.7}$$

When Eq. 2.6 holds, the work done is dependent only upon the position of the points A and B, that is it does not depend upon the path traced out by the point in moving from A to B. V is known as potential energy.

Dynamics

If a vector operator $\{\nabla\}_1$ (del) is defined by

$$\{\nabla\}_1 = \begin{bmatrix} \dfrac{\partial}{\partial x} \\ \dfrac{\partial}{\partial y} \\ \dfrac{\partial}{\partial z} \end{bmatrix}, \qquad (2.8)$$

then for the case in which the work done is not path dependent,

$$-\{F_p\}_1^T d\{R_p\}_{1/1} = dV = \{\nabla\}_1^T V d\{R_p\}_{1/1}$$

or

$$\{F_p\}_1 = -\{\nabla\}_1 V . \qquad (2.9)$$

If both sides of Eq. 2.9 are premultiplied by

$$[\nabla]_1 = \begin{bmatrix} 0 & -\dfrac{\partial}{\partial z} & \dfrac{\partial}{\partial y} \\ \dfrac{\partial}{\partial z} & 0 & -\dfrac{\partial}{\partial x} \\ -\dfrac{\partial}{\partial y} & \dfrac{\partial}{\partial x} & 0 \end{bmatrix}$$

then

$$[\nabla]_1 \{F_p\}_1 = -[\nabla]_1 \{\nabla\}_1 V. \qquad (2.10)$$

The right hand side of Eq. 2.10 is

$$\begin{bmatrix} 0 & \dfrac{\partial}{\partial z} & -\dfrac{\partial}{\partial y} \\ -\dfrac{\partial}{\partial z} & 0 & \dfrac{\partial}{\partial x} \\ \dfrac{\partial}{\partial y} & -\dfrac{\partial}{\partial x} & 0 \end{bmatrix} \begin{bmatrix} \dfrac{\partial V}{\partial x} \\ \dfrac{\partial V}{\partial y} \\ \dfrac{\partial V}{\partial z} \end{bmatrix} = \begin{bmatrix} \dfrac{\partial^2 V}{\partial z \partial y} - \dfrac{\partial^2 V}{\partial y \partial z} \\ -\dfrac{\partial^2 V}{\partial z \partial x} + \dfrac{\partial^2 V}{\partial z \partial z} \\ \dfrac{\partial^2 V}{\partial y \partial x} - \dfrac{\partial^2 V}{\partial x \partial y} \end{bmatrix} = \begin{bmatrix} -\dfrac{\partial F_y}{\partial z} + \dfrac{\partial F_z}{\partial y} \\ \dfrac{\partial F_x}{\partial z} - \dfrac{\partial F_y}{\partial x} \\ -\dfrac{\partial F_x}{\partial y} + \dfrac{\partial F_y}{\partial x} \end{bmatrix}$$

which is a null vector since

$$\dfrac{\partial^2 V}{\partial q_1 \partial q_2} = \dfrac{\partial^2 V}{\partial q_2 \partial q_1} .$$

Hence

$$[\nabla]_1 \{F_p\}_1 = 0 \qquad (2.11)$$

when the work done by $\{F_p\}_1$ is independent of the path which its point of application traces out.

As a slight digression, it is interesting to note that

$$[\nabla]\{B\} ,$$

where $\{B\}$ is any vector, is known as the curl or rotation (rot.) of $\{B\}$. The reason for the use of this term is somewhat obscure, but if the reader cares to find

$$[\nabla]_1\{V_{BA}\}_{1/1} = [\nabla]_1[\omega_2]_{1/1}\{V_{BA}\}_{1/1}$$

where $\{\omega_2\}_{1/1}$ is treated as a constant, then it will be found that

$$[\nabla]_1\{V_{BA}\}_{1/1} = 2\{\omega_2\}_{1/1},$$

when the reason for the use of the terms curl and rot will seem more understandable.

It is also of interest to note that the vector

$$\{\nabla\}_1\phi ,$$

where ϕ is any scalar, is known as the gradient of ϕ or grad ϕ.

Since

$$\{F_P\}_1^T = m_P\{A_P\}_{1/1}^T$$

Eq. 2.5 can be written

$$W_{A \to B} = m_P \int_A^B \{A_P\}_{1/1}^T \frac{d}{dt}\{R_P\}_{1/1} dt = m_P \int_A^B \{A_P\}_{1/1}^T \{V_P\}_{1/1} dt$$

$$= m_P \int_A^B \{V_P\}_{1/1}^T \{A_P\}_{1/1} dt = m_P \int_A^B \{V_P\}_{1/1}^T \frac{d}{dt}\{V_P\}_{1/1} dt$$

$$= m_P \int_A^B (v_x dv_x + v_y dv_y + v_z dv_z)$$

$$= \frac{m_P}{2}(v_x^2 + v_y^2 + v_z^2)\Big|_A^B = \frac{m_P}{2}\left\{|V_P|_1^2\Big|_B - |V_P|_1^2\Big|_A\right\}$$

$$= T_P|_B - T_P|_A \qquad (2.12)$$

where

$$T_P = \frac{m_P}{2}|V_P|_1^2 = \frac{m_P}{2}\{V_P\}_{1/1}^T\{V_P\}_{1/1} = \frac{m_P}{2}\{V_P\}_{1/3}^T\{V_P\}_{1/3} \qquad (2.13)$$

is the kinetic energy of the particle. Combining Eqs. 2.6 and 2.12 gives

Dynamics

$$W_{A \to B} = -(V_B - V_A) = T_P|_B - T_P|_A$$

or

$$T_P|_A + V_A = T_P|_B + V_B = \text{constant} \qquad (2.14)$$

when the reason for the introduction of the negative sign in Eq. 2.6 becomes clear. Thus, for the case in which potential energy is defined, that is Eq. 2.6 is valid, the sum of kinetic energy and potential energy is constant throughout the motion and the force $\{F_P\}$ is said to be conservative.

2.4. The Activity of a Force and its Relationship to the Rate of Change of Kinetic Energy

The activity, power or rate of working of a force is defined as the scalar product of the force and the velocity of its point of application. Hence, for the case in which $\{F_P\}$ acts on a particle of mass m_P

$$\begin{aligned}
\text{Activity} &= \{F_P\}_1^T \{V_P\}_{1/1} = m_P \{A_P\}_{1/1}^T \{V_P\}_{1/1} \\
&= m_P \{V_P\}_{1/1}^T \{A_P\}_{1/1} = m_P \{V_P\}_{1/1}^T \{\dot{V}_P\}_{1/1} \\
&= m_P \{V_P\}_{1/3}^T \{\dot{V}_P\}_{1/3} .
\end{aligned} \qquad (2.15)$$

Now

$$T_P = \frac{m_P}{2} \{V_P\}_{1/1}^T \{V_P\}_{1/1}$$

and

$$\begin{aligned}
\dot{T}_P &= \frac{m_P}{2} \left\{ \{\dot{V}_P\}_{1/1}^T \{V_P\}_{1/1} + \{V_P\}_{1/1}^T \{\dot{V}_P\}_{1/1} \right\} \\
&= m_P \{V_P\}_{1/1}^T \{\dot{V}_P\}_{1/1} .
\end{aligned} \qquad (2.16)$$

Hence

$$\begin{aligned}
\dot{T}_P &= m_P \{V_P\}_{1/1}^T \{\dot{V}_P\}_{1/1} = \{F_P\}_3^T \{V_P\}_{1/1} \\
&= m_P \{V_P\}_{1/3}^T \{\dot{V}_P\}_{1/3} = \{F_P\}_3^T \{V_P\}_{1/3} .
\end{aligned} \qquad (2.17)$$

2.5. Impulse and Momentum

Now

$$\{F_P\}_1 = m_P \frac{d}{dt} \{V_P\}_{1/1} = \frac{d}{dt} \{G_P\}_{1/1}$$

and therefore

$$\int_{t_2}^{t_1} \{F_P\}_1 dt = m_P \left\{ \{V_P\}_{1/1}\big|_{t_2} - \{V_P\}_{1/1}\big|_{t_1} \right\}$$

$$= \{G_P\}_{1/1}\big|_{t_2} - \{G_P\}_{1/1}\big|_{t_1} \qquad (2.18)$$

where the time integral of the force $\{F_P\}$ is known as the impulse of the force.

2.6. Centre of Mass

If a rigid body, body 2, is composed of n particles of mass m_i, then the total mass of the body is given by

$$m_2 = \sum_1^n m_i . \qquad (2.19)$$

Also, if the body is positioned relative to some reference frame, frame 3, then the position of the centre of mass of body 2, C_2 is given by

$$\{R_{C_2O_3}\}_{3/3} = \{R_2\}_{3/3} = \frac{1}{m_2} \sum_1^n m_i \{R_i\}_{3/3} \qquad (2.20)$$

where $\{R_i\}_{3/3}$ is the position of m_i measured in frame 3 and referred to frame 3. In particular, if the origin of frame 3 is at C_2, then

$$\sum_1^n m_i \{R_i\}_{3/3} \qquad (2.21)$$

is a null vector.

Similarly, for a system of bodies 2, 3 and 4 making the composite body 5 positioned relative to frame 6, the position of the centre of mass of the composite body 5, C, is given by

$$\{R_{C_5O_6}\}_{6/6} = \{R_5\}_{6/6} = \frac{m_2\{R_2\}_{6/6} + m_3\{R_3\}_{6/6} + m_4\{R_3\}_{6/6}}{m_2 + m_3 + m_4} \qquad (2.22)$$

In particular, if the origin of frame 6 is at C, then

$$m_2\{R_2\}_{6/6} + m_2\{R_3\}_{6/6} + m_4\{R_4\}_{6/6} \qquad (2.23)$$

is a null vector.

Refer to Problems 4.1 to 4.15 and Problems 4.44 to 4.47.

Dynamics

2.7. Force Moment, Moment of Momentum and Moment of Rate of Change of Momentum

A force has, in addition to its capacity to cause or tend to cause a change of the state of the translational motion of a body, a capacity to cause or tend to cause a change of the rotational motion of a body. The magnitude of this turning effect or moment about a point is the product of the magnitude of the force and the distance from the point to the line of action of the force measured along a line at right angles to the line of action of the force as shown in Fig. 2.1. Hence

$$|M_A| = |R_{NA}||F| \qquad (2.24)$$

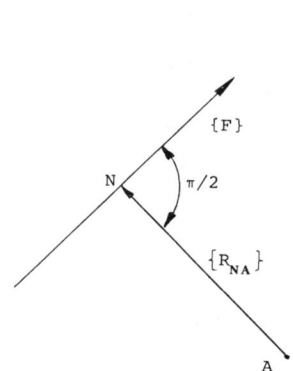 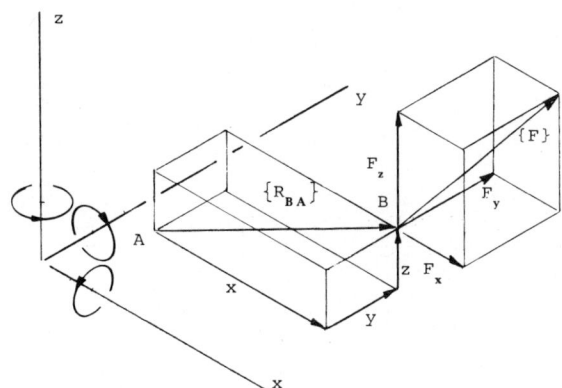

Fig. 2.1. Fig. 2.2.

In the more general situation of Fig. 2.2, where a point B on the line of action of $\{F\}$ is given relative to A, the point about which the moment is to be determined, using $\{R_{BA}\}$, the moment in each of the directions x, y and z is given by

$$\begin{aligned} M_A^x &= yF_z - zF_y \\ M_A^y &= zF_x - xF_z \\ M_A^z &= xF_y - yF_x \end{aligned} \qquad (2.25)$$

This set of equations can be arranged in vector and matrix form as

$$\begin{bmatrix} M_A^x \\ M_A^y \\ M_A^z \end{bmatrix} = \begin{bmatrix} 0 & -z & y \\ z & 0 & -x \\ -y & x & 0 \end{bmatrix} \begin{bmatrix} F_x \\ F_y \\ F_z \end{bmatrix} = \begin{bmatrix} -zF_y + yF_z \\ zF_x - xF_z \\ -yF_x + xF_y \end{bmatrix}$$

or

$$\{M_A\} = [R_{BA}]\{F\} . \qquad (2.26)$$

Where it is necessary to specify the frames to which the vectors are referred, then, for example

$$\{M_A\}_2 = [R_{BA}]_{2/2}\{F\}_2 \ . \tag{2.27}$$

That the moment vector $\{M_A\}$ is perpendicular to the plane containing the vectors $\{R_{BA}\}$ and $\{F\}$ is readily shown by the fact that

$$\{M_A\}^T\{R_{BA}\} \quad \text{and} \quad \{M_A\}^T\{F\}$$

are zero.

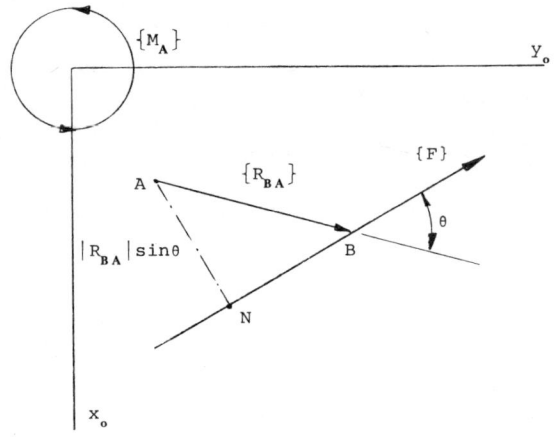

Fig. 2.3.

The magnitude of the moment vector is given by

$$|M_A| = |R_{BA}||F|\sin\theta$$

as can be seen from Fig. 2.3 which shows the $x_o y_o$ plane in which the $\{R_{BA}\}$ and $\{F\}$ vectors lie.

The concept of force moment can usefully be extended to that of moment of momentum or angular momentum and the rate of change of moment of momentum or the rate of change of angular momentum. Thus, for example, the moment of momentum of a particle is written

$$\{H_P\}_{1/1} = [R_{BA}]_{1/1}\{G_P\}_{1/1} = m_P[R_{BA}]_{1/1}\{V_P\}_{1/1} \ . \tag{2.28}$$

Similarly, the moment of the rate of change of momentum is written

$$[R_{BA}]_{1/1}\{\dot{G}_P\}_{1/1} = m_P[R_{BA}]_{1/1}\{A_P\}_{1/1} \ . \tag{2.29}$$

Refer to Problems 4.16 to 4.22 and Problems 4.48 to 4.53.

Dynamics

2.8. The Linear Momentum of a Rigid Body

The linear momentum of a particle of body 2 of mass m_i at P, measured with respect to frame 1 and referred to frame 1, is given by

$$\{G_P\}_{1/1} = m_i \{V_P\}_{1/1} .$$

The total linear momentum of body 2 is thus

$$\{G_2\}_{1/1} = \sum m_i \{V_P\}_{1/1}$$

where the summation is effected over the whole of body 2. Now

$$\{V_P\}_{1/1} = \{V_2\}_{1/1} + [\omega_2]_{1/1} \{R_{PC}\}_{1/1}$$

and therefore the linear momentum of body 2, measured with respect to frame 1 and referred to frame 1 is given by

$$\{G_2\}_{1/1} = \sum m_i \{V_2\}_{1/1} + \sum m_i [\omega_2]_{1/1} \{R_{PC}\}_{1/1}$$

and therefore the linear momentum of body 2, measured with respect to frame 1 and referred to frame 1, is given by

$$\{G_2\}_{1/1} = \sum m_i \{V_2\}_{1/1} + \sum m_i [\omega_2]_{1/1} \{R_{PC}\}_{1/1}$$

The factors $\{V_2\}_{1/1}$ and $[\omega_2]_{1/1}$ can be taken outside the summation sign since they are characteristics of the body rather than the particle. Thus

$$\{G_2\}_{1/1} = \{V_2\}_{1/1} \sum m_i + [\omega_2]_{1/1} \sum m_i \{R_{PC}\}_{1/1}$$

$$= m_2 \{V_2\}_{1/1} \quad (2.30)$$

since

$$m_2 = \sum m_i \quad \text{and} \quad \sum m_i \{R_{PC}\}_{1/1} = \{0\}$$

being the first moment of mass of the body about its centre of mass. By differentiation of Eq. 2.30

$$\{\dot{G}_2\}_{1/1} = m_2 \{\dot{V}_2\}_{1/1} = m_2 \{A_2\}_{1/1} \quad (2.31)$$

and this can be equated to the total external force on body 2 referred to frame 1 since the internal forces on the particles of the body sum to zero.

2.9. The Moment of Momentum of a Rigid Body About its Centre of Mass

Refer to Fig. 2.4 which shows body 2 moving relative to an inertial frame 1. The angular velocity of the body is $\{\omega_2\}_1$ and the velocity of its centre of mass $\{V_2\}_1$. The momentum of a particle of the body at P, which has a mass m_i, measured in frame 1 and referred to frame 1 is

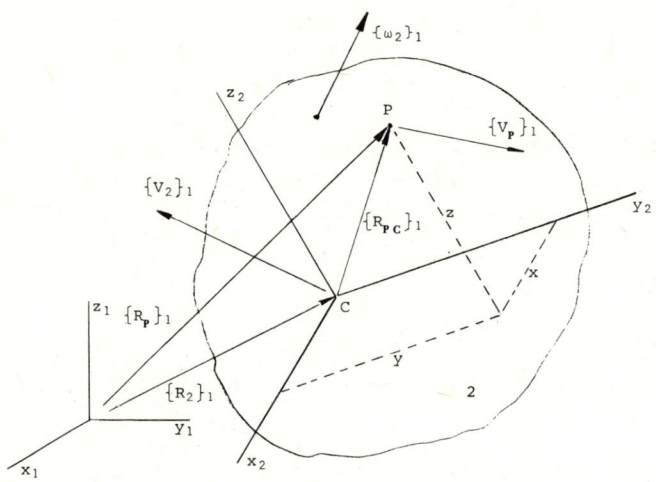

Fig. 2.4.

given by

$$[R_{PC}]_{1/1} m_i \{V_P\}_{1/1} = [R_{PC}]_{2/1} m_i \{V_P\}_{1/1}$$

and therefore the moment of momentum of body 2 about its centre of mass, measured in frame 1 and referred to frame 1, is given by

$$\{H_2\}_{1/1} = \sum m_i [R_{PC}]_{2/1} \{V_P\}_{1/1}$$

where the summation is effected over the whole of body 2. Since

$$\{V_P\}_{1/1} = \{V_2\}_{1/1} + [\omega_2]_{1/1} \{R_{PC}\}_{2/1}$$
$$= \{V_2\}_{1/1} - [R_{PC}]_{2/1} [\omega_2]_{1/1}$$

the moment of momentum can be written

$$\{H_2\}_{1/1} = \sum m_i [R_{PC}]_{2/1} \left(\{V_2\}_{1/1} - [R_{PC}]_{2/1} \{\omega_2\}_{1/1} \right)$$

$$= - \sum m_i [V_2]_{1/1} \{R_{PC}\}_{2/1}$$

$$+ \sum -m_i [R_{PC}]_{2/1} [R_{PC}]_{2/1} \{\omega_2\}_{1/1}$$

$$= -[V_2]_{1/1} \left\{ \sum m_i \{R_{PC}\}_{2/1} \right\}$$

$$+ \left[\sum -m_i [R_{PC}]^2_{2/1} \right] \{\omega_2\}_{1/1}$$

Dynamics

The factors $\{V_2\}_{1/1}$ and $\{\omega_2\}_{1/1}$ can be taken outside the summation sign since they are characteristics of the body rather than the particle. Now

$$\sum m_i \{R_{PC}\}_{2/1} = \{0\}$$

since this term is the first moment of the body about its centre of mass. Hence

$$\{H_2\}_{1/1} = \left[\sum -m_i [R_{PC}]^2_{2/1} \right] \{\omega_2\}_{1/1} \qquad (2.32)$$

and on defining

$$[I_2]_{2/1} = \sum -m_i [R_{PC}]^2_{2/1}, \qquad (2.33)$$

which is the inertia matrix for body 2 measured with respect to frame 2 and referred to frame 1,

$$\{H_2\}_{1/1} = [I_2]_{2/1} \{\omega_2\}_{1/1} . \qquad (2.34)$$

Since

$$[R]^2_{2/2} = [\ell_1]_2 [R]_{2/1} [\ell_1]^T_2 [\ell_1]_2 [R]_{2/1} [\ell_1]^T_2$$

$$= [\ell_1]_2 [R]^2_{2/1} [\ell_1]^T_2 ,$$

$$[I_2]_{2/2} = \sum -m_i [R_{PC}]^2_{2/2} = \sum -m_i [\ell_1]_2 [R_{PC}]^2_{2/2} [\ell_1]^T_2$$

$$= [\ell_1]_2 \left[\sum -m_i [R_{PC}]_{2/1} \right] [\ell_1]^T_2$$

$$= [\ell_1]_2 [I_2]_{2/1} [\ell_1]^T_2 \qquad (2.35)$$

which is the inertia matrix for body 2, measured with respect to frame 2 and referred to frame 2. Similarly

$$[I_2]_{2/1} = [\ell_1]_2 [I_2]_{2/2} [\ell_1]^T_2 .$$

If

$$\{R_{PC}\}_{2/2} = \begin{bmatrix} x \\ y \\ z \end{bmatrix}$$

then

$$[R_{PC}]^2_{2/2} = \begin{bmatrix} 0 & -z & y \\ z & 0 & -x \\ -y & x & 0 \end{bmatrix} \begin{bmatrix} 0 & -z & y \\ z & 0 & -x \\ -y & x & 0 \end{bmatrix}$$

$$= \begin{bmatrix} -(y^2 + z^2) & xy & xz \\ xy & -(x^2 + z^2) & zy \\ xz & zy & -(x^2 + y^2) \end{bmatrix}$$

and

$$[I_2]_{2/2} =$$

$$\begin{bmatrix} \sum m_i(y^2+z^2) & \sum -m_i xy & \sum -m_i xz \\ \sum -m_i xy & \sum m_i(x^2+z^2) & \sum -m_i zy \\ \sum -m_i xz & \sum -m_i zy & \sum m_i(x^2+y^2) \end{bmatrix}$$

$$= \begin{bmatrix} I_{x_2 x_2} & I_{x_2 y_2} & I_{x_2 z_2} \\ I_{x_2 y_2} & I_{y_2 y_2} & I_{z_2 y_2} \\ I_{x_2 z_2} & I_{y_2 z_2} & I_{z_2 z_2} \end{bmatrix} \qquad (2.36)$$

Thus, in general, the inertia matrix is symmetric. In particular, for example

$$I_{x_2 x_2} = \sum m_i (y^2+z^2)$$

is the moment of inertia of body 2 about the x axis (see Fig. 2.5), while

$$I_{x_2 y_2} = \sum -m_i xy$$

is the xy product of inertia.

The statement of Eq. 2.36 is tedious to write out and it can be abbreviated to, for example,

$$[I_2]_{2/2} = \begin{bmatrix} A & D & E \\ D & B & F \\ E & F & C \end{bmatrix}, \quad [I_2]_{2/2} = \begin{bmatrix} I_x & I_{xy} & I_{xz} \\ I_{xy} & I_y & I_{yz} \\ I_{xz} & I_{yz} & I_z \end{bmatrix}$$

or any other such convenient contraction.

2.10. The Relationship Between Moments of Inertia Measured in Different Frames

By reference to Fig. 2.6 it can be seen that the inertia matrix for body 2, measured in frame 3 and referred to frame 3 is given by

$$[I_2]_{3/3} = \sum -m_i [R_P]^2_{3/3} = \sum -m_i \left[[R_{PC}]_{3/3} + [R_2]_{3/3} \right]^2$$

$$= \sum -m_i \left[[R_{PC}]_{2/3} + [R_2]_{3/3} \right]^2$$

$$= \sum -m_i \Big[[R_{PC}]^2_{2/3} + [R_2]_{3/3}[R_{PC}]_{2/3}$$
$$\qquad\qquad + [R_{PC}]_{2/3}[R_2]_{3/3} + [R_2]^2_{3/3} \Big].$$

$$I_{x_2x_2} = \sum m_i (y^2 + z^2)$$

$$I_{y_2y_2} = \sum m_i (x^2 + z^2)$$

$$I_{z_2z_2} = \sum m_i (x^2 + y^2)$$

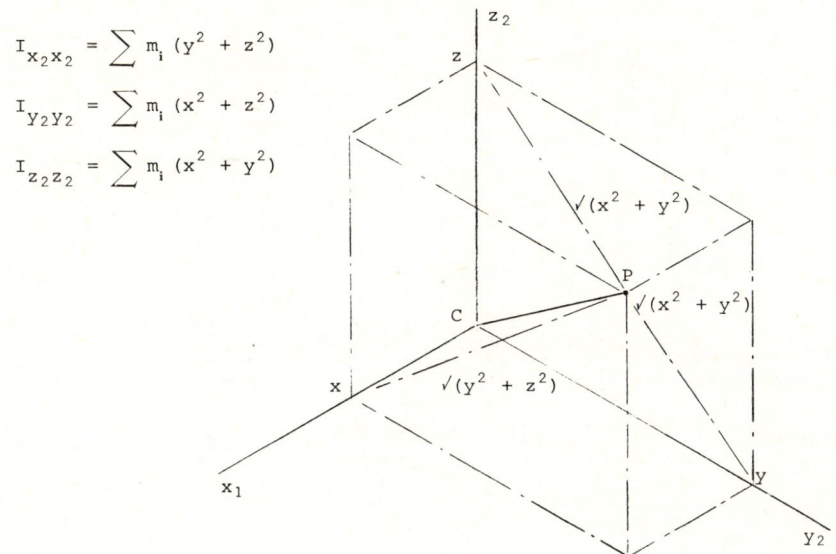

Fig. 2.5.

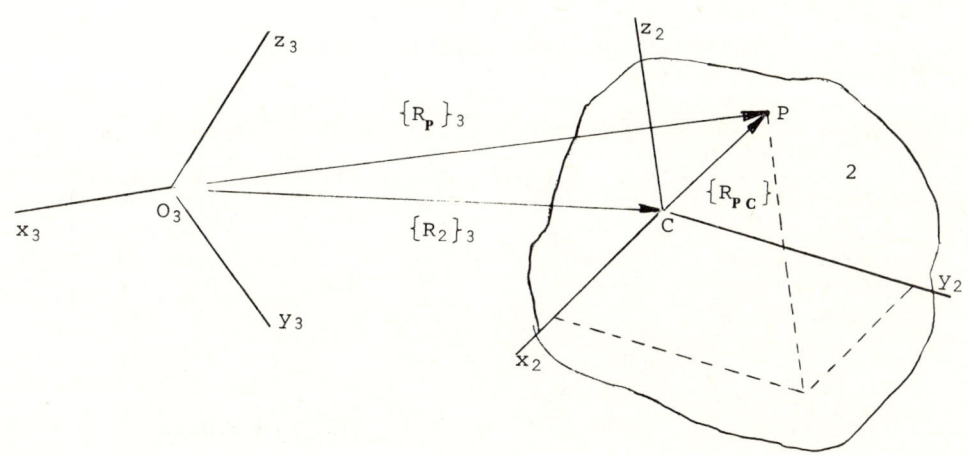

Fig. 2.6.

Now

$$\sum -m_i [R_2]_{3/3} [R_{PC}]_{2/3} = -[R_2]_{2/3} \sum m_i [R_{PC}]_{2/3}$$

and

$$\sum -m_i [R_{PC}]_{2/3} [R_2]_{3/3} = \left[\sum -m_i [R_{PC}]_{2/3} \right] [R_2]_{3/3}$$

are both null matrices since the origin of frame 2 is at the centre of mass of body 2. Hence

$$[I_2]_{3/3} = \sum -m_i [R_{PC}]^2_{2/3} - \sum m_i [R_2]^2_{3/3}$$

$$= [I_2]_{2/3} - m_2 [R_2]^2_{3/3}$$

and by Eq. 2.35

$$[I_2]_{3/3} = [\ell_2]_3 [I_2]_{2/2} [\ell_2]^T_3 - m_2 [R_2]^2_{3/3} . \qquad (2.36)$$

Refer to Problems 4.23 to 4.25 and Problems 4.54 to 4.56.

2.11. The Rate of Change of Angular Momentum of a Rigid Body About its Centre of Mass

Equation 2.34 gives the angular momentum of body 2 about its centre of mass, measured in frame 1 and referred to frame 1, as

$$\{H_2\}_{1/1} = [I_2]_{2/1} \{\omega_2\}_{1/1}$$

and this can be written

$$\{H_2\}_{1/1} = [\ell_2]_1 [I_2]_{2/2} [\ell_2]^T_1 \{\omega_2\}_{1/1}$$

by Eq. 2.35. This expression for angular momentum is readily differentiated because

$$[I_2]_{2/2}$$

the inertia matrix for body 2, measured in frame 2 and referred to frame 2, is a constant since frame 2 is fixed in body 2. Hence

$$\{\dot{H}_2\}_{1/1} = [\omega_2]_{1/1} [I_2]_{2/1} \{\omega_2\}_{1/1}$$
$$+ [\ell_2]_1 [I_2]_{2/2} [\omega_1]_{2/2} [\ell_1]_2 \{\omega_2\}_{1/1}$$
$$+ [I_2]_{2/1} \{\dot{\omega}_2\}_{1/1} .$$

The middle term on the right hand side of this equation can be written

$$-[\ell_2]_1 [I_2]_{2/2} [\ell_1]_2 [\omega_2]_{1/1} [\ell_1]^T_2 [\ell_1]_2 \{\omega_2\}_{1/1}$$

$$= -[\ell_2]_1 [I_2]_{2/2} [\ell_1]_2 [\omega_2]_{1/1} \{\omega_2\}_{1/1} = \{0\}$$

since

$$[\omega_1]_{2/2} = -[\omega_2]_{1/2} = -[\ell_1]_2 [\omega_2]_{1/1} [\ell_1]^T_2$$

and

$$[\omega_2]_{1/1} \{\omega_2\}_{1/1} = \{0\} .$$

Hence

Dynamics 53

$$\{\dot{H}_2\}_{1/1} = \lfloor \omega_2 \rfloor_{1/1} [I_2]_{2/1} \{\omega_2\}_{1/1} + [I_2]_{2/1} \{\dot{\omega}_2\}_{1/1} \quad (2.38)$$

and
$$\{\dot{H}_2\}_{1/n} = \lfloor \omega_2 \rfloor_{1/n} [I_2]_{2/n} \{\omega_2\}_{1/n} + [I_2]_{2/n} \{\dot{\omega}_2\}_{1/n}. \quad (2.39)$$

This rate of change of angular momentum is identically equal to the sum of the moment of the external forces about C and the external couples acting on body 2.

Refer to Problems 4.26 to 4.32 and Problems 4.57 to 4.61.

2.12. The Moment of Momentum of a Rigid Body About Any Point Q and the Rate of Change of Moment of Momentum About That Point

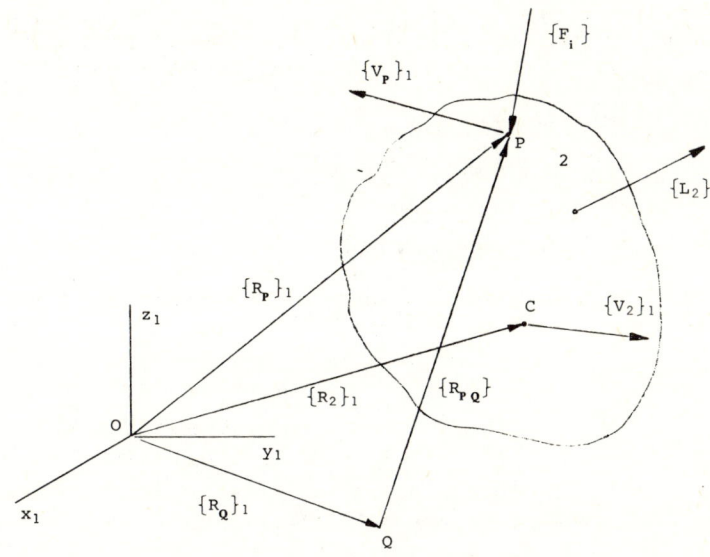

Fig. 2.7.

The moment of momentum of body 2 about any point Q, measured in frame 1 and referred to frame 1, is given by (Fig. 2.7)

$$\{H_{2Q}\}_{1/1} = \sum m_i [R_{PQ}]_{1/1} \{V_P\}_{1/1}$$
$$= \sum m_i \left[[R_{CQ}]_{1/1} + [R_{PC}]_{2/1} \right] \Big\{ \{V_2\}_{1/1}$$
$$+ \lfloor \omega_2 \rfloor_{1/1} \{R_{PC}\}_{2/2} \Big\}$$

$$= \sum m_i [R_{CQ}]_{1/1} \{V_2\}_{1/1} + \left[\sum m_i [R_{PC}]_{2/1}\right] \{V_2\}_{1/1}$$

$$+ [R_{CQ}]_{1/1} [\omega_2]_{1/1} \sum m_i \{R_{PC}\}_{2/1}$$

$$+ \left[\sum -m_i [R_{PC}]_{2/1}\right] \{\omega_2\}_{1/1}$$

and since the second and third terms on the right hand side of this equation are null matrices

$$\{H_{2Q}\}_{1/1} = m_2 [R_{CQ}]_{1/1} \{V_2\}_{1/1} + \{H_2\}_{1/1} \ . \tag{2.40}$$

Hence, by differentiation,

$$\{\dot{H}_{2Q}\}_{1/1} = m_2 [V_{CQ}]_{1/1} \{V_2\}_{1/1} + m_2 [R_{CQ}]_{1/1} \{A_2\}_{1/1} + \{\dot{H}_2\}_{1/1}$$

$$\tag{2.41}$$

2.13. The Relationship Between the Moment of the External Forces and Couples on a Rigid Body About Any Point Q and the Rate of Change of Moment of Momentum About That Point

Let $\{F_i\}$ be the external force on a particle of body 2 of mass m_i at P. In the case of a rigid body the internal forces cancel out in pairs and therefore

$$\sum \{F_i\} = \{F_2\}$$

where $\{F_2\}$ is the total external force on body 2. Let the body also be acted upon by external couples which reduce to $\{L_2\}$. The moment of the external forces and couples about Q is given by

$$\{M_{2Q}\}_1 = \sum [R_{PQ}]_{1/1} \{F_i\}_1 + \{L_2\}_1$$

$$= \sum [R_P]_{1/1} \{F_i\}_1 + \{L_2\}_1 - \sum [R_Q]_{1/1} \{F_i\}_1$$

$$= \{M_{2O}\}_1 - m_2 [R_Q]_{1/1} \{A_2\}_{1/1} \tag{2.42}$$

since

$$\{F_i\}_1 = \{F_2\}_1 = \{\dot{G}_2\}_{1/1} = m_2 \{A_2\}_{1/1} \ .$$

Now the rate of change of angular momentum about O, the origin of an inertial frame, is equal to the moment of the external forces and couples about that point. Thus

$$\{M_{2O}\}_1 = \{\dot{H}_{2O}\}_{1/1} \tag{2.43}$$

Dynamics

and since
$$\{H_{2Q}\}_{1/1} = \sum m_i [R_{PQ}]_{1/1}\{V_P\}_{1/1}$$
$$= \sum m_i [R_P]_{1/1}\{V_P\}_{1/1} - [R_Q]_{1/1}\sum m_i \{V_P\}_{1/1}$$
$$= \{H_{2O}\}_{1/1} - m_2[R_Q]_{1/1}\{V_2\}_{1/1}$$

then, by differentiation,
$$\{\dot{H}_{2Q}\}_{1/1} = \{\dot{H}_{2O}\}_{1/1} - m_2[V_Q]_{1/1}\{V_2\}_{1/1} - m_2[R_Q]_{1/1}\{A_2\}_{1/1}$$
$$(2.44)$$

By Eq. 2.43 in Eq. 2.42
$$\{\dot{H}_{2O}\}_{1/1} = \{M_{2Q}\}_1 + m_2[R_Q]_{1/1}\{A_2\}_{1/1}$$

and this result in Eq. 2.44 gives
$$\{M_{2Q}\}_1 = \{\dot{H}_{2Q}\}_{1/1} + m_2[V_Q]_{1/1}\{V_2\}_{1/1}$$

Substituting for $\{\dot{H}_{2Q}\}_{1/1}$ from Eq. 2.41 gives
$$\{M_{2Q}\}_1 = m_2[V_{CQ}]_{1/1}\{V_2\}_{1/1} + m_2[R_{CQ}]_{1/1}\{A_2\}_{1/1}$$
$$+ m_2[V_Q]_{1/1}\{V_2\}_{1/1} + \{\dot{H}_2\}_{1/1}$$

and since
$$\{V_Q\}_{1/1} + \{V_{CQ}\}_{1/1} = \{V_2\}_{1/1}$$

this reduces to
$$\{M_{2Q}\}_{1/1} = m_2[R_{CQ}]_{1/1}\{A_2\}_{1/1} + \{\dot{H}_2\}_{1/1}. \qquad (2.45)$$

Thus, for the particular case in which Q is at C,
$$\{M_2\}_1 = \{\dot{H}_2\}_{1/1} \qquad (2.46)$$

and for the case in which Q is at O
$$\{M_{2O}\}_1 = m_2[R_2]_{1/1}\{A_2\}_{1/1} + \{\dot{H}_2\}_{1/1} = \{\dot{H}_{2O}\}_{1/1}. \qquad (2.47)$$

2.14. The Kinetic Energy of a Rigid Body

The kinetic energy of body 2 is given by
$$2T_2 = \sum m_i |V_P|_1^2 = \sum m_i \{V_P\}_{1/n}^T \{V_P\}_{1/n} \qquad (2.48)$$

where n is any reference frame. Now, by reference to Fig. 2.4,

$$\{V_p\}_{1/n} = \{V_2\}_{1/n} + [\omega_2]_{1/n}\{R_{PC}\}_{2/n}$$

and therefore, using an abbreviated form of the statements

$$\{V_p\}^T_{1/n}\{V_p\}_{1/n} = \{\{V\} + [\omega]\{R\}\}^T\{\{V\} + [\omega]\{R\}\}$$
$$\text{①}\text{②}\text{③}$$

$$= \{V\}^T\{V\} + \{[\omega]\{R\}\}^T\{V\} + \{V\}^T[\omega]\{R\}$$
$$\text{④}$$

$$+ \{[\omega]\{R\}\}^T[\omega]\{R\}.$$

The summation involving term 3 will be zero since

$$\{V_2\}_{1/n} \quad \text{and} \quad \{\omega_2\}_{1/n}$$

are constants for the body. Term 2 can be written

$$\{-[R]\{\omega\}\}^T\{V\} = -\{\omega\}^T[R]^T\{V\} = \{\omega\}^T[R]\{V\}$$

so that the summation involving this term is also zero. Term 4 can be written

$$\{-[R]\{\omega\}\}^T\{-[R]\{\omega\}\} = \{\{\omega\}^T[R]^T\}\{[R]\{\omega\}\} = -\{\omega\}^T[R]^2\{\omega\}$$

Equation 2.48 thus reduces to

$$2T_2 = \sum m_i\{V_2\}^T_{1/n}\{V_2\}_{1/n} + \sum -m_i\{\omega_2\}^T_{1/n}[R_{PC}]^2_{2/n}\{\omega_2\}_{1/n}$$

$$= m_2\{V_2\}^T_{1/n}\{V_2\}_{1/n} + \{\omega_2\}^T_{1/n}[I_2]_{2/n}\{\omega_2\}_{1/n}$$

$$= \{V_2\}^T_{1/n}\{G_2\}_{1/n} + \{\omega_2\}^T_{1/n}\{H_2\}_{1/n}$$

$$= 2T_{2_{translation}} + 2T_{2_{rotation}} \qquad (2.49)$$

2.15. The Rate of Change of Kinetic Energy of a Rigid Body

Equation 2.49 can be written in the abbreviated form

$$2T_2 = m\{V\}^T\{V\} + \{\omega\}^T[I]\{\omega\}.$$

Hence, by differentiation

$$2\dot{T}_2 = m\{A\}^T\{V\} + m\{V\}^T\{A\} + \{\dot{\omega}\}^T[I]\{\omega\}$$
$$+ \{\omega\}^T\{[\omega][I]\{\omega\} + [I]\{\dot{\omega}\}\}.$$

Now
$$\{\omega\}^T[\omega] = \{[\omega]^T\{\omega\}\}^T = \{-[\omega]\{\omega\}\}^T$$

which is zero. Also

Dynamics

$$\{\omega\}^T[I]\{\dot{\omega}\} = \{[I]\{\omega\}\}^T\{\dot{\omega}\} = \{\dot{\omega}\}^T[I]\{\omega\} .$$

Hence, since

$$\{A\}^T\{V\} = \{V\}^T\{A\} ,$$

$$\dot{T}_2 = m_2\{A_2\}^T_{1/n}\{V_2\}_{1/n} + \{\dot{\omega}_2\}^T_{1/n}[I_2]_{2/n}\{\omega_2\}_{1/n}$$

$$= \{A_2\}^T_{1/n}\{G_2\}_{1/n} + \{\dot{\omega}_2\}^T_{1/n}\{H_2\}_{1/n} . \qquad (2.50)$$

2.16. The Special Case of the Motion of a Solid of Revolution

Links in three dimensional mechanisms can frequently be treated as solids of revolution, which are constrained by connections to them that exert no moment about the axis of generation, and consequently, if at some time the motion about this axis is zero, then it will always be zero.

Consider the case of a solid of revolution, body 4, which has its axis of generation along the x_4 axis. If $\omega_x = 0$, then

$$\{H_4\}_{1/4} = [I_4]_{4/4}\{\omega_4\}_{1/4} = \begin{bmatrix} J & 0 & 0 \\ 0 & I & 0 \\ 0 & 0 & I \end{bmatrix}\begin{bmatrix} 0 \\ \omega_y \\ \omega_z \end{bmatrix}$$

$$= I\begin{bmatrix} 0 \\ \omega_y \\ \omega_z \end{bmatrix} = I\{\omega_4\}_{1/4}$$

or

$$\{H_4\}_{1/1} = I\{\omega_4\}_{1/1} \qquad (2.51)$$

and the angular momentum vector is parallel to the angular velocity vector. Also

$$\{\dot{H}_4\}_{1/4} = \lfloor\omega_4\rfloor_{1/4}\lfloor I_4\rfloor_{4/4}\{\omega_4\}_{1/4} + \lfloor I_4\rfloor_{4/4}\{\dot{\omega}_4\}_{1/4}$$

$$= \begin{bmatrix} 0 & -\omega_z & \omega_y \\ \omega_z & 0 & 0 \\ -\omega_y & 0 & 0 \end{bmatrix}\begin{bmatrix} J & 0 & 0 \\ 0 & I & 0 \\ 0 & 0 & I \end{bmatrix}\begin{bmatrix} 0 \\ \omega_y \\ \omega_z \end{bmatrix}$$

$$+ \begin{bmatrix} J & 0 & 0 \\ 0 & I & 0 \\ 0 & 0 & I \end{bmatrix}\begin{bmatrix} 0 \\ \dot{\omega}_y \\ \dot{\omega}_z \end{bmatrix} = I\{\dot{\omega}\}_{1/4}$$

or

$$\{\dot{H}_4\}_{1/1} = I\{\dot{\omega}_4\}_{1/1} \qquad (2.52)$$

and the rate of change of angular momentum vector is parallel to the angular velocity vector. Further, the rotational kinetic enegy and its rate of change are given by

$$2T_{4_{rot}} = \{\omega_4\}^T_{1/1}\{H_4\}_{1/1} = I\{\omega_4\}^T_{1/1}\{\omega_4\}_{1/1} = I|\omega_4|^2_1 \quad (2.53)$$

and

$$\dot{T}_{4_{rot}} = \{\dot{\omega}_4\}^T_{1/1}\{H_4\}_{1/1} = I\{\dot{\omega}_4\}^T_{1/1}\{\omega_4\}_{1/1} \quad (2.54)$$

2.17. Rotation About a Fixed Axis

Consider the motion of body 4 about the z_4 axis which remains parallel to and fixed with respect to an inertial axis z_1. Then

$$\{\dot{H}_4\}_{1/4} = [\omega_4]_{1/4}[I_4]_{4/4}\{\omega_4\}_{1/4} + [I_4]_{4/4}\{\dot{\omega}_4\}_{1/4}$$

$$= \begin{bmatrix} 0 & -\omega_z & 0 \\ \omega_z & 0 & 0 \\ 0 & 0 & 0 \end{bmatrix} \begin{bmatrix} A & D & E \\ D & B & F \\ E & F & C \end{bmatrix} \begin{bmatrix} 0 \\ 0 \\ \omega_z \end{bmatrix}$$

$$+ \begin{bmatrix} A & D & E \\ D & B & F \\ E & F & C \end{bmatrix} \begin{bmatrix} 0 \\ 0 \\ \dot{\omega}_z \end{bmatrix} = \begin{bmatrix} E\dot{\omega}_z - F\omega_z^2 \\ F\dot{\omega}_z + E\omega_z^2 \\ C\dot{\omega}_z \end{bmatrix}. \quad (2.55)$$

Refer to Problems 4.33 to 4.40 and Problems 4.62 to 4.83.

2.18. Principal Axes and Principal Moments of Inertia of a Rigid Body With a Plane of Symmetry

Let y_2z_2 plane be the plane of symmetry of body 2 as shown in Fig. 2.8. Since

$$\sum xy = 0 \quad \text{and} \quad \sum zx = 0$$

the inertia matrix will be of the form

$$[I_2]_{2/2} = \begin{bmatrix} A_2 & 0 & 0 \\ 0 & B_2 & F_2 \\ 0 & F_2 & C_2 \end{bmatrix}.$$

Thus, for frame 3 positioned as shown in Fig. 2.8,

$$[I_2]_{2/3} = [\ell_2]_3[I_2]_{2/2}[\ell_2]^T_3$$

or

$$[\ell_3]^T_2[I_2]_{2/3} = [I_2]_{2/2}[\ell_2]^T_3$$

Dynamics

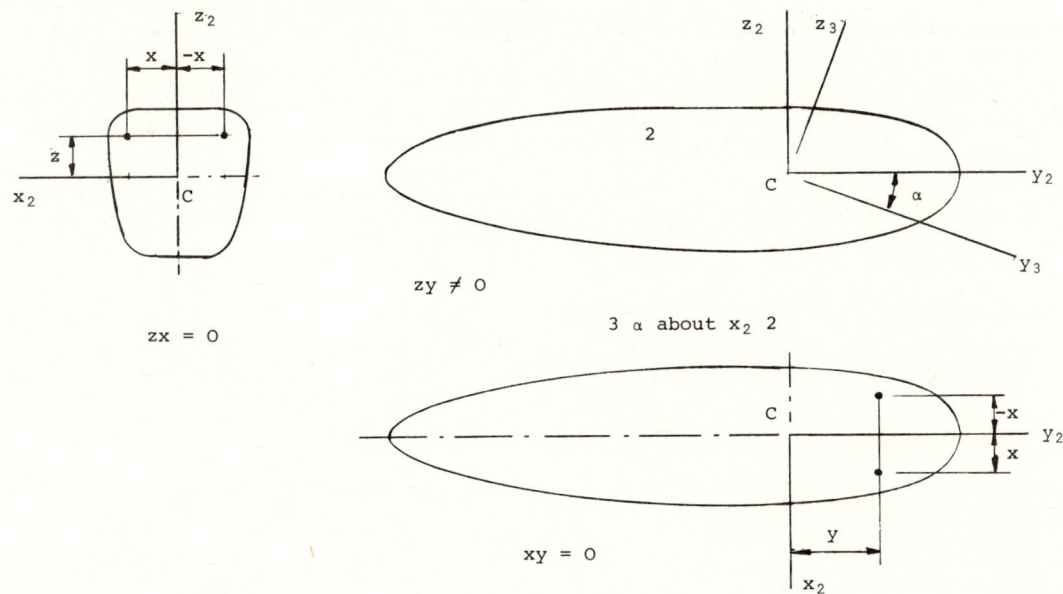

Fig. 2.8.

$$\begin{bmatrix} 1 & 0 & 0 \\ 0 & c\alpha & s\alpha \\ 0 & -s\alpha & c\alpha \end{bmatrix} \begin{bmatrix} A_3 & 0 & 0 \\ 0 & B_3 & F_3 \\ 0 & F_3 & C_3 \end{bmatrix} = \begin{bmatrix} A_2 & 0 & 0 \\ 0 & B_2 & F_2 \\ 0 & F_2 & C_2 \end{bmatrix} \begin{bmatrix} 1 & 0 & 0 \\ 0 & c\alpha & s\alpha \\ 0 & -s\alpha & c\alpha \end{bmatrix}$$

$$\begin{bmatrix} A_3 & 0 & 0 \\ 0 & B_3 c\alpha + F_3 s\alpha & F_3 c\alpha + C_3 s\alpha \\ 0 & -B_3 s\alpha + F_3 c\alpha & -F_3 s\alpha + C_3 c\alpha \end{bmatrix}$$

$$= \begin{bmatrix} A_2 & 0 & 0 \\ 0 & B_2 c\alpha - F_2 s\alpha & B_2 s\alpha + C_2 c\alpha \\ 0 & F_2 c\alpha + C_2 s\alpha & F_2 s\alpha + C_2 c\alpha \end{bmatrix} .$$

In particular

$$B_3 \cos\alpha + F_3 \sin\alpha = B_2 \cos\alpha - F_2 \sin\alpha$$

and

$$-B_3 \sin\alpha + F_3 \cos\alpha = F_2 \cos\alpha - C_2 \sin\alpha$$

giving

$$\tan\alpha = \frac{B_2 - B_3}{F_2 + F_3} = \frac{F_2 - F_3}{C_2 - B_3} \qquad (2.56)$$

and

$$B_3^2 + (C_2 + B_2)B_3 + F_3^2 + B_2C_2 - F_2^2 = 0 . \qquad (2.57)$$

Now the equation to a circle drawn on the xy plane, centred at (a,0) and of radius r is

$$x^2 - 2ax + a^2 + y^2 - r^2 = 0.$$

Comparing corresponding terms in this equation and those in Eq. 2.57

$$x \equiv B_3, \; y \equiv F_3, \; 2a \equiv C_2 + B_2, \; a^2 - r^2 \equiv B_2C_2 - F_2^2$$

and therefore

$$r^2 = \frac{(C_2 - B_2)^2}{4} + F_2^2 .$$

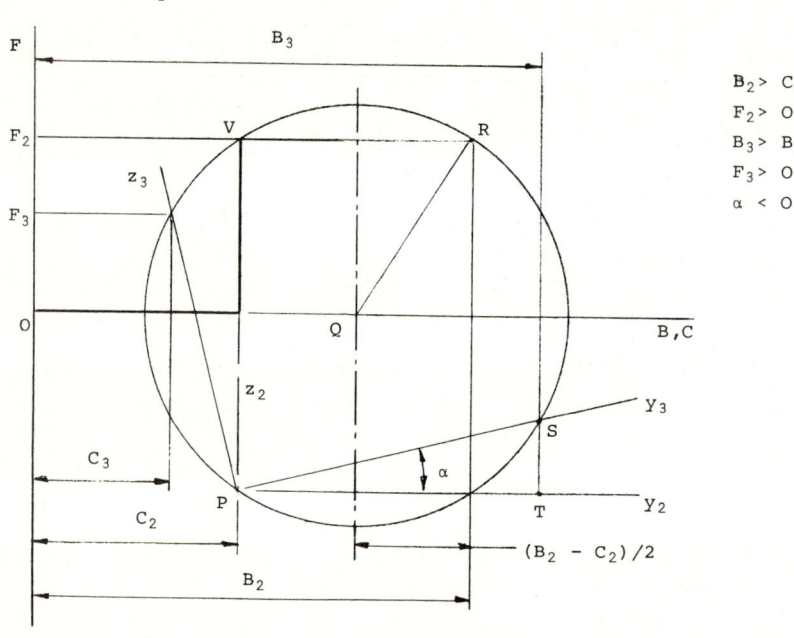

Fig.2.9.

The circle corresponding to Eq. 2.57, for

$$B_2 > C_2 \quad \text{and} \quad F_2 > 0$$

is shown in Fig. 2.9. B_2 and C_2 have been set off along an axis parallel to the y_2 axis and F_2 has been set off along an axis parallel to the z_2 axis to define the position of points V and R. A circle through these points, with its centre on the line OQ, centred at Q where

$$OQ = (C_2 + B_2)/2,$$

has a radius

Dynamics

$$RQ = \sqrt{\{(B_2 - C_2)^2/4 + F_2^2\}}$$

as required. For frame 3, obtained by rotating frame 2 through the angle α about x_2, i.e. $[\ell_2]_3$ corresponding to a negative rotation, $B_2 > B_3$, $F_3 < F_2$ and

$$\tan\alpha = \frac{TS}{PT} = \frac{F_2 - F_3}{B_3 - C_2} = -\frac{F_2 - F_3}{C_2 - B_3}.$$

The reader is invited to draw the circle for other cases, e.g.

(a) $B_2 > C_2$, $F_3 > 0$, $B_3 < B_2$,

(b) $B_2 < C_2$, $F_3 > 0$, $B_3 > C_2$ and

(c) $B_2 < C_2$, $F_3 < 0$, $B_3 > C_2$.

and indentify the position of the point P in relation to V(C ,F) from which frames 2 and 3 can be drawn in their correct relative positions with the correct angle α between them.

There will clearly be an angle α given by

$$\tan\alpha = \frac{B_2 - B_3}{F_2} = \frac{F_2}{C_2 - B_3}$$

for which the product of inertia term F_3 is zero. The moments of inertia in this case are the principal moments of inertia of the body. Also, the axes of frame 3 for this condition are the set of principal axes corresponding to the point C in body 2. There is a set of principal axes and principal moments of inertia for every point in the plane of symmetry.

2.19. Principal Axes and Principal Moments of Inertia For Any Rigid Body

The inertia matrix for body 2, measured with respect to frame 3 and referred to frame 3, is written

$$[I_2]_{3/3}$$

and it will, in general, be of the symmetric form

$$\begin{bmatrix} A & D & E \\ D & B & F \\ E & F & C \end{bmatrix}.$$

It will always be possible to find a frame 4, with the same origin as that of frame 3, such that

$$[I_2]_{3/4}$$

is of the form

$$\begin{bmatrix} \lambda_1 & 0 & 0 \\ 0 & \lambda_2 & 0 \\ 0 & 0 & \lambda_3 \end{bmatrix}$$

where λ_1, λ_2 and λ_3 are the principal moments of inertia of body 2 for the point corresponding to the origin of frame 3. The set of principal axes through a given point are such as to make the product of inertia terms in the inertia matrix referred to them equal to zero.

It is necessary to find, for the given inertia matrix

$$[I_2]_{3/3}, [I_2]_{3/4} \text{ and } [\ell_3]_4$$

such that

$$[I_2]_{3/4} = [\ell_3]_4 [I_2]_{3/3} [\ell_3]_4^T \tag{2.58}$$

is a diagonal matrix. On premultiplying Eq. 2.58 by $[\ell_3]_4^T$ it becomes

$$[\ell_3]_4^T [I_2]_{3/4} = [I_2]_{3/3} [\ell_3]_4^T$$

$$\begin{bmatrix} a_1 & b_1 & c_1 \\ a_2 & b_2 & c_2 \\ a_3 & b_3 & c_3 \end{bmatrix} \begin{bmatrix} \lambda_1 & 0 & 0 \\ 0 & \lambda_2 & 0 \\ 0 & 0 & \lambda_3 \end{bmatrix} =$$

$$\begin{bmatrix} A & D & E \\ D & B & F \\ E & F & C \end{bmatrix} \begin{bmatrix} a_1 & b_1 & c_1 \\ a_2 & b_2 & c_2 \\ a_3 & b_3 & c_3 \end{bmatrix}$$

$$\begin{bmatrix} a_1\lambda_1 & b_1\lambda_2 & c_1\lambda_3 \\ a_2\lambda_1 & b_2\lambda_2 & c_2\lambda_3 \\ a_3\lambda_1 & b_3\lambda_2 & c_3\lambda_3 \end{bmatrix} =$$

$$\begin{bmatrix} a_1A + a_2D + a_3E & b_1A + b_2D + b_3E \\ a_1D + a_2B + a_3F & b_1D + b_2B + b_3F & \text{etc.} \\ a_1E + a_2F + a_3C & b_1E + b_2F + b_3C \end{bmatrix}$$

Equating the first columns of these matrices gives

$$\begin{aligned} a_1\lambda_1 &= a_1A + a_2D + a_3E \\ a_2\lambda_1 &= a_1D + a_2B + a_3F \\ a_3\lambda_1 &= a_1E + a_2F + a_3C \end{aligned} \tag{2.59}$$

and this set of equations can be written in the form

$$(A - \lambda_1)a_1 + Da_2 + Ea_3 = 0$$
$$Da_1 + (B - \lambda_1)a_2 + Fa_3 = 0 \quad (2.60)$$
$$Ea_1 + Fa_2 + (C - \lambda_1)a_3 = 0$$

or

$$\begin{bmatrix} A - \lambda_1 & D & E \\ D & B - \lambda_1 & F \\ E & F & C - \lambda_1 \end{bmatrix} \begin{bmatrix} a_1 \\ a_2 \\ a_3 \end{bmatrix} = \begin{bmatrix} 0 \\ 0 \\ 0 \end{bmatrix} \quad (2.61)$$

or

$$\begin{bmatrix} A & D & E \\ D & B & F \\ E & F & C \end{bmatrix} \begin{bmatrix} a_1 \\ a_2 \\ a_3 \end{bmatrix} = \lambda_1 \begin{bmatrix} a_1 \\ a_2 \\ a_3 \end{bmatrix} \quad (2.62)$$

which can be abbreviated to

$$[I_2]_{3/3}\{a\} = \lambda_1\{a\} . \quad (2.63)$$

The set of homogeneous equations Eqs. 2.60 require that the determinant of the square matrix in Eqs. 2.61 is zero. Thus

$$\begin{vmatrix} A - \lambda_1 & D & E \\ D & B - \lambda_1 & F \\ E & F & C - \lambda_1 \end{vmatrix} = 0$$

which leads to the cubic equation

$$\lambda_1^3 - (A + B + C)\lambda_1^2 + (AB + BC + AC - F^2 - E^2 - D^2)\lambda_1$$
$$+ AF^2 + BE^2 + CD^2 - ABC - 2DEF = 0 . \quad (2.64)$$

This equation has three positive real roots corresponding to the three principal moments of inertia λ_1, λ_2 and λ_3. If each of these three roots is substituted, in turn, in Eq. 2.62 three separate $\{a\}$'s are determined which correspond to the three columns

$$\begin{bmatrix} a_1 \\ a_2 \\ a_3 \end{bmatrix} \quad \begin{bmatrix} b_1 \\ b_2 \\ b_3 \end{bmatrix} \quad \text{and} \quad \begin{bmatrix} c_1 \\ c_2 \\ c_3 \end{bmatrix}$$

of

$$[\ell_3 \; l_4^T] .$$

Equations of the form of Eqs. 2.62 occur frequently in the solution of physical problems so that computer programmes to solve them are readily available, especially for the case in which the square matrix is symmetric. Their solution longhand, even with a calculator, is tedious.

The quantities λ_1 λ_2 and λ_3, the principal moments of inertia, are the *eigenvalues* of the inertia matrix

$$[I_2]_{3/3}$$

and the three columns $\{a\}$ are its *eigenvectors*.

That the three columns $\{a\}$ are orthogonal (as they must be, being columns of a transformation matrix) is readily shown as follows. On writing

$$[I_2]_{3/3} = [I]$$

and
$$[I]\{a\} = \lambda_1\{a\} \qquad (2.65)$$
$$[I]\{b\} = \lambda_2\{b\} . \qquad (2.67)$$

Also

$$\{[I]\{b\}\}^T = \lambda_2\{b\}^T$$

or

$$\{b\}^T[I]^T = \{b\}^T[I] = \lambda_2\{b\}^T$$

and thus, on postmultiplication by $\{a\}$

$$\{b\}^T[I]\{a\} = \lambda_2\{b\}^T\{a\} .$$

Futher, on postmultiplying Eq. 2.65 by $\{b\}^T$,

$$\{b\}^T[I]\{a\} = \lambda_1\{b\}^T\{a\} . \qquad (2.68)$$

On subtracting Eq. 2.68 from Eq. 2.67

$$(\lambda_2 - \lambda_1)\{b\}^T\{a\} = 0$$

and this requires that the columns $\{b\}$ and $\{a\}$ be orthogonal since their scalar product is zero.

Refer to Problems 4.41 and 4.43 and Problems 4.84 and 4.85.

Chapter 3

Solution of Kinematics Problems

3.1 Solved Problems

Problem 3.1. A point P moves in a circular path of radius a so that the angle θ it subtends at the centre O of the circle increases uniformly with time. Find for the frames chosen in Fig. 3.1a

$$\frac{d}{dt}\{R_{PO}\}_{2/2}, \quad \frac{d}{dt}\{R_{PO}\}_{1/2}, \quad \frac{d}{dt}\{R_{PO}\}_{1/1} = \{V_{PO}\}_{1/1},$$

$$\{V_{PO}\}_{1/2} = \{\dot{R}_{PO}\}_{1/2} = \underset{\Delta t \to 0}{\text{Limit}} \frac{\{\Delta R_{PO}\}_{1/2}}{\Delta t},$$

$$\frac{d^2}{dt^2}\{R_{PO}\}_{1/2}, \quad \frac{d}{dt}\{V_{PO}\}_{1/2}, \quad \frac{d}{dt}\{V_{PO}\}_{1/1} = \{A_{PO}\}_{1/1}$$

and

$$\{\dot{V}_{PO}\}_{1/2} = \{A_{PO}\}_{1/2} = \underset{\Delta t \to 0}{\text{Limit}} \frac{\{\Delta V_{PO}\}_{1/2}}{\Delta t}.$$

Solution. Now

$$\{R_{PO}\}_{2/2} = \begin{bmatrix} a \\ 0 \\ 0 \end{bmatrix} \quad \text{and therefore} \quad \frac{d}{dt}\{R_{PO}\}_{2/2} = \begin{bmatrix} 0 \\ 0 \\ 0 \end{bmatrix}.$$

Also

$$\{R_{PO}\}_{1/1} = \begin{bmatrix} a\cos\theta \\ a\sin\theta \\ 0 \end{bmatrix} \quad \text{and therefore} \quad \{R_{PO}\}_{1/2} = \begin{bmatrix} a \\ 0 \\ 0 \end{bmatrix} = \{R_{PO}\}_{2/2}.$$

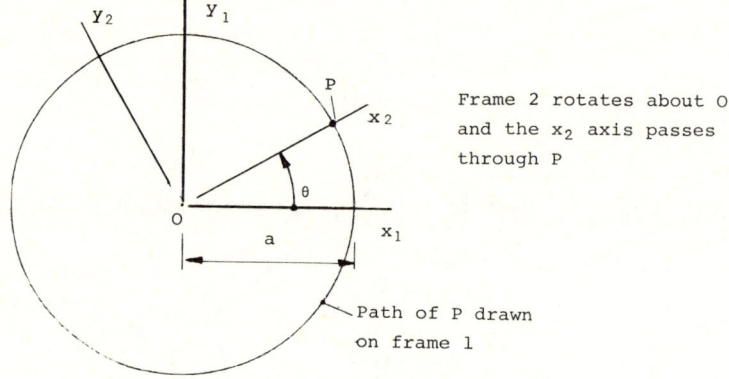

Frame 2 rotates about O and the x_2 axis passes through P

Path of P drawn on frame 1

Fig. 3.1a.

(Draw the components of $a\cos\theta$ and $a\sin\theta$ along the x_2 and y_2 axes). Thus

$$\frac{d}{dt}\{R_{PO}\}_{1/2} = \begin{bmatrix} 0 \\ 0 \\ 0 \end{bmatrix}.$$

By reference to Fig. 3.1b it can be seen that

$$\{\Delta R_{PO}\}_{1/2} = \begin{bmatrix} -a(1 - \cos\Delta\theta) \\ a\sin\Delta\theta \\ 0 \end{bmatrix} \tag{1}$$

and

$$\{\Delta R_{PO}\}_{1/1} = \begin{bmatrix} -a(1 - \cos\Delta\theta)\cos\theta - a\sin\Delta\theta\sin\theta \\ a\sin\Delta\theta\cos\theta - a(1 - \cos\Delta\theta)\sin\theta \\ 0 \end{bmatrix}. \tag{2}$$

Hence

$$\frac{d}{dt}\{R_{PO}\}_{1/1} = a\dot\theta \begin{bmatrix} -\sin\theta \\ \cos\theta \\ 0 \end{bmatrix} = \underset{\Delta t\to 0}{\text{Limit}}\frac{\{\Delta R_{PO}\}_{1/1}}{\Delta t} = \{V_{PO}\}_{1/1}$$

and

$$\{V_{PO}\}_{1/2} = \{\dot R_{PO}\}_{1/2} = \underset{\Delta t\to 0}{\text{Limit}}\frac{\{\Delta R_{PO}\}_{1/2}}{\Delta t} = a\dot\theta \begin{bmatrix} 0 \\ 1 \\ 0 \end{bmatrix}.$$

Also

Solution of Kinematics Problems

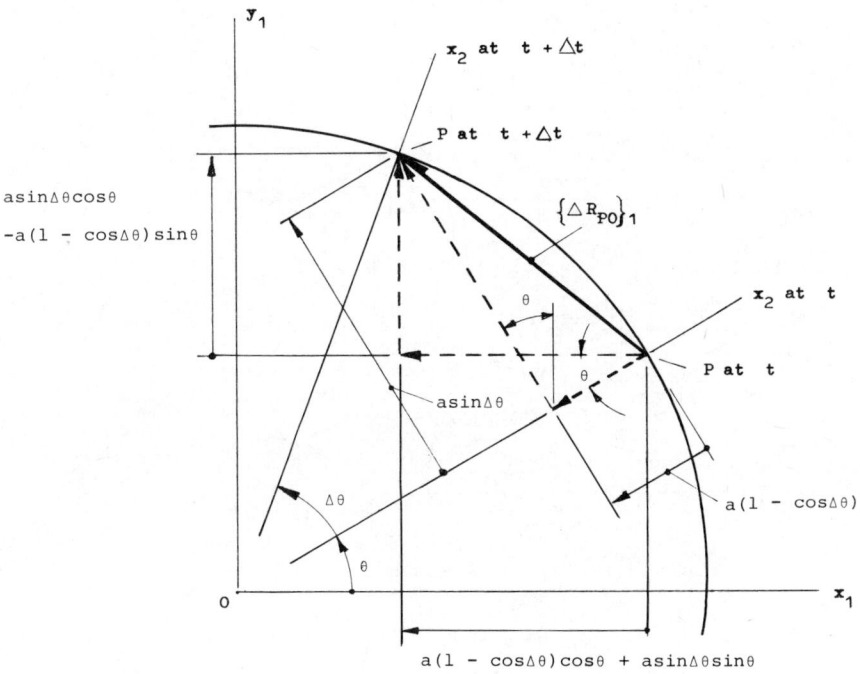

Fig. 3.1b.

$$\frac{d^2}{dt^2}\{R_{PO}\}_{1/2} = \begin{bmatrix} 0 \\ 0 \\ 0 \end{bmatrix}, \quad \frac{d}{dt}\{V_{PO}\}_{1/2} = \begin{bmatrix} 0 \\ 0 \\ 0 \end{bmatrix}$$

and

$$\frac{d^2}{dt^2}\{R_{PO}\}_{1/1} = \frac{d}{dt}\{V_{PO}\}_{1/1} = \{A_{PO}\}_{1/1} = a\dot{\theta}^2 \begin{bmatrix} -\cos\theta \\ -\sin\theta \\ 0 \end{bmatrix}.$$

By reference to Fig. 3.1c

$$\{\Delta V_{PO}\}_{1/2} = \begin{bmatrix} -a\sin\Delta\theta \\ -2a\sin^2(\Delta\theta/2) \\ 0 \end{bmatrix}$$

and therefore

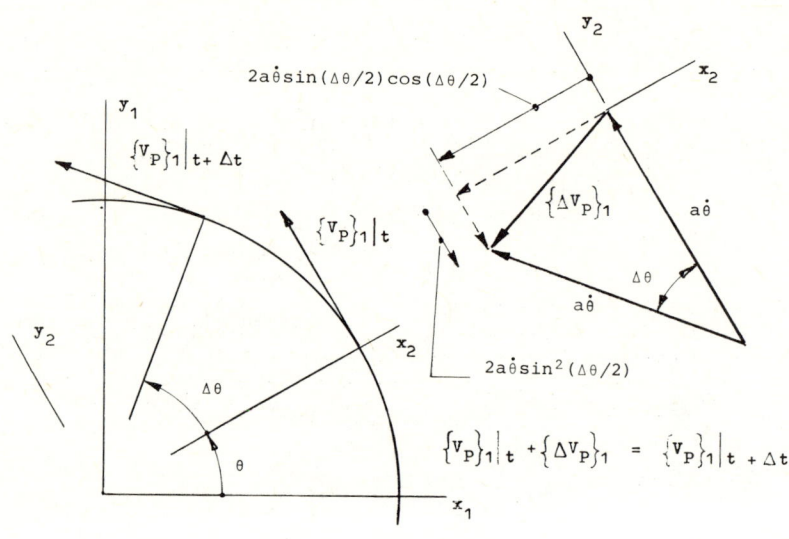

Fig. 3.1c.

$$\{\dot{V}_{PO}\}_{1/2} = \{A_{PO}\}_{1/2} = \underset{\Delta t \to 0}{\text{Limit}} \frac{\{\Delta V_{PO}\}_{1/2}}{\Delta t} = a\dot{\theta}^2 \begin{bmatrix} -1 \\ 0 \\ 0 \end{bmatrix}.$$

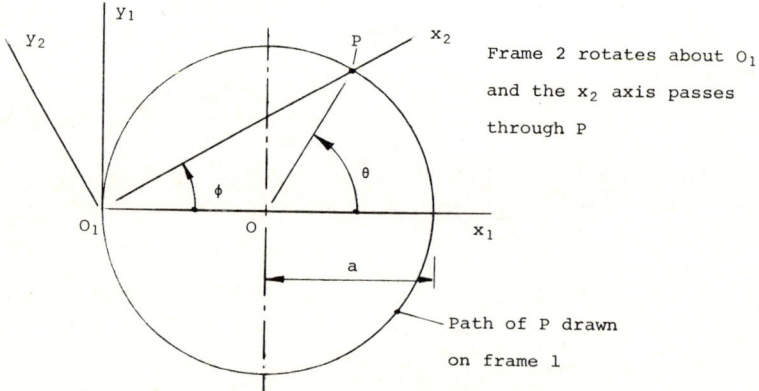

Frame 2 rotates about O_1 and the x_2 axis passes through P

Path of P drawn on frame 1

Fig. 3.1d.

It should be noted that the results obtained are peculiar to the axes chosen in Fig. 3.1a, and that if, for example, the axes of Fig. 3.1d were chosen, then different results would have been obtained for the given motion. The importance of specifying the axes being used is

Solution of Kinematics Problems

thus clear. The reader is invited to solve this problem for the axes of Fig. 3.1d.

Problem 3.2. The position of frame 4 relative to frame 1 is specified by

$$[\ell_4]_1 = \begin{bmatrix} 0.741516 & 0.45315 & -0.494731 \\ -0.595012 & 0.78485 & -0.172904 \\ 0.309955 & 0.4226 & 0.85165 \end{bmatrix}.$$

Find the consecutive rotations α, β and γ necessary to align frame 1 with frame 4 when they are performed in the following order:

(a) $1 \; \gamma$ about $z_1 \; 2 \; \alpha$ about $x_2 \; 3 \; \beta$ about $y_3 \; 4$,

(b) $1 \; \gamma$ about $z_1 \; 2 \; \beta$ about $y_2 \; 3 \; \alpha$ about $x_3 \; 4$,

(c) $1 \; \alpha$ about $x_1 \; 2 \; \beta$ about $y_2 \; 3 \; \gamma$ about $z_3 \; 4$,

(d) $1 \; \alpha$ about $x_1 \; 2 \; \gamma$ about $z_2 \; 3 \; \beta$ about $y_3 \; 4$.

Solution. (a)

$$[\ell_4]_1 = \begin{bmatrix} c\gamma & -s\gamma & 0 \\ s\gamma & c\gamma & 0 \\ 0 & 0 & 1 \end{bmatrix} \begin{bmatrix} 1 & 0 & 0 \\ 0 & c\alpha & -s\alpha \\ 0 & s\alpha & c\alpha \end{bmatrix} \begin{bmatrix} c\beta & 0 & s\beta \\ 0 & 1 & 0 \\ -s\beta & 0 & c\beta \end{bmatrix}$$

$$= \begin{bmatrix} c\gamma c\beta - s\gamma s\alpha s\beta & -s\gamma c\alpha & c\gamma s\beta + s\gamma s\alpha c\beta \\ s\gamma c\beta - c\gamma s\alpha s\beta & c\gamma c\alpha & s\gamma s\beta - c\gamma s\alpha c\beta \\ -c\alpha s\beta & s\alpha & c\alpha c\beta \end{bmatrix}.$$

Referring to the terms in the given and derived matrices using the usual a_{ij} notation

$a_{32} = 0.4226 = \sin\alpha$, $\alpha = 25°$.

$a_{12}/a_{22} = 0.45315/0.78485 = -\tan\gamma$, $\gamma = -30°$.

$a_{31}/a_{33} = 0.309955/0.85165 = -\tan\beta$, $\beta = -20°$.

(b)
$$[\ell_4]_1 = \begin{bmatrix} c\gamma & -s\gamma & 0 \\ s\gamma & c\gamma & 0 \\ 0 & 0 & 1 \end{bmatrix} \begin{bmatrix} c\beta & 0 & s\beta \\ 0 & 1 & 0 \\ -s\beta & 0 & c\beta \end{bmatrix} \begin{bmatrix} 1 & 0 & 0 \\ 0 & c\alpha & -s\alpha \\ 0 & s\alpha & c\alpha \end{bmatrix}$$

$$= \begin{bmatrix} c\gamma c\beta & c\gamma s\beta s\alpha - s\gamma c\alpha & c\gamma s\beta c\alpha + s\gamma s\alpha \\ s\gamma c\beta & s\gamma s\beta s\alpha + c\gamma c\alpha & s\gamma s\beta c\alpha - c\gamma s\alpha \\ -s\beta & c\beta s\alpha & c\beta c\alpha \end{bmatrix}.$$

$a_{31} = 0.309955 = -\sin\beta$, $\beta = -18.06°$.

$a_{21}/a_{11} = -0.59012/0.741516 = \tan\gamma$, $\gamma = -38.74°$.

$a_{32}/a_{33} = 0.4226/0.85165 = \tan\alpha$, $\alpha = 26.4°$.

(c)
$$[\ell_4]_1 = \begin{bmatrix} 1 & 0 & 0 \\ 0 & c\alpha & -s\alpha \\ 0 & s\alpha & c\alpha \end{bmatrix} \begin{bmatrix} c\beta & 0 & s\beta \\ 0 & 1 & 0 \\ -s\beta & 0 & c\beta \end{bmatrix} \begin{bmatrix} c\gamma & -s\gamma & 0 \\ s\gamma & c\gamma & 0 \\ 0 & 0 & 1 \end{bmatrix}$$

$$= \begin{bmatrix} c\beta c\gamma & -c\beta s\gamma & s\beta \\ s\alpha s\beta c\gamma & -s\alpha s\beta s\gamma & -s\alpha c\beta \\ -c\alpha s\beta c\gamma & c\alpha s\beta s\gamma & c\alpha c\beta \end{bmatrix}.$$

$a_{13} = -0.494731 = \sin\beta$, $\beta = -29.65°$.

$a_{12}/a_{11} = 0.45315/0.75151 = \tan\gamma$, $\gamma = -34.43°$.

$a_{23}/a_{33} = -0.172904/0.85165 = -\tan\alpha$, $\alpha = 11.48°$.

(d)
$$[\ell_4]_1 = \begin{bmatrix} 1 & 0 & 0 \\ 0 & c\alpha & -s\alpha \\ 0 & s\alpha & c\alpha \end{bmatrix} \begin{bmatrix} c\gamma & -s\gamma & 0 \\ s\gamma & c\gamma & 0 \\ 0 & 0 & 1 \end{bmatrix} \begin{bmatrix} c\beta & 0 & s\beta \\ 0 & 1 & 0 \\ -s\beta & 0 & c\beta \end{bmatrix}$$

$$= \begin{bmatrix} c\gamma c\beta & -s\gamma & c\gamma s\beta \\ c\alpha s\gamma c\beta + s\alpha s\beta & c\alpha c\gamma & c\alpha s\gamma s\beta - s\alpha c\beta \\ s\alpha s\gamma c\beta - c\alpha s\beta & s\alpha c\gamma & s\alpha s\gamma s\beta - c\alpha c\beta \end{bmatrix}.$$

$a_{12} = 0.45315 = -\sin\gamma$, $\gamma = -26.95°$.

$a_{32}/a_{22} = 0.4226/0.78485 = \tan\alpha$, $\alpha = 38.13°$.

$a_{13}/a_{11} = -0.494731/0.741516 = \tan\beta$, $\beta = -49.04°$.

A summary of these results is as follows:

Solution of Kinematics Problems

	$\alpha°$	$\beta°$	$\gamma°$
(a)	25	-20	-30
(b)	26.3	-18.06	-38.74
(c)	11.48	-29.65	-31.43
(d)	38.13	-33.71	-49.04 .

Problem 3.3. Body 2 rotates about an axis fixed in body 1 and body 3 slides on body 2 as shown in Fig. 3.3a. Obtain expressions for

$$\{V_A\}_{1/1}, \quad \{V_A\}_{1/2}, \quad \{A_A\}_{1/1} \quad \text{and} \quad \{A_A\}_{1/2}$$

where A is a point fixed in body 3.

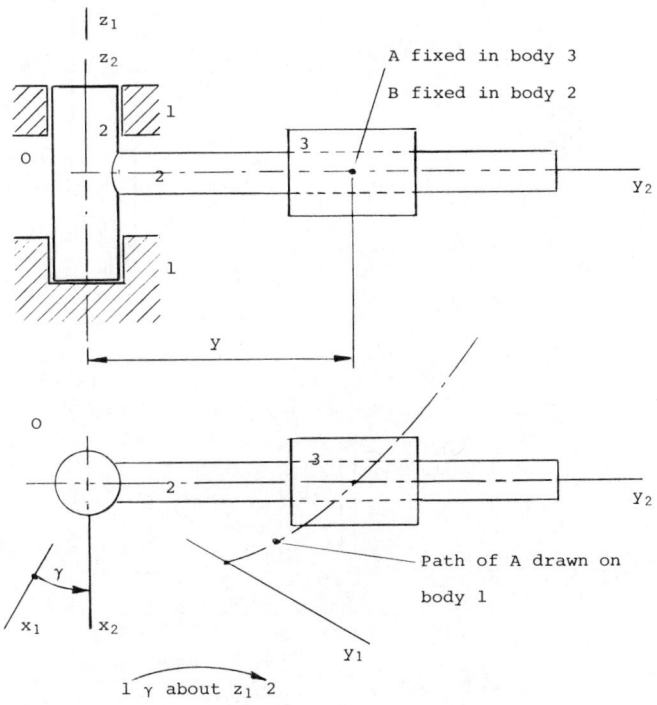

Fig. 3.3a.

Solution. Now

$$\{R_A\}_{1/1} = [\ell_2]_1 \{R_A\}_{2/2} = \begin{bmatrix} c\gamma & -s\gamma & 0 \\ s\gamma & c\gamma & 0 \\ 0 & 0 & 1 \end{bmatrix} \begin{bmatrix} 0 \\ y \\ 0 \end{bmatrix}$$

and therefore, by differentiation with respect to time

$$\{V_A\}_{1/1} = \dot{\gamma} \begin{bmatrix} -s\gamma & -c\gamma & 0 \\ c\gamma & -s\gamma & 0 \\ 0 & 0 & 0 \end{bmatrix} \begin{bmatrix} 0 \\ y \\ 0 \end{bmatrix} + \begin{bmatrix} c\gamma & -s\gamma & 0 \\ s\gamma & c\gamma & 0 \\ 0 & 0 & 1 \end{bmatrix} \begin{bmatrix} 0 \\ \dot{y} \\ 0 \end{bmatrix}$$

$$= -\dot{\gamma}y \begin{bmatrix} c\gamma \\ s\gamma \\ 0 \end{bmatrix} + \dot{y} \begin{bmatrix} -s\gamma \\ c\gamma \\ 0 \end{bmatrix} = \begin{bmatrix} -\dot{\gamma}yc\gamma + \dot{y}s\gamma \\ -\dot{\gamma}ys\gamma + \dot{y}c\gamma \\ 0 \end{bmatrix} . \quad (1)$$

The acceleration is obtained by a further differentiation with respect to time

$$\{A_A\}_{1/1} = -\dot{\gamma}\dot{y} \begin{bmatrix} c\gamma \\ s\gamma \\ 0 \end{bmatrix} - \ddot{\gamma}y \begin{bmatrix} c\gamma \\ s\gamma \\ 0 \end{bmatrix} - \dot{\gamma}^2 y \begin{bmatrix} -s\gamma \\ c\gamma \\ 0 \end{bmatrix} + \ddot{y} \begin{bmatrix} -s\gamma \\ c\gamma \\ 0 \end{bmatrix} + \dot{\gamma}\dot{y} \begin{bmatrix} -c\gamma \\ s\gamma \\ 0 \end{bmatrix}$$

$$= -2\dot{\gamma}\dot{y} \begin{bmatrix} c\gamma \\ s\gamma \\ 0 \end{bmatrix} - \ddot{\gamma}y \begin{bmatrix} c\gamma \\ s\gamma \\ 0 \end{bmatrix} - \dot{\gamma}^2 y \begin{bmatrix} -s\gamma \\ c\gamma \\ 0 \end{bmatrix} + \ddot{y} \begin{bmatrix} -s\gamma \\ c\gamma \\ 0 \end{bmatrix} . \quad (2)$$

From Eq. 1

$$\{V_A\}_{1/2} = [\ell_1]_2 \{V_A\}_{1/1}$$

$$= \begin{bmatrix} c\gamma & s\gamma & 0 \\ -s\gamma & c\gamma & 0 \\ 0 & 0 & 1 \end{bmatrix} \left[-\dot{\gamma}y \begin{bmatrix} c\gamma \\ s\gamma \\ 0 \end{bmatrix} + \dot{y} \begin{bmatrix} -s\gamma \\ c\gamma \\ 0 \end{bmatrix} \right] = \begin{bmatrix} -\dot{\gamma}y \\ \dot{y} \\ 0 \end{bmatrix} . \quad (3)$$

Similarly, from Eq. 2

$$\{A_A\}_{1/2} = [\ell_1]_2 \{A_A\}_{1/1}$$

$$= -2\dot{\gamma}\dot{y} \begin{bmatrix} 1 \\ 0 \\ 0 \end{bmatrix} - \ddot{\gamma}y \begin{bmatrix} 1 \\ 0 \\ 0 \end{bmatrix} - \dot{\gamma}^2 y \begin{bmatrix} 0 \\ 1 \\ 0 \end{bmatrix} + \ddot{y} \begin{bmatrix} 0 \\ 1 \\ 0 \end{bmatrix} . \quad (4)$$

Vectors corresponding to Eqs. 3 and 4 are shown in Fig. 3.3b. It should be noted that the component velocities and accelerations are independent of γ when the vectors are referred to frame 2. When applying the results of kinematic analysis to dynamics problems it will be found convenient to 'work in' a frame for which the

Solution of Kinematics Problems

components of as many vectors as possible are as simple as possible. Skill in solving problems is thus largely a matter of the judicious choice of a system of axes.

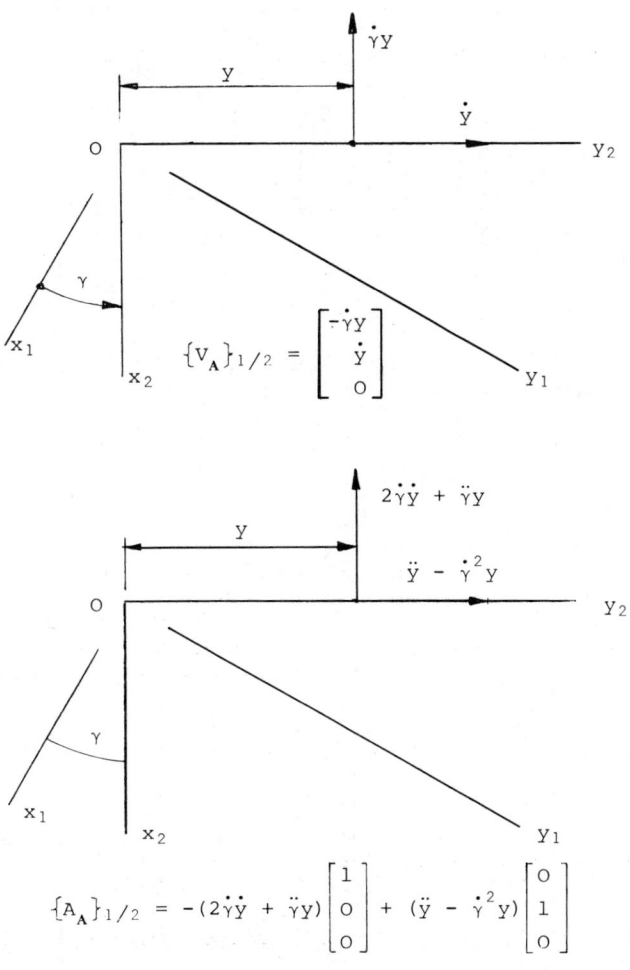

Fig. 3.3b.

The corresponding expressions for the velocity and acceleration of a point B fixed in body 2 with which A is coincident are obtained from Eqs. 3 and 4 by setting $\dot{y}$ and $\ddot{y}$ equal to zero. Hence

$$\{V_B\}_{1/2} = \dot{\gamma}y \begin{bmatrix} -1 \\ 0 \\ 0 \end{bmatrix} \quad \text{and} \quad \{A_B\}_{1/2} = \ddot{\gamma}y \begin{bmatrix} -1 \\ 0 \\ 0 \end{bmatrix} + \dot{\gamma}^2 y \begin{bmatrix} 0 \\ -1 \\ 0 \end{bmatrix}.$$

Problem 3.4. Body 2 is moving relative to body 1 such that, at a given instant, the velocities of points B and C fixed in body 2 relative to A fixed in body 2 are given by

$$\{V_{BA}\}_{1/1} = \begin{bmatrix} 3 \\ 4 \\ 5 \end{bmatrix} \text{ m/s} \quad \text{and} \quad \{V_{CA}\}_{1/1} = \begin{bmatrix} 1 \\ v \\ w \end{bmatrix} \text{ m/s}.$$

The position of B and C relative to A are

$$\{R_{BA}\}_{1/1} = \begin{bmatrix} 2 \\ 1 \\ -2 \end{bmatrix} \text{ m} \quad \text{and} \quad \{R_{CA}\}_{1/1} = \begin{bmatrix} 8 \\ 5 \\ 6 \end{bmatrix} \text{ m}.$$

Find
$\{\omega_2\}_{1/1}$, v and w.

Solution. Let

$$\{\omega_2\}_{1/1} \quad \begin{bmatrix} \omega_x \\ \omega_y \\ \omega_z \end{bmatrix}.$$

Then

$$\{V_{BA}\}_{1/1} = \begin{bmatrix} 3 \\ 4 \\ 5 \end{bmatrix} = \begin{bmatrix} 0 & -\omega_z & \omega_y \\ \omega_z & 0 & -\omega_x \\ -\omega_y & \omega_x & 0 \end{bmatrix} \begin{bmatrix} 2 \\ 1 \\ -2 \end{bmatrix} = \begin{bmatrix} -\omega_z - 2\omega_y \\ 2\omega_z + 2\omega_x \\ -2\omega_y + \omega_x \end{bmatrix}, \quad (1)$$

$$\{V_{CA}\}_{1/1} = \begin{bmatrix} 1 \\ v \\ w \end{bmatrix} = \begin{bmatrix} 0 & -\omega_z & \omega_y \\ \omega_z & 0 & -\omega_x \\ -\omega_y & \omega_x & 0 \end{bmatrix} \begin{bmatrix} 8 \\ 5 \\ 6 \end{bmatrix} = \begin{bmatrix} -5\omega_z + 6\omega_y \\ 8\omega_z - 6\omega_x \\ -8\omega_y + 5\omega_x \end{bmatrix}, \quad (2)$$

$$\{V_{BA}\}^T_{1/1}\{R_{BA}\}_{1/1} = \begin{bmatrix} 3 & 4 & 5 \end{bmatrix} \begin{bmatrix} 2 \\ 1 \\ -2 \end{bmatrix} = 6 + 4 - 10 = 0 \quad (3)$$

(which means that the data are correct) and

$$\{V_{CA}\}^T_{1/1}\{R_{CA}\}_{1/1} = \begin{bmatrix} 1 & v & w \end{bmatrix} \begin{bmatrix} 8 \\ 5 \\ 6 \end{bmatrix} = 8 + 5v + 6w = 0. \quad (4)$$

There are five unknown quantities (ω_x, ω_y, ω_z, v and w) and therefore five independent equations must be selected. Only two of Eqs. 1 and Eqs. 2 are independent and by selecting two from each, together with

Solution of Kinematics Problems

Eq. 4, the five independent equations become

$$
\begin{aligned}
-2\omega_y - \omega_z &= 3 \\
\omega_x + \omega_z &= 2 \\
6\omega_y - 5\omega_z &= 1 \\
-6\omega_x + 8\omega_z - v &= 0 \\
5v + 6w &= -8
\end{aligned}
\qquad (5)
$$

These equations have the solution (see the Appendix)

$\omega_x = 3.25$ rad/s, $\omega_y = -0.875$ rad/s, $\omega_z = -1.25$ rad/s, $v = -29.5$ m/s and $w = 23.25$ m/s.

As a check on the work it should be found that

$$\{V_{BC}\}_{1/1}^T \{R_{BC}\}_{1/1} = 0 .$$

Now

$$\{V_{BC}\}_{1/1} = \{V_{BA}\}_{1/1} - \{V_{CA}\}_{1/1} = \begin{bmatrix} 3 \\ 4 \\ 5 \end{bmatrix} - \begin{bmatrix} 1 \\ -29.5 \\ 23.25 \end{bmatrix}$$

$$= \begin{bmatrix} 2 \\ 33.5 \\ -18.25 \end{bmatrix} \text{m/s}$$

and

$$\{R_{BC}\}_{1/1} = \{R_{BA}\}_{1/1} - \{R_{CA}\}_{1/1} = \begin{bmatrix} 2 \\ 1 \\ -2 \end{bmatrix} - \begin{bmatrix} 8 \\ 5 \\ 6 \end{bmatrix} = \begin{bmatrix} -6 \\ -4 \\ -8 \end{bmatrix} \text{m}$$

giving for the required scalar product of relative velocity and position

$$\begin{bmatrix} 2 & 23.5 & -18.25 \end{bmatrix} \begin{bmatrix} -6 \\ -4 \\ -8 \end{bmatrix} = -12 - 134 + 146 = 0$$

as it should be.

Appendix.

Solution of Eqs. 5.

These equations can be written in the matrix form

$$\begin{bmatrix} 0 & -2 & -1 & 0 & 0 \\ 1 & 0 & 1 & 0 & 0 \\ 0 & 6 & -5 & 0 & 0 \\ -6 & 0 & 8 & -1 & 0 \\ 0 & 0 & 0 & 5 & 6 \end{bmatrix} \begin{bmatrix} \omega_x \\ \omega_y \\ \omega_z \\ v \\ w \end{bmatrix} = \begin{bmatrix} 3 \\ 2 \\ 1 \\ 0 \\ -8 \end{bmatrix}$$

and solved by computer. However, with a limited number of equations the work involved in reducing the square matrix to echlon form, or even diagonalising it, is tolerable. The procedure, no more than a systematic elimination process, is set out in detail below. The first stage, if necessary, is to make the first term of the first column non-zero by equation interchange.

```
i        1    0    1    0    0    2
ii       0   -2   -1    0    0    3
iii      0    6   -5    0    0    1
iv      -6    0    8   -1    0    0
v        0    0    0    5    6   -8

i x 6    6    0    6    0    0   12
iv      -6    0    8   -1    0    0
    ∑    0    0   14   -1    0   12  ─┐
                                       │
i        1    0    1    0    0    2   │
ii       0   -2   -1    0    0    3   │
iii      0    6   -5    0    0    1   │
iv       0    0   14   -1    0   12  ◄┘
v        0    0    0    5    6   -8

ii x 3        -6   -3    0    0    9
iii            6   -5    0    0    1
         ∑     0   -8    0    0   10  ─┐
                                        │
i        1    0    1    0    0    2    │
ii       0   -2   -1    0    0    3    │
iii      0    0   -8    0    0   10   ◄┘
iv       0    0   14   -1    0   12
v        0    0    0    5    6   -8

iii x 14          -112    0    0   140
iv  x 8            112   -8    0    96
            ∑        0   -8    0   236  ─┐
                                          │
i        1    0    1    0    0     2     │
ii       0   -2   -1    0    0     3     │
iii      0    0   -8    0    0    10     │
iv       0    0    0   -8    0   236    ◄┘
v        0    0    0    5    6    -8

iv x 5                  -40    0  1180
v  x 8                   40   48  - 64
                  ∑       0   48  1116
```

so that the equations finally become

Solution of Kinematics Problems

$$\begin{bmatrix} 1 & 0 & 1 & 0 & 0 \\ 0 & -2 & -1 & 0 & 0 \\ 0 & 0 & -8 & 0 & 0 \\ 0 & 0 & 0 & -8 & 0 \\ 0 & 0 & 0 & 0 & 48 \end{bmatrix} \begin{bmatrix} \omega_x \\ \omega_y \\ \omega_z \\ v \\ w \end{bmatrix} = \begin{bmatrix} 2 \\ 3 \\ 10 \\ 236 \\ 1116 \end{bmatrix}$$

from which the unknowns are readily determined.

Problem 3.5. Figure 3.5 shows a schematic arrangement of an offset crank and connecting rod mechanism. The frame of the mechanism is body 1, the crank OA body 2 and the connecting rod AB body 3. Reference frame 1 is arranged with its z_1 axis along the axis about which the crank rotates and so that OA lies in the $x_1 y_1$ plane. The end B of the connecting rod is constrained to move along a straight line PQ in the $y_1 z_1$ plane and parallel to the x_1 axis.

Formulate equations which will permit a determination of the velocities

$$\{V_B\}_{1/1}, \quad \{V_{BA}\}_{1/1}, \quad \{\omega_3^n\}_{1/1} \quad \text{and} \quad \{V_C\}_{1/1}$$

for any position of the mechanism. In particular, evaluate the above velocities for the case in which the angular velocity of the crank is 10 rad/s and $\alpha = 30°$.

Solution. All vectors will be measured in and referred to frame 1 and therefore the 1/1 suffix can be omitted throughout the solution. Now

$$\{R_{BA}\} = \{R_B\} - \{R_A\}$$

$$\begin{bmatrix} x_3 \\ y_3 \\ a \end{bmatrix} = \begin{bmatrix} x \\ 0 \\ a \end{bmatrix} - \begin{bmatrix} x_2 \\ y_2 \\ 0 \end{bmatrix} \qquad (1)$$

where a = 50 mm and

$$\{V_{BA}\} = \{V_B\} - \{V_A\}$$

$$\begin{bmatrix} u_3 \\ v_3 \\ w_3 \end{bmatrix} = \begin{bmatrix} V \\ 0 \\ 0 \end{bmatrix} - \begin{bmatrix} u_2 \\ v_2 \\ w_2 \end{bmatrix}. \qquad (2)$$

In this case $w_2 = 0$ and therefore $w_3 = 0$, but they are retained for the sake of generality. Also, since

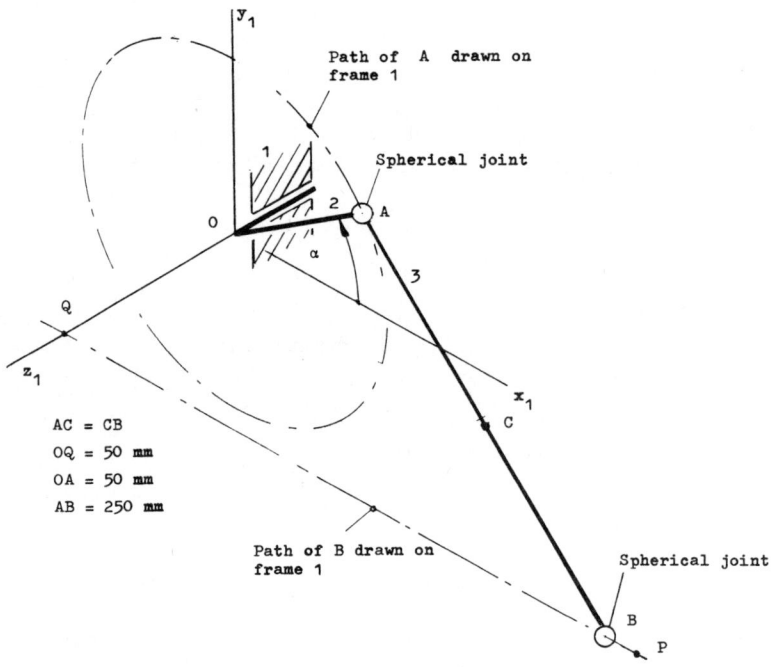

Fig. 3.5.

$\{V_{BA}\}$ is perpendicular to $\{R_{BA}\}$,

$$\{R_{BA}\}^T\{V_{BA}\} = 0$$

$$\begin{bmatrix} x_3 & y_3 & a \end{bmatrix} \begin{bmatrix} u_3 \\ v_3 \\ w_3 \end{bmatrix} = 0$$

$$u_3 x_3 + v_3 y_3 + w_3 a = 0 . \tag{3}$$

Equations 2 and 3 can be combined to give the single set of equations

$$\begin{bmatrix} 1 & 0 & 0 & -1 \\ 0 & 1 & 0 & 0 \\ 0 & 0 & 1 & 0 \\ x_3 & y_3 & a & 0 \end{bmatrix} \begin{bmatrix} u_3 \\ v_3 \\ w_3 \\ V \end{bmatrix} = \begin{bmatrix} -u_2 \\ -v_2 \\ -w_2 \\ 0 \end{bmatrix} - \{V_A\} \tag{4}$$

$\{R_{BA}\}^T$ $\{R_{QP}\}/|R_{QP}|$

Solution of Kinematics Problems

Equations 4 are of the form

$$[A]\{V_o\} = \{V_i\} \tag{5}$$

where the A matrix is a characteristic of the mechanism, the V_i matrix is a column of 'knowns' and the V_o matrix is a column of 'unknowns'. The set of equations has the solution

$$\{V_o\} = [A]^{-1}\{V_i\} . \tag{6}$$

The angular velocity of the connecting rod and the linear velocity of the point C on it can thus be found from

$$\{\omega_3^n\} = \frac{[R_{BA}]\{V_{BA}\}}{|R_{BA}|^2} = \frac{1}{r^2}\begin{bmatrix} 0 & -z_3 & y_3 \\ z_3 & 0 & -x_3 \\ -y_3 & x_3 & 0 \end{bmatrix}\begin{bmatrix} u_3 \\ v_3 \\ w_3 \end{bmatrix} = \begin{bmatrix} \omega_x \\ \omega_y \\ \omega_z \end{bmatrix} \tag{7}$$

where $r = |R_{EA}|$ and

$$\{V_C\} = \{V_A\} + [\omega_3^n]\{R_{CA}\}$$

$$\begin{bmatrix} u_2 \\ v_2 \\ w_2 \end{bmatrix} + \begin{bmatrix} 0 & -\omega_z & \omega_y \\ \omega_z & 0 & -\omega_x \\ -\omega_y & \omega_x & 0 \end{bmatrix}\begin{bmatrix} x_3/2 \\ y_3/2 \\ z_3/2 \end{bmatrix} . \tag{8}$$

In the above equations

$$y_3 = -a\sin\alpha , \quad x_3 = \sqrt{(b^2 - c^2 - a^2\sin^2\alpha)} ,$$

$$x = a\cos\alpha + x_3 , \quad u_2 = \dot{x}_2 = \frac{d}{dt}a\cos\alpha = -a\dot{\alpha}\sin\alpha$$

and

$$v_2 = \dot{y}_2 = \frac{d}{dt}a\sin\alpha = a\dot{\alpha}\cos\alpha$$

where

$$b = 250 \text{ mm} \quad \text{and} \quad c = 50 \text{ mm}.$$

The above relationships could be embodied in a computer programme to evaluate the required velocities.

For the case in which $\dot{\alpha} = 10$ rad/s and $\alpha = 30°$

$$y_3 = -25 \text{ mm} , \quad x_3 = 243.7 \text{ mm} , \quad u_2 = -250 \text{ mm/s}$$

and

$$v_3 = 433 \text{ mm/s} ,$$

when Eqs. 4 become

80   Matrix Methods in Engineering Mechanics

$$\begin{bmatrix} 1 & 0 & 0 & -1 \\ 0 & 1 & 0 & 0 \\ 0 & 0 & 1 & 0 \\ 243.7 & -25 & 50 & 0 \end{bmatrix} \begin{bmatrix} u_3 \\ v_3 \\ w_3 \\ V \end{bmatrix} = \begin{bmatrix} 250 \\ -433 \\ 0 \\ 0 \end{bmatrix}$$

and their solution gives

$$\{V_{BA}\} = \begin{bmatrix} -44.43 \\ -433 \\ 0 \end{bmatrix} \text{ mm/s} \quad \text{and} \quad \{V_B\} = \begin{bmatrix} -294 \\ 0 \\ 0 \end{bmatrix} \text{ mm/s}.$$

Also, by Eqs. 7 and 8

$$\{\omega_3^n\} = \begin{bmatrix} 0.346 \\ -0.035 \\ -1.706 \end{bmatrix} \text{ rad/s} \quad \text{and} \quad \{V_C\} = \begin{bmatrix} -272 \\ 216 \\ 0 \end{bmatrix} \text{ mm/s}.$$

Problem 3.6. A circular disc, body 2, of radius a rolls and slips on a plane, body 1. The motion is such that the plane of the disc is in the $z_1 x_1$ plane and the path of the point of contact is the x_1 axis. for points specified as follows:

- C   the centre of the disc,
- A   a point fixed on the periphery of the disc which is at some instant at the point of contact,
- Z   a point on the line joining A and C,
- B   a point fixed on the periphery of the disc which is instantaneously at the point of contact,
- D   a point fixed in frame 3 as shown in Fig. 3.6a, where $[\ell_3]_1$ is unit matrix,
- E   a point which moves along the x_1 axis such that it remains coincident with D
- F   a point fixed in frame 1 which is coincident with B,

and find

$$\{R_Z\}_{1/1}, \{R_C\}_{1/1}, \{R_A\}_{1/1}, \{R_{AC}\}_{1/1}, \{R_D\}_{1/1}, \{R_E\}_{1/1},$$

$$\{V_Z\}_{1/1}, \{V_C\}_{1/1}, \{V_A\}_{1/1}, \{V_{AC}\}_{1/1}, \{V_D\}_{1/1}, \{V_E\}_{1/1},$$

$$\{V_B\}_{1/1}, \{V_{BC}\}_{1/1}, \{A_Z\}_{1/1}, \{A_C\}_{1/1}, \{A_A\}_{1/1}, \{A_B\}_{1/1}$$

and

$$\{A_{BC}\}_{1/1}.$$

Use the expression for

$$\{V_Z\}_{1/1}$$

Solution of Kinematics Problems

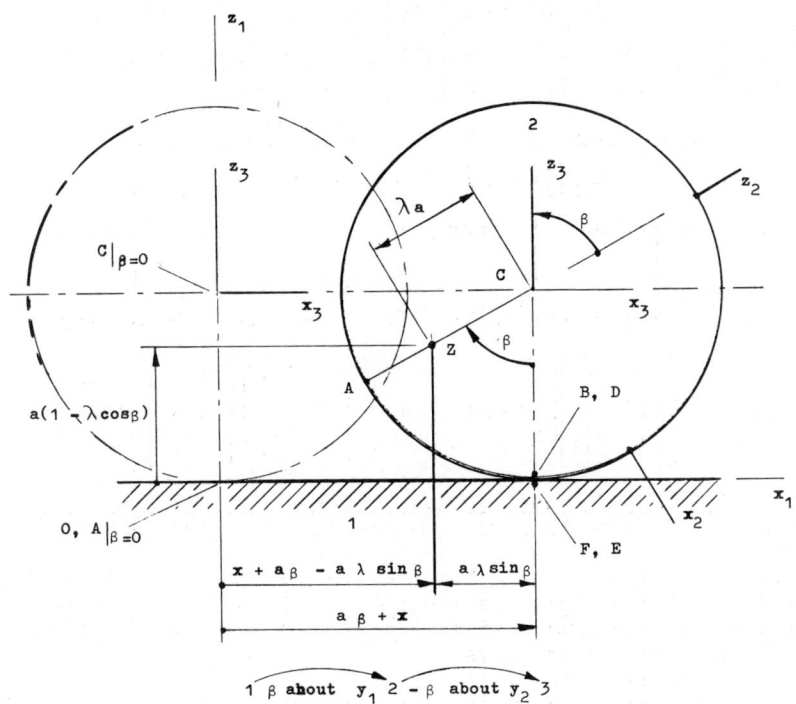

Fig. 3.6a.

to locate the instantaneous axis for the motion of the disc relative to the plane. Also locate the central axis for the motion of the disc relative to the plane using both

$$\{V_A\}_{1/1} \quad \text{and} \quad \{V_C\}_{1/1} ,$$

and show, in each case, that it is an instantaneous axis.

Solution. Let the points C, Z and A be initially on the z_1 axis. When the disc turns through the angle β the centre of the disc, C, moves through

$$a\beta + x$$

$a\beta$ being that part of the displacement due to the rotation of the disc and x that part of the displacement due to slip at the point of contact. The position of the point Z is thus given by

$$\{R_z\}_{1/1} = \begin{bmatrix} x + a\beta - \lambda as\beta \\ 0 \\ a - \lambda ac\beta \end{bmatrix} \quad (1)$$

as can by seen by reference to Fig. 3.6a. Therefore

$$\{R_C\}_{1/1} = \{R_z\}_{1/1}\big|_{\lambda=0} = \begin{bmatrix} x + a\beta \\ 0 \\ a \end{bmatrix},$$

$$\{R_A\}_{1/1} = \{R_z\}_{1/1}\big|_{\lambda=0} = \begin{bmatrix} x + a(\beta - s\beta) \\ 0 \\ a(1 - c\beta) \end{bmatrix}$$

and

$$\{R_{AC}\}_{1/1} = \{R_A\}_{1/1} - \{R_C\}_{1/1} = a\begin{bmatrix} -s\beta \\ 0 \\ -c\beta \end{bmatrix}.$$

Differentiating Eq. 1 with respect to time gives

$$\{V_z\}_{1/1} = \begin{bmatrix} \dot{x} + a\dot{\beta} - \lambda a\dot{\beta}c\beta \\ 0 \\ \lambda a\dot{\beta}s\beta \end{bmatrix} \quad (2)$$

and therefore

$$\{V_C\}_{1/1} = \{V_z\}_{1/1}\big|_{\lambda=0} = \begin{bmatrix} \dot{x} + a\dot{\beta} \\ 0 \\ 0 \end{bmatrix} = \frac{d}{dt}\{R_C\}_{1/1},$$

$$\{V_A\}_{1/1} = \{V_z\}_{1/1}\big|_{\lambda=1} = \begin{bmatrix} \dot{x} + a\dot{\beta}(1 - c\beta) \\ 0 \\ a\dot{\beta}s\beta \end{bmatrix} = \frac{d}{dt}\{R_A\}_{1/1},$$

$$\{V_{AC}\}_{1/1} = \{V_A\}_{1/1} - \{V_C\}_{1/1} = a\dot{\beta}\begin{bmatrix} -c\beta \\ 0 \\ s\beta \end{bmatrix} = \frac{d}{dt}\{R_{AC}\}_{1/1}$$

and

$$\{V_B\}_{1/1} = \{V_A\}_{1/1}\big|_{\beta=0} = \begin{bmatrix} \dot{x} \\ 0 \\ 0 \end{bmatrix}.$$

It can thus be seen that when rolling without slip is taking place the point B is at rest relative to frame 1. Also

Solution of Kinematics Problems

$$\{V_{BC}\}_{1/1} = \{V_B\}_{1/1} - \{V_C\}_{1/1} = a\beta \begin{bmatrix} -1 \\ 0 \\ 0 \end{bmatrix}.$$

Now

$$\{R_D\}_{1/1} = \{R_C\}_{1/1} + \{R_{DC}\}_{1/1} = \begin{bmatrix} x + a\beta \\ 0 \\ 0 \end{bmatrix} = \{R_E\}_{1/1}$$

and therefore

$$\{V_D\}_{1/1} = \{V_E\}_{1/1} = \begin{bmatrix} \dot{x} + a\dot{\beta} \\ 0 \\ 0 \end{bmatrix}.$$

Figure 3.6b shows a velocity diagram which illustrates the above results.

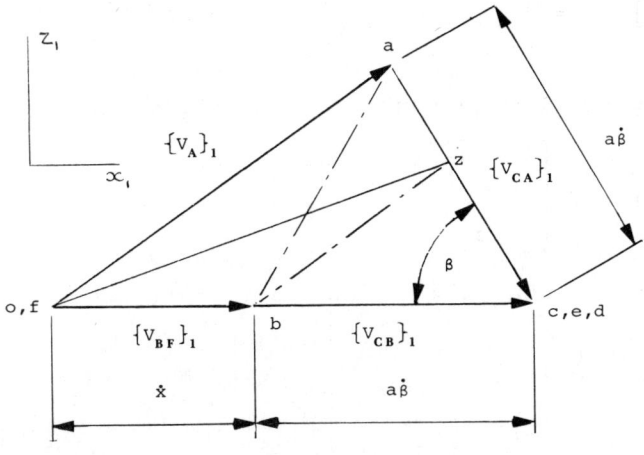

Fig. 3.6b.

Differentiating Eq. 2 with respect to time gives

$$\{A_Z\}_{1/1} = \begin{bmatrix} \ddot{x} + a\ddot{\beta}(1 - \lambda c\beta) + \lambda a\dot{\beta}^2 s\beta \\ 0 \\ \lambda a(\ddot{\beta}s\beta + \dot{\beta}^2 c\beta) \end{bmatrix}$$

and therefore

$$\{A_C\}_{1/1} = \{A_Z\}_{1/1}\big|_{\lambda=0} = \begin{bmatrix} \ddot{x} + a\ddot{\beta} \\ 0 \\ 0 \end{bmatrix},$$

$$\{A_A\}_{1/1} = \{A_Z\}_{1/1}\big|_{\lambda=1} = \begin{bmatrix} \ddot{x} + a\ddot{\beta}(1 - c\beta) + a\dot{\beta}^2 s\beta \\ 0 \\ a(\ddot{\beta}s\beta + \dot{\beta}^2 c\beta) \end{bmatrix},$$

$$\{A_B\}_{1/1} = \{A_A\}_{1/1}\big|_{\beta=0} = \begin{bmatrix} \ddot{x} \\ 0 \\ a\dot{\beta}^2 \end{bmatrix} \neq \frac{d}{dt}\{V_B\}_{1/1} = \begin{bmatrix} \ddot{x} \\ 0 \\ 0 \end{bmatrix}$$

and

$$\{A_{BC}\}_{1/1} = \{A_B\}_{1/1} - \{A_C\}_{1/1} = \begin{bmatrix} -a\ddot{\beta} \\ 0 \\ a\dot{\beta}^2 \end{bmatrix} \neq \frac{d}{dt}\{V_{BC}\}_{1/1} = \begin{bmatrix} -a\ddot{\beta} \\ 0 \\ 0 \end{bmatrix}.$$

The instantaneous axis for the motion of the disc relative to the plane will intersect the line CB at some point for which

$$\{V_Z\}_{1/1}\big|_{\beta=0} = \begin{bmatrix} \dot{x} + a\dot{\beta} - \lambda a\dot{\beta} \\ 0 \\ 0 \end{bmatrix} = \begin{bmatrix} 0 \\ 0 \\ 0 \end{bmatrix}$$

as shown in Fig. 3.6c. The position of the instantaneous axis is therefore given by

$$\lambda = 1 + \dot{x}/a\dot{\beta}.$$

If $\dot{x} > 0$ then $\lambda > 1$ and the instantaneous axis is below B.
If $\dot{x} < 0$ and $\dot{x} < a\dot{\beta}$ then $\lambda < 1$ and the instantaneous axis is between B and C.
If $\dot{x} < 0$ and $\dot{x} > a\dot{\beta}$ then $\lambda < 0$ and the instantaneous axis is above C.

Let Q be a point on the central axis for the motion of the disc relative to the plane. Then by Eq. 1.48

$$\{R_{QA}\}_{1/1} = \frac{[\omega_2]_{1/1}\{V_A\}_{1/1}}{|\omega_2|_1^2} = \frac{1}{\dot{\beta}}\begin{bmatrix} 0 & 0 & 1 \\ 0 & 0 & 0 \\ -1 & 0 & 0 \end{bmatrix}\begin{bmatrix} \dot{x} + a\dot{\beta}(1 - c\beta) \\ 0 \\ a\dot{\beta}s\beta \end{bmatrix}$$

$$= \begin{bmatrix} as\beta \\ 0 \\ -\dot{x}/\dot{\beta} - a(1 - c\beta) \end{bmatrix}.$$

Now

$$\{R_{BA}\}_{1/1} = \{R_B\}_{1/1} - \{R_A\}_{1/1} = \begin{bmatrix} as\beta \\ 0 \\ -a(1 - c\beta) \end{bmatrix}$$

and therefore

$$|R_{QC}| = a + \dot{x}/\dot{\beta} \quad \text{or} \quad \lambda = 1 + \dot{x}/a\dot{\beta}$$

which means that Q is at rest, making the central axis an instantaneous axis. It is left as an exercise for the reader to find a point S relative to C on the central axis and show that S is at rest relative to the plane.

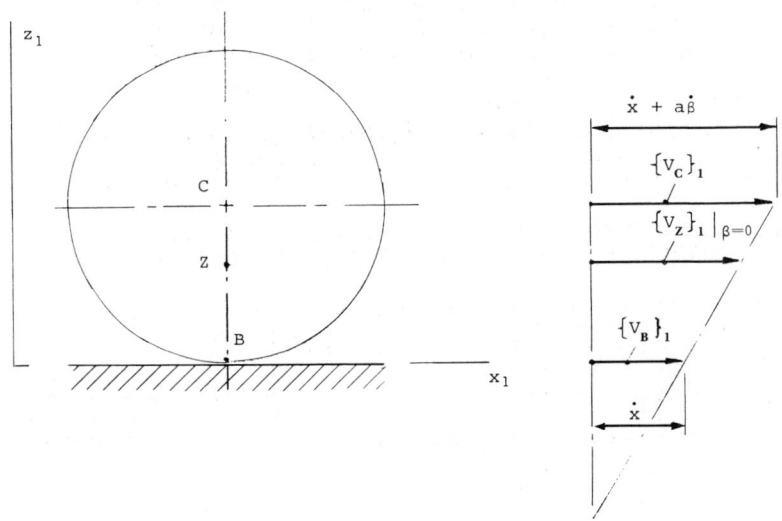

Fig. 3.6c.

Problem 3.7. Body 4 consists of a circular disc of radius na, a straight rod and a small sphere. The axis of the rod lies along the polar axis of the disc and one end of the rod is attached to a plane face of the disc. The small sphere is attached to the other end of the rod as shown in Fig. 3.7a. The sphere fits in a hemi-spherical socket in body 1 such that its centre lies in the plane on which the circumference of the disc rolls without slip.

For the case in which body 4 rolls at a constant rate and the point of contact of the disc completes one circuit on body 1 in time T, find, for the frames shown

$$\{V_C\}_{1/3}, \quad \{V_A\}_{1/3} \quad \text{and} \quad \{A_A\}_{1/3},$$

where A is a point fixed on the periphery of the disc at the point of contact between the disc and body 1 at some instant of time.

Draw an appropriate angular velocity vector diagram abd use it to find

$$\{\omega_2\}_{1/3}, \quad \{\omega_4\}_{3/3} \quad \text{and} \quad \{\omega_4\}_{1/3}.$$

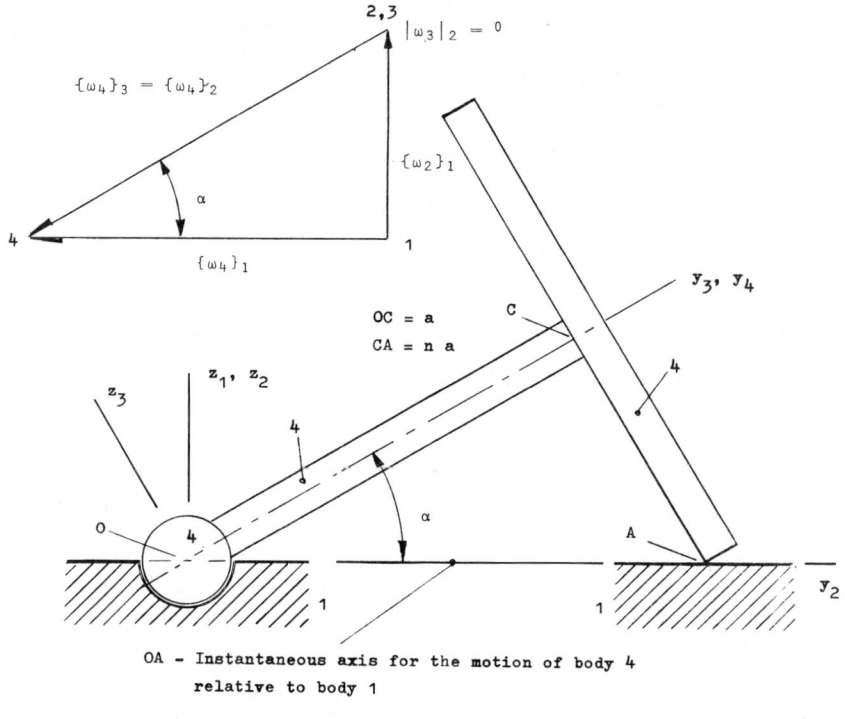

Fig. 3.7a.

Solution. The reference frames are located as follows:

Frame 1 has its x_1 and y_1 axes fixed in the plane on which the disc rolls and its z_1 axis perpendicular to this plane.

Frame 2 has its z_2 axis coincident with the z_1 axis and rotates about the z_1 axis so that the y_2 axis always passes through the point of contact between the disc and the plane.

Frame 3 is positioned relative to frame 2 by rotating it from coincidence with frame 2 through the angle α about the x_2 axis when the y_3 axis lies along the axis of the rod. The angle α is a fixed angle.

Frame 4 is fixed in body 4 such that the y_3 and y_4 axes coaxial and the $x_4 z_4$ axes lie in the face of the disc to which the rod is attached. Frame 4 thus rotates relative to frame 3 about the y_3 axis during the motion.

Solution of Kinematics Problems

Now
$$\{R_c\}_{1/1} = [\ell_2]_1 [\ell_3]_2 \{R_c\}_{3/3}$$

and therefore
$$\{V_c\}_{1/1} = \{\dot{R}_c\}_{1/1} = [\omega_2]_{1/1}[\ell_2]_1[\ell_3]_2\{R_c\}_{3/3}$$
$$+ [\ell_2]_1[\omega_3]_{2/2}[\ell_3]_2\{R_c\}_{3/3}$$
$$+ [\ell_2]_1[\ell_3]_2\{\dot{R}_c\}_{3/3} \, . \tag{1}$$

Since $\{\omega_3\}_2$ is a null vector, α being a constant, and $\{\dot{R}_c\}_3$ is a null vector, C being a fixed point in frame 3, Eq. 1 reduces to

$$\{V_c\}_{1/1} = [\omega_2]_{1/1}[\ell_2]_1[\ell_3]_2\{R_c\}_{3/3} \, . \tag{2}$$

The y_2 axis completes one revolution in time T and therefore
$$|\omega_2|_1 = 2\pi/T = \Omega = \dot{\gamma}$$

giving
$$\{\omega_2\}_{1/1} = \Omega \begin{bmatrix} 0 \\ 0 \\ 1 \end{bmatrix} \, .$$

Also, since
$$[\ell_2]_1 = \begin{bmatrix} c\gamma & -s\gamma & 0 \\ s\gamma & c\gamma & 0 \\ 0 & 0 & 1 \end{bmatrix}, \quad [\ell_3]_2 = \begin{bmatrix} 1 & 0 & 0 \\ 0 & c\alpha & -s\alpha \\ 0 & s\alpha & c\alpha \end{bmatrix}$$

and
$$\{R_c\}_{3/3} = \begin{bmatrix} 0 \\ a \\ 0 \end{bmatrix}$$

Eq. 2 reduces to
$$\{V_c\}_{1/1} = -\Omega a c\alpha \begin{bmatrix} c\gamma \\ s\gamma \\ 0 \end{bmatrix} = -\frac{2\pi a \cos\alpha}{T} \begin{bmatrix} \cos\gamma \\ \sin\gamma \\ 0 \end{bmatrix} \, . \tag{3}$$

Therefore
$$\{V_c\}_{1/3} = [\ell_2]_3[\ell_1]_2\{V_c\}_{1/1} = -\Omega a \cos\alpha \begin{bmatrix} 1 \\ 0 \\ 0 \end{bmatrix} \, . \tag{4}$$

Now $\{A_c\}_{1/1}$ is readily found from Eq. 3

$$\{A_c\}_{1/1} = \{\dot{V}_c\}_{1/1} = -\Omega^2 a \cos\alpha \begin{bmatrix} -\sin\gamma \\ \cos\gamma \\ 0 \end{bmatrix} \tag{5}$$

or more circuitously from Eq. 2, remembering that Ω is a constant,

$$\{A_C\}_{1/1} = \{\dot{V}_C\}_{1/1} = \lfloor \omega_2 \rfloor^2_{1/1} \lfloor \ell_2 \rfloor_1 \lfloor \ell_3 \rfloor_1 \{R_C\}_{3/3}.$$

The reader is asked to evaluate this expression for the acceleration of C to establish that it does give the result of Eq. 5. Hence

$$\{A_C\}_{1/3} = \lfloor \ell_2 \rfloor_3 \lfloor \ell_1 \rfloor_2 \{A_C\}_{1/1} = -\Omega^2 a \cos\alpha \begin{bmatrix} 0 \\ -\cos\alpha \\ \sin\alpha \end{bmatrix}$$

The position of A, where A is fixed in body 4 on the periphery of the disc as shown in Fig. 3.7b, is given by

$$\{R_A\}_{1/1} = \{R_C\}_{1/1} + \{R_{AC}\}_{1/1}$$
$$\{R_C\}_{1/1} + \lfloor \ell_2 \rfloor_1 \lfloor \ell_3 \rfloor_2 \lfloor \ell_4 \rfloor_3 \{R_{AC}\}_{4/4} \qquad (6)$$

where

$$\{R_{AC}\}_{4/4} = \begin{bmatrix} 0 \\ 0 \\ -na \end{bmatrix},$$

a constant, and therefore, since $\{\omega_3\}_2$ is a null vector

$$\{V_A\}_{1/1} = \{V_C\}_{1/1} + \Big[\lfloor \omega_2 \rfloor_{1/1} \lfloor \ell_2 \rfloor_1 \lfloor \ell_3 \rfloor_2 \lfloor \ell_4 \rfloor_3$$
$$+ \lfloor \ell_2 \rfloor_1 \lfloor \ell_3 \rfloor_2 \lfloor \omega_4 \rfloor_{3/3} \lfloor \ell_4 \rfloor_3 \Big] \{R_{AC}\}_{4/4}. \qquad (7)$$

Therefore

$$\{V_A\}_{1/3} = \lfloor \ell_1 \rfloor_3 \{V_A\}_{1/1} = \lfloor \ell_2 \rfloor_3 \lfloor \ell_1 \rfloor_2 \{V_A\}_{1/1}$$
$$= \{V_C\}_{1/3} + \Big[\lfloor \ell_2 \rfloor_3 \lfloor \ell_1 \rfloor_2 \lfloor \omega_2 \rfloor_{1/1} \lfloor \ell_2 \rfloor_1 \lfloor \ell_3 \rfloor_2$$
$$+ \lfloor \ell_2 \rfloor_3 \lfloor \ell_1 \rfloor_2 \lfloor \ell_2 \rfloor_1 \lfloor \ell_3 \rfloor_2 \lfloor \omega_4 \rfloor_{3/3} \Big] \lfloor \ell_4 \rfloor_3 \{R_{AC}\}_{4/4}.$$

Now

$$\lfloor \ell_2 \rfloor_3 \lfloor \ell_1 \rfloor_2 \lfloor \omega_2 \rfloor_{1/1} \lfloor \ell_2 \rfloor_1 \lfloor \ell_3 \rfloor_2 = \lfloor \ell_1 \rfloor_3 \lfloor \omega_2 \rfloor_{1/1} \lfloor \ell_1 \rfloor_3^T = \lfloor \omega_2 \rfloor_{1/3}$$

and

$$\lfloor \ell_2 \rfloor_3 \lfloor \ell_1 \rfloor_2 \lfloor \ell_2 \rfloor_1 \lfloor \ell_3 \rfloor_2 \lfloor \omega_4 \rfloor_{3/3} = \lfloor \ell_1 \rfloor_3 \lfloor \ell_3 \rfloor_1 \lfloor \omega_4 \rfloor_{3/3} = \lfloor \omega_4 \rfloor_{3/3}$$

giving

$$\{V_A\}_{1/3} = \{V_C\}_{1/3} + \Big[\lfloor \omega_2 \rfloor_{1/3} + \lfloor \omega_4 \rfloor_{3/3}\Big] \lfloor \ell_4 \rfloor_3 \{R_{AC}\}_{4/4}$$
$$= \{V_C\}_{1/3} + \lfloor \omega_4 \rfloor_{1/3} \lfloor \ell_4 \rfloor_3 \{R_{AC}\}_{4/4} \qquad (8)$$

since

$$\{\omega_4\}_1 = \{\omega_2\}_1 + \{\omega_4\}_3, \quad \{\omega_3\}_2 \text{ being a null vector.}$$

Solution of Kinematics Problems

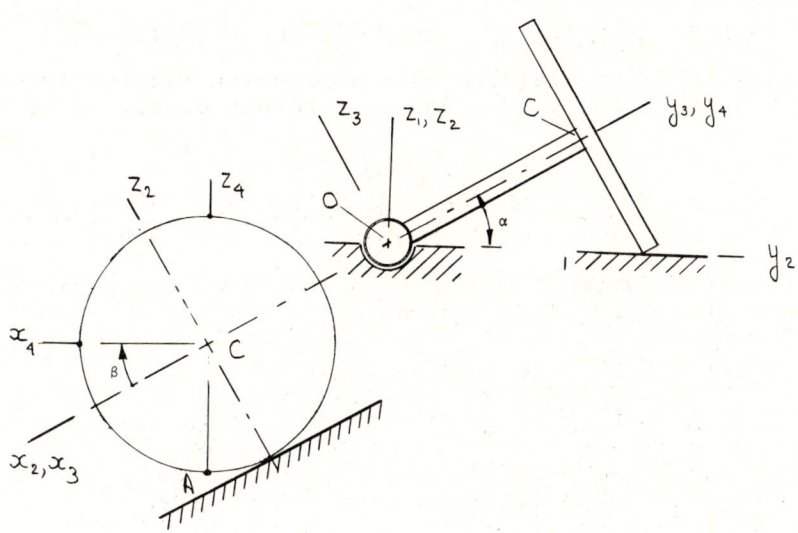

Fig. 3.7b.

On writing

$$\{\omega_2\}_{1/3} = [\ell_1]_3 \{\omega_2\}_{1/1} = \Omega \begin{bmatrix} 0 \\ s\alpha \\ c\alpha \end{bmatrix}, \quad \{\omega_4\}_{3/3} = \begin{bmatrix} 0 \\ \omega_y \\ 0 \end{bmatrix}$$

and

$$[\ell_4]_3 = \begin{bmatrix} c\beta & 0 & s\beta \\ 0 & 1 & 0 \\ -s\beta & 0 & c\beta \end{bmatrix}$$

Eq. 8 becomes

$$\{V_A\}_{1/3} = \Omega a c\alpha \begin{bmatrix} -1 \\ 0 \\ 0 \end{bmatrix} + \Omega \begin{bmatrix} 0 & -c\alpha & s\alpha \\ c\alpha & 0 & 0 \\ -s\alpha & 0 & 0 \end{bmatrix}$$

$$+ \omega_y \begin{bmatrix} 0 & 0 & 1 \\ 0 & 0 & 0 \\ -1 & 0 & 0 \end{bmatrix} \begin{bmatrix} c\beta & 0 & s\beta \\ 0 & 1 & 0 \\ -s\beta & 0 & c\beta \end{bmatrix} \begin{bmatrix} 0 \\ 0 \\ -na \end{bmatrix}$$

and this reduces to

$$\{V_A\}_{1/3} = \begin{bmatrix} -\Omega a \cos\alpha - na\cos\beta(\Omega\sin\alpha + \omega_y) \\ -\Omega na\cos\alpha\sin\beta \\ na\sin\beta(\Omega\sin\alpha + \omega_y) \end{bmatrix}.$$

When $\beta = 0$, the velocity of A is zero since it is then at the point of contact between the plane and the disc. Hence

$$\{V_A\}_{1/3}\big|_{\beta=0} = \begin{bmatrix} 0 \\ 0 \\ 0 \end{bmatrix} = \begin{bmatrix} -\Omega ac\alpha - na(\Omega s\alpha + \omega_y) \\ 0 \\ 0 \end{bmatrix}$$

which gives $\omega_y = \dot{\beta} = -\Omega/\sin\alpha$ since $n = \tan\alpha$.
Therefore

$$\{\omega_4\}_{3/3} = \begin{bmatrix} 0 \\ -\Omega/\sin\alpha \\ 0 \end{bmatrix} , \quad [\ell_4]_3 = \begin{bmatrix} c\beta & 0 & -s\beta \\ 0 & 1 & 0 \\ s\beta & 0 & c\beta \end{bmatrix}$$

where $\beta = \Omega t/\sin\alpha$.
The velocity of A, measured in frame 1 and referred to frame 3, is thus given by

$$\{V_A\}_{1/3} = \Omega a\cos\alpha \begin{bmatrix} \cos\beta - 1 \\ \tan\alpha\sin\beta \\ \sin\beta \end{bmatrix}. \tag{9}$$

Now

$$\{V_A\}_{1/1} = [\ell_3]_1 \{V_A\}_{1/3}$$

differentiates to

$$\{A_A\}_{1/1} = [\omega_3]_{1/1}[\ell_3]_1\{V_A\}_{1/3} + [\ell_3]_1 \frac{d}{dt}\{V_A\}_{1/3}$$

and since $\{\omega_3\}_2$ is a null vector

$$\{A_A\}_{1/3} = [\ell_1]_3\{A_A\}_{1/1} = [\omega_2]_{1/3}\{V_A\}_{1/3} + \frac{d}{dt}\{V_A\}_{1/3}$$

$$= \Omega^2 ac\alpha \begin{bmatrix} 0 & -c\alpha & s\alpha \\ c\alpha & 0 & 0 \\ -s\alpha & 0 & 0 \end{bmatrix} \begin{bmatrix} c\beta - 1 \\ \tan\beta s\alpha \\ s\beta \end{bmatrix}$$

$$+ \Omega\dot{\beta}ac\alpha \begin{bmatrix} -s\beta \\ \tan\alpha c\beta \\ c\beta \end{bmatrix}$$

$$= \Omega^2 a \begin{bmatrix} -\sin\beta/\tan\alpha \\ \cos\beta(1 + \cos^2\alpha) - \cos^2\alpha \\ \cos\beta\cos^2\alpha/\tan\alpha \end{bmatrix} \tag{10}$$

The angular velocity vector diagram shown in Fig. 3.7a is constructed as follows:

$\{\omega_2\}_1$ is represented by the vector 1 ⟶ 2 of magnitude Ω parallel

Solution of Kinematics Problems

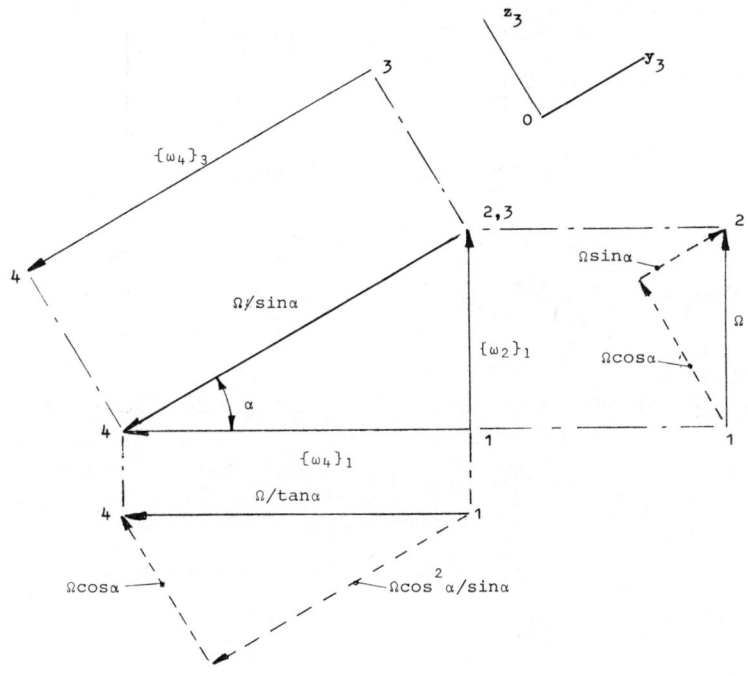

Fig. 3.7c.

to and in the direction of the z_1 axis.
$\{\omega_3\}_2$ is a null vector and the points 2 and 3 are therefore co-incident.
$\{\omega_4\}_3$ is represented by the vector 3 ⟶ 4 parallel to OC or the y_3 axis. Point 4 is not yet located on this vector.
Point O is fixed in both body 4 and body 1. Also, the point in the disc at the point of contact with body 1 is instantaneously at rest with respect to body 1. The y_2 axis is thus the instantaneous axis for the relative motion of bodies 4 and 1. The vector diagram can thus be completed by drawing a line parallel to to the y_2 axis through point 1. Point 4 is given by the intersection of this vector with the vector representing $\{\omega_4\}_3$ on which point 4 was to be located. The vector representing $\{\omega_4\}_3$ is thus directed negatively along the y_3 axis while the vector representing $\{\omega_4\}_1$ (1 ⟶ 4) is directed negatively along the y_2 axis.

It is therefore possible, having drawn the angular velocity vector diagram, to obtain the magnitude of the vectors and hence deduce their column matrix form. If

$$|\omega_2|_1 = \Omega \quad \text{then} \quad |\omega_4|_3 = \Omega/\sin\alpha \quad \text{and} \quad |\omega_4|_1 = \Omega/\tan\alpha .$$

Since the angular velocity vector diagram is 'fixed' in frames 2 and 3, it is easy to obtain the matrices which describe the various angular velocities by reference to Fig. 3.7c. Hence

$$\{\omega_2\}_{1/3} = \Omega \begin{bmatrix} 0 \\ \sin\alpha \\ \cos\alpha \end{bmatrix}, \quad \{\omega_4\}_{3/3} = -\Omega/\sin\alpha \begin{bmatrix} 0 \\ 1 \\ 0 \end{bmatrix}$$

and

$$\{\omega_4\}_{1/3} = \Omega/\tan\alpha \begin{bmatrix} 0 \\ -\cos\alpha \\ \sin\alpha \end{bmatrix}.$$

Problem 3.8. Figure 3.8 shows the schematic arrangement of a mechanism. Body 2 rotates at a constant rate of 6 rad/s about the y_1 axis of a frame fixed in body 1 so that point B moves in a circular path, centred at O, in the x_1z_1 plane. Body 4 rotates about an axis fixed in body 1 which is parallel to the z_1 axis so that point A moves in a circular path, centred at C, in the x_1y_1 plane. Body 3 couples bodies 2 and 4 by means of spherical joints at A and B.

Find, for the given configuration of the mechanism,

$$\{\omega_4\}_{1/1}, \quad \{\omega_3^n\}_{1/1}, \quad \{\dot{\omega}_4\}_{1/1} \quad \text{and} \quad \{\dot{\omega}_3^n\}_{1/1}.$$

Solution. All vectors will be measured in and referred to frame 1 and therefore the 1/1 suffix can be omitted throughout the solution. Now

$$\{R_{AB}\} = \{R_A\} - \{R_B\}$$

$$= \begin{bmatrix} 2 \\ 4 \\ 0 \end{bmatrix} - \begin{bmatrix} 0 \\ 0 \\ -4 \end{bmatrix} = \begin{bmatrix} 2 \\ 4 \\ 4 \end{bmatrix} \text{ cm} \tag{1}$$

and

$$\{V_B\} = [\omega_2]\{R_B\} = \begin{bmatrix} 0 & 0 & -6 \\ 0 & 0 & 0 \\ 6 & 0 & 0 \end{bmatrix} \begin{bmatrix} 0 \\ 0 \\ -4 \end{bmatrix} = \begin{bmatrix} 24 \\ 0 \\ 0 \end{bmatrix} \text{ cm/s} \tag{2}$$

giving

$$\{V_{AC}\} = \{V_B\} + \{V_{AB}\}$$

$$\begin{bmatrix} 0 \\ V \\ 0 \end{bmatrix} = \begin{bmatrix} 24 \\ 0 \\ 0 \end{bmatrix} + \begin{bmatrix} u \\ v \\ w \end{bmatrix} \text{ cm/s}. \tag{3}$$

Also, since

$$\{V_{AB}\}^T\{R_{AB}\} = 0, \quad 2u + 4v + 4w = 0. \tag{4}$$

Solution of Kinematics Problems

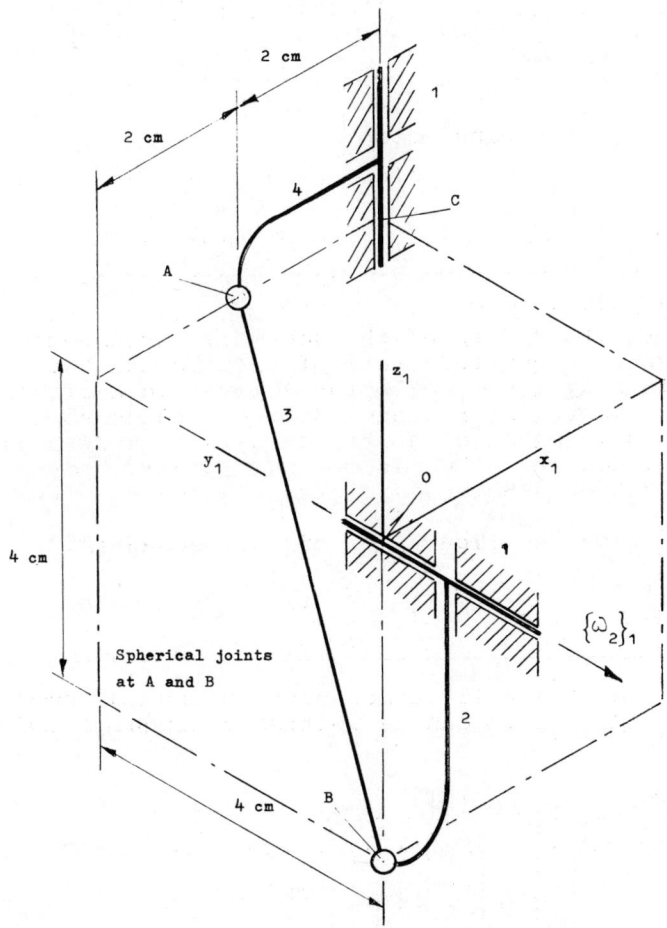

Fig. 3.8.

Equations 3 and 4 can be combined to give the single set of equations

$$\begin{bmatrix} 1 & 0 & 0 & 0 \\ 0 & 1 & 0 & -1 \\ 0 & 0 & 1 & 0 \\ \boxed{2 \quad 4 \quad 4} & 0 \end{bmatrix} \begin{bmatrix} u \\ v \\ w \\ V \end{bmatrix} = \begin{bmatrix} -24 \\ 0 \\ 0 \\ 0 \end{bmatrix} \leftarrow -\{V_B\}$$

$\{R_{AB}\}^T \qquad -\{R_{CA}\}/|R_{CA}|$

(5)

and they have the solution

$$[u \quad v \quad w \quad V] = [-24 \quad 12 \quad 0 \quad 12] \text{ cm/s}.$$

Now

$$\{V_{AC}\} = [\omega_4]\{R_{AC}\}$$

$$\begin{bmatrix} 0 \\ 12 \\ 0 \end{bmatrix} = \begin{bmatrix} 0 & -\omega_z & 0 \\ \omega_z & 0 & 0 \\ 0 & 0 & 0 \end{bmatrix} \begin{bmatrix} -2 \\ 0 \\ 0 \end{bmatrix} = \begin{bmatrix} 0 \\ -2\omega_z \\ 0 \end{bmatrix}$$

giving

$$\{\omega_4\} = \begin{bmatrix} 0 \\ 0 \\ -6 \end{bmatrix} \text{ rad/s }.$$

Also

$$\{\omega_3^n\} = \frac{[R_{AB}]\{V_{AB}\}}{|R_{AB}|^2} = \frac{1}{36}\begin{bmatrix} 0 & -4 & 4 \\ 4 & 0 & -2 \\ -4 & 2 & 0 \end{bmatrix} \begin{bmatrix} -24 \\ 12 \\ 0 \end{bmatrix} = -\frac{2}{3}\begin{bmatrix} 2 \\ 4 \\ -5 \end{bmatrix} \text{ rad/s}.$$

The acceleration of the points A, B and C are related by

$$\{A_{AC}\} = \{A_B\} + \{A_{AB}\}$$

$$\{A_{AC}^P\} + \{A_{AC}^n\} = \{A_B^P\} + \{A_{AB}^P\} + \{A_{AB}^n\} . \tag{6}$$

The directions of the normal components of acceleration are specified by

$$\{A_{AC}^n\}^T\{R_{AC}\} = 0 , \quad \begin{bmatrix} r_4 & s_4 & t_4 \end{bmatrix} \begin{bmatrix} -2 \\ 0 \\ 0 \end{bmatrix} = 0$$

or

$$-2r_4 + 0s_4 + 0t_4 = 0 \tag{7}$$

and

$$\{A_{AB}^n\}^T\{R_{AB}\} = 0 , \quad \begin{bmatrix} r_3 & s_3 & t_3 \end{bmatrix} \begin{bmatrix} 2 \\ 4 \\ 4 \end{bmatrix} = 0$$

or

$$2r_3 + 4s_3 + 4t_3 = 0 . \tag{8}$$

The parallel component acceleration terms of Eq. 6 are evaluated as follows.

$$|V_{AB}|^2 = 24^2 + 12^2 + 0 = 720 \text{ cm}^2/\text{s}^2,$$

$$\{A_{AB}^P\} = \frac{|V_{AB}|^2\{R_{BA}\}}{|R_{AB}|^2} = -\frac{720}{36}\begin{bmatrix} 2 \\ 4 \\ 4 \end{bmatrix} = -\begin{bmatrix} 40 \\ 80 \\ 80 \end{bmatrix} \text{ cm/s}^2 ,$$

Solution of Kinematics Problems

$$\{A_B^P\} = -|V_B|^2 \frac{\{R_B\}}{|R_B|^2} = -\frac{576}{16}\begin{bmatrix} 0 \\ 0 \\ -4 \end{bmatrix} = \begin{bmatrix} 0 \\ 0 \\ 144 \end{bmatrix} \text{cm/s}^2,$$

and

$$\{A_{AC}^P\} = |V_{AC}|^2 \frac{\{R_{CA}\}}{|R_{AC}|^2} = \frac{144}{4}\begin{bmatrix} 2 \\ 0 \\ 0 \end{bmatrix} = \begin{bmatrix} 72 \\ 0 \\ 0 \end{bmatrix} \text{cm/s}^2.$$

These results in Eq. 6 give

$$\begin{bmatrix} 72 \\ 0 \\ 0 \end{bmatrix} + \begin{bmatrix} r_4 \\ s_4 \\ t_4 \end{bmatrix} = \begin{bmatrix} 0 \\ 0 \\ 144 \end{bmatrix} - \begin{bmatrix} 40 \\ 80 \\ 80 \end{bmatrix} + \begin{bmatrix} r_3 \\ s_3 \\ t_3 \end{bmatrix}. \tag{9}$$

Equations 7, 8 and 9 can be combined into a single set of equations

$$\begin{bmatrix} 1 & 0 & 0 & -1 & 0 & 0 \\ 0 & 1 & 0 & 0 & -1 & 0 \\ 0 & 0 & 1 & 0 & 0 & -1 \\ 2 & 4 & 4 & 0 & 0 & 0 \\ 0 & 0 & 0 & -2 & 0 & 0 \\ 0 & 0 & 0 & 0 & 0 & 0 \end{bmatrix} \begin{bmatrix} r_3 \\ s_3 \\ t_3 \\ r_4 \\ s_4 \\ t_4 \end{bmatrix} = \begin{bmatrix} \begin{bmatrix} 0 \\ 0 \\ 144 \end{bmatrix} + \begin{bmatrix} 40 \\ 80 \\ 80 \end{bmatrix} + \begin{bmatrix} 72 \\ 0 \\ 0 \end{bmatrix} \\ 0 \\ 0 \\ 0 \end{bmatrix} \tag{10}$$

$\{R_{BA}\}^T \quad -\{R_{CA}\}^T$

and they have the solution

$$\begin{bmatrix} r_3 & s_3 & t_3 & r_4 & s_4 & t_4 \end{bmatrix} = \begin{bmatrix} 112 & 8 & -64 & 0 & -72 & 0 \end{bmatrix} \text{cm/s}^2.$$

Hence

$$\{\dot{\omega}_3^n\} = [R_{AB}] \frac{\{A_{AB}^n\}}{|R_{AB}|^2} = \frac{1}{36}\begin{bmatrix} 0 & -4 & 4 \\ 4 & 0 & -2 \\ -4 & 2 & 0 \end{bmatrix} \begin{bmatrix} 112 \\ 8 \\ -64 \end{bmatrix} = \begin{bmatrix} -8 \\ 16 \\ -12 \end{bmatrix} \text{rad/s}^2$$

and

$$\{\dot{\omega}_4\} = [R_{AC}] \frac{\{A_{AC}^n\}}{|R_{AC}|^2} = \frac{1}{4}\begin{bmatrix} 0 & 0 & 0 \\ 0 & 0 & 2 \\ 0 & -2 & 0 \end{bmatrix} \begin{bmatrix} 0 \\ -72 \\ 0 \end{bmatrix} = \begin{bmatrix} 0 \\ 0 \\ 36 \end{bmatrix} \text{rad/s}^2.$$

If the velocity determining equations had been written more generally then they would have appeared thus,

$$\begin{bmatrix} u_4 \\ v_4 \\ w_4 \end{bmatrix} = \begin{bmatrix} u_2 \\ v_2 \\ 0 \end{bmatrix} + \begin{bmatrix} u_3 \\ v_3 \\ w_3 \end{bmatrix},$$

$$\{V_{AB}\}^T \{R_{AB}\} = 0, \quad x_3 u_3 + y_3 v_3 + z_3 w_3 = 0$$

and

$$\{V_{AC}\}^T \{R_{AC}\} = 0, \quad x_4 u_4 + y_4 v_4 + 0 \cdot w_4 = 0,$$

in which case their combined matrix form would be

$$\begin{bmatrix} 1 & 0 & 0 & -1 & 0 & 0 \\ 0 & 1 & 0 & 0 & -1 & 0 \\ 0 & 0 & 1 & 0 & 0 & -1 \\ x_3 & y_3 & z_3 & 0 & 0 & 0 \\ 0 & 0 & 0 & -x_4 & -y_4 & -z_4 \\ 0 & 0 & 0 & 0 & 0 & 0 \end{bmatrix} \begin{bmatrix} u_3 \\ v_3 \\ w_3 \\ u_4 \\ v_4 \\ w_4 \end{bmatrix} = \begin{bmatrix} -u_2 \\ -v_2 \\ 0 \\ 0 \\ 0 \\ 0 \end{bmatrix}. \quad (11)$$

On comparing the accleration determining equations, Eqs. 10, with the above velocity determining equations it will be clear that their form is the same. It is

$$\begin{bmatrix} \text{Matrix characteristic} \\ \text{of mechanism} \\ \text{geometry.} \end{bmatrix} \begin{bmatrix} \text{Matrix} \\ \text{of} \\ \text{unknowns.} \\ \text{'Outputs'} \end{bmatrix} = \begin{bmatrix} \text{Matrix} \\ \text{of} \\ \text{knowns.} \\ \text{'Inputs'} \end{bmatrix}.$$

These matrices can be partitioned as follows. The 'characteristic' matrix becomes

$$\begin{bmatrix} \begin{bmatrix} \ddots & & \\ & 1 & \\ & & \ddots \end{bmatrix} & \begin{bmatrix} \ddots & -1 & \\ & & \ddots \\ & & \end{bmatrix} \\ \begin{bmatrix} \{R_{AB}\}^T \end{bmatrix} & \begin{bmatrix} 0 \end{bmatrix} \\ \begin{bmatrix} 0 \end{bmatrix} & \begin{bmatrix} -\{R_{AC}\}^T \end{bmatrix} \\ & \begin{bmatrix} 0 \end{bmatrix} \end{bmatrix}.$$

The 'output' matrices for velocity and acceleration determination are respectively

$$\begin{bmatrix} \{V_{AB}\} \\ \{V_{AC}\} \end{bmatrix} \quad \text{and} \quad \begin{bmatrix} \{A^n_{AB}\} \\ \{A^n_{AC}\} \end{bmatrix}.$$

Solution of Kinematics Problems

The 'input' matrices for velocity and acceleration determination are respectively

$$\begin{bmatrix} -\{V_B\} \\ \{0\} \end{bmatrix}$$

and

$$\begin{bmatrix} -\{A_B^P\} - \{A_{AB}^P\} - \{A_{AC}^P\} \\ \{0\} \end{bmatrix}.$$

A recognition of these characteristics in the formulation of the problem is particularly useful when a computer is employed to solve the equations.

Problem 3.9. In the linkage shown in Fig. 3.9 a rod, body 2, has its end A constrained to move along the x_1 axis of frame 1 fixed in body 1. The rod is also constrained so that it slides through a diametral hole in a sphere, body 3, at C. The sphere is free to rotate in a block which constrains the centre of the sphere to move along a path DE, which is fixed in body 1, in the $y_1 z_1$ plane and parallel to the z_1 axis.

In the given position of the linkage the velocity and acceleration of A and C, the centre of the sphere, are respectively

$$\{V_A\}_{1/1} = \begin{bmatrix} 2 \\ 0 \\ 0 \end{bmatrix} \text{ m/s}, \quad \{V_C\}_{1/1} = \begin{bmatrix} 0 \\ 0 \\ 1 \end{bmatrix} \text{ m/s}$$

$$\{A_A\}_{1/1} = \frac{100}{81}\begin{bmatrix} 5 \\ 0 \\ 0 \end{bmatrix} \text{ m/s}^2 \quad \text{and} \quad \{A_C\}_{1/1} = \frac{100}{81}\begin{bmatrix} 0 \\ 0 \\ 6 \end{bmatrix} \text{ m/s}^2.$$

Find

$$\{\overset{n}{\omega}_2\}_{1/1}, \ \{V_{BC}\}_{1/1}, \ \{A_{BC}\}_{2/1} \text{ and } \{\overset{\cdot n}{\omega}_2\}_{1/1}.$$

Solution. Since the suffixes are, with one exception 1/1 throughout the solution they can be omitted in other than the exceptional case. The exception is the vector

$$\{A_{BC}\}_2,$$

(See paragraph 1.13 and in particular Eqs. 1.69 and 1.70).

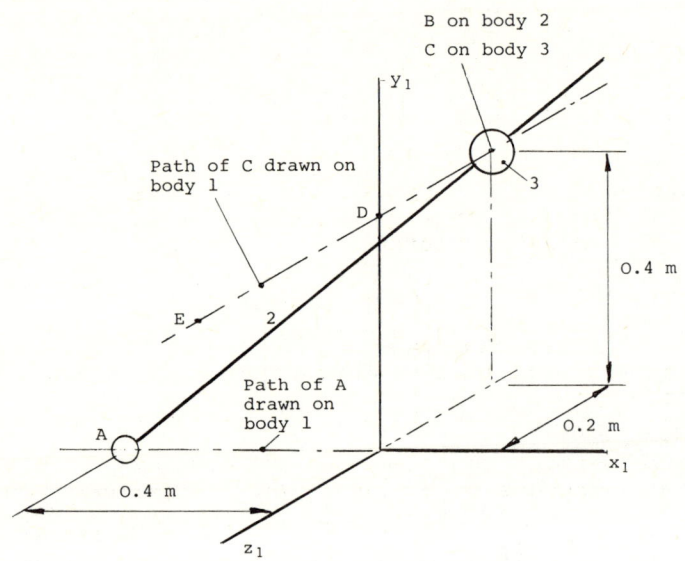

Fig. 3.9.

Now

$$\{R_A\} = \begin{bmatrix} -0.4 \\ 0 \\ 0 \end{bmatrix} \text{ m} \;, \quad \{R_B\} = \begin{bmatrix} 0 \\ 0.4 \\ -0.2 \end{bmatrix} \text{ m}$$

and therefore

$$\{R_{AB}\} = \begin{bmatrix} 0 \\ 0.4 \\ -0.2 \end{bmatrix} - \begin{bmatrix} -0.4 \\ 0 \\ 0 \end{bmatrix} = 0.2 \begin{bmatrix} 2 \\ 2 \\ -1 \end{bmatrix} \text{ m} \;. \tag{1}$$

Also

$$\{V_C\} = \{V_A\} + \{V_{BA}\} + \{V_{BC}\} \tag{2}$$

and

$$\{R_{BA}\}^T \{V_{BA}\} = 0 \;. \tag{3}$$

If

$$\{V_{BA}\} = \begin{bmatrix} u \\ v \\ w \end{bmatrix} \text{ and } \{V_{BC}\} = \frac{V\{R_{BA}\}}{|R_{BA}|}$$

then Eqs. 2 and 3 become

$$\begin{bmatrix} u \\ v \\ w \end{bmatrix} + \frac{V}{3}\begin{bmatrix} 2 \\ 2 \\ -1 \end{bmatrix} = \begin{bmatrix} 0 \\ 0 \\ 1 \end{bmatrix} - \begin{bmatrix} 2 \\ 0 \\ 0 \end{bmatrix} \qquad (4)$$

and

$$2u + 2v - w = 0 \qquad (5)$$

Equations 4 and 5 can be written

$$\begin{bmatrix} 1 & 0 & 0 & 2/3 \\ 0 & 1 & 0 & 2/3 \\ 0 & 0 & 1 & -1/3 \\ 2 & 2 & -1 & 0 \end{bmatrix} \begin{bmatrix} u \\ v \\ w \\ V \end{bmatrix} = \begin{bmatrix} -2 \\ 0 \\ 1 \\ 0 \end{bmatrix} . \qquad (6)$$

To determine velocities for a series of configurations of the linkage it would be advantageous to express Eqs. 6 in the more general form

$$\begin{bmatrix} 1 & 0 & 0 & \dfrac{\{R_{BA}\}}{|R_{BA}|} \\ 0 & 1 & 0 & \\ 0 & 0 & 1 & \\ \{R_{BA}\}^T & & 0 \end{bmatrix} \begin{bmatrix} u \\ v \\ w \\ V \end{bmatrix} = \begin{bmatrix} \{V_C\} - \{V_A\} \\ \\ 0 \end{bmatrix} . \qquad (7)$$

The solution of Eqs. 6 give

$$\{V_{BA}\} = \frac{2}{9}\begin{bmatrix} -4 \\ 5 \\ 2 \end{bmatrix} \text{ m/s} \quad \text{and} \quad \{V_{BC}\} = \frac{5}{9}\begin{bmatrix} -2 \\ -2 \\ 1 \end{bmatrix} \text{ m/s}.$$

The angular velocity of the rod, body 2, is given by

$$\{\omega_2^n\} = \frac{[R_{BA}]\{V_{BA}\}}{|R_{BA}|^2} = \frac{0.2 \times 2}{0.36 \times 9}\begin{bmatrix} 0 & 1 & 2 \\ -1 & 0 & -2 \\ -2 & 2 & 0 \end{bmatrix}\begin{bmatrix} -4 \\ 5 \\ 2 \end{bmatrix} = \frac{10}{9}\begin{bmatrix} 1 \\ 0 \\ 2 \end{bmatrix} \text{ rad/s}.$$

The acceleration of B relative to A parallel to $\{R_{BA}\}$ is given by

$$\{A_{BA}^P\} = [\omega_2^n]\{V_{BA}\} = \frac{10 \times 2}{9 \times 9}\begin{bmatrix} 0 & -2 & 0 \\ 2 & 0 & -1 \\ 0 & 1 & 0 \end{bmatrix}\begin{bmatrix} -4 \\ 5 \\ 2 \end{bmatrix} = \frac{100}{81}\begin{bmatrix} -2 \\ -2 \\ 1 \end{bmatrix} \text{ m/s}^2$$

and the accleration of C relative to B which is normal to $\{R_{BA}\}$ is given by

$$2\left[\omega_2^n\right]\{V_{BC}\} = \frac{2 \times 10 \times 5}{9 \times 9}\begin{bmatrix} 0 & -2 & 0 \\ 2 & 0 & -1 \\ 0 & 1 & 0 \end{bmatrix}\begin{bmatrix} -2 \\ -2 \\ 1 \end{bmatrix} = \frac{100}{81}\begin{bmatrix} 4 \\ -5 \\ -2 \end{bmatrix} \text{ m/s}^2.$$

For the relative acclerations

$$\{A_C\} = \{A_A\} + \{A_{BA}\} + \{A_{CB}\}$$
$$= \{A_A\} + \{A_{BA}^n\} + \{A_{BA}^p\} + 2\left[\omega_2^n\right]\{V_{CB}\} + \{A_{CB}\}_{2/1} \qquad (8)$$

or

$$\{A_{BA}^n\} + \{A_{CB}\}_{2/1} = \{A_C\} - \{A_A\} - \{A_{BA}^p\} - 2\left[\omega_2^n\right]\{V_{CB}\}$$

$$\begin{bmatrix} r \\ s \\ t \end{bmatrix} + \frac{A\{R_{BA}\}}{|R_{BA}|} = \frac{100}{81}\begin{bmatrix} 0 - 5 + 2 - 4 \\ 0 - 0 + 2 + 5 \\ 6 - 0 - 1 + 2 \end{bmatrix}. \qquad (9)$$

Also, since

$$\{R_{BA}\}^T\{A_{BA}^n\} = 0,$$

it follows that the acceleration determining equations can be written in the combined form

$$\begin{bmatrix} 1 & 0 & 0 & 2/3 \\ 0 & 1 & 0 & 2/3 \\ 0 & 0 & 1 & -1/3 \\ 2 & 2 & -1 & 0 \end{bmatrix}\begin{bmatrix} r \\ s \\ t \\ A \end{bmatrix} = \frac{700}{81}\begin{bmatrix} -1 \\ 1 \\ 1 \\ 0 \end{bmatrix}. \qquad (10)$$

The solution of Eqs. 10 give

$$\{A_{BA}^n\} = \frac{700}{81}\begin{bmatrix} 1 \\ 3 \\ -1 \end{bmatrix} \text{ m/s}^2 \quad \text{and} \quad \{A_{CB}\}_{2/1} = \frac{700}{81}\begin{bmatrix} -2 \\ -2 \\ 1 \end{bmatrix} \text{ m/s}^2.$$

The angular acceleration of body 2 is given by

$$\{\dot{\omega}_2^n\} = \frac{\left[R_{BA}\right]\{A_{BA}^n\}}{|R_{BA}|^2} = \frac{0.2 \times 700}{0.36 \times 81}\begin{bmatrix} 0 & 1 & 2 \\ -1 & 0 & -2 \\ -2 & 2 & 0 \end{bmatrix}\begin{bmatrix} 1 \\ 3 \\ -1 \end{bmatrix} = \begin{bmatrix} -14.4 \\ 4.8 \\ 19.2 \end{bmatrix} \text{ rad/s}^2.$$

Problem 3.10. The position of a point P is measured in a reference frame 4 which is fixed in the earth as shown in Fig. 3.10. Obtain an expression for

$$\{A_{PO}\}_{1/4}$$

where frame 1 is fixed in the stars (an inertial frame of reference).

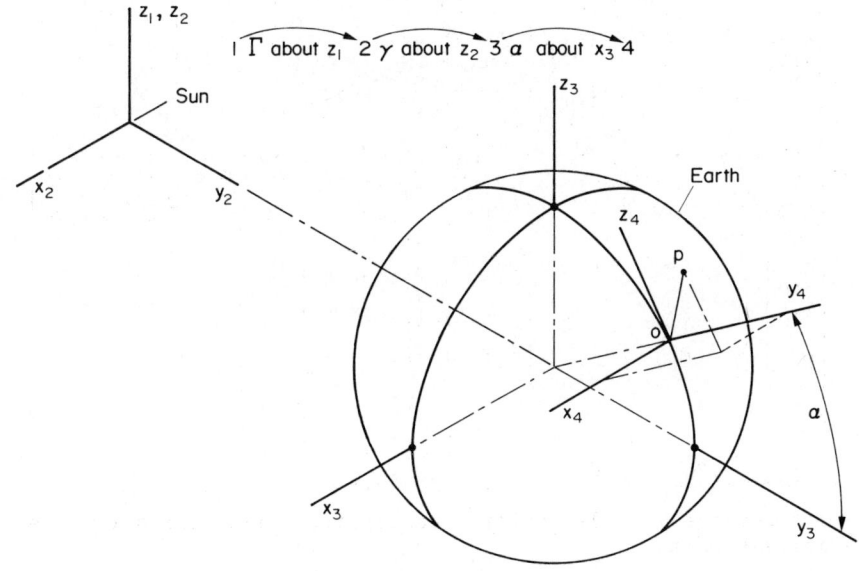

Fig. 3.10.

Solution. Introduce a frame 2 with its origin at the centre of the sun such that the y_2 axis passes through the centre of the earth. Also introduce a frame 3 which is fixed in the earth with its z_3 axis parallel to the z_2 axis of frame 2. Hence

$$\{\omega_4\}_1 = \{\omega_2\}_1 + \{\omega_3\}_2 + \{\omega_4\}_3 = \{\omega_3\}_1$$

since $\{\omega_4\}_3$ is a null vector. Therefore

$$\{\omega_3\}_{1/4} = \{\omega_4\}_{1/4} = [\ell_3]_4 \{\{\omega_2\}_{1/1} + \{\omega_3\}_{2/2}\}$$

$$= \begin{bmatrix} 1 & 0 & 0 \\ 0 & c\alpha & s\alpha \\ 0 & -s\alpha & c\alpha \end{bmatrix} \left[\begin{bmatrix} 0 \\ 0 \\ \Omega \end{bmatrix} + \begin{bmatrix} 0 \\ 0 \\ \omega_e \end{bmatrix} \right] = (\Omega + \omega_e) \begin{bmatrix} 0 \\ s\alpha \\ c\alpha \end{bmatrix}$$

since $[\ell_1]_2$ and $[\ell_2]_3$ have no effect of the transformation of $\{\omega_2\}_{1/1}$

and $\{\omega_3\}_{2/2}$. Now

$$\{R_{PO}\}_{1/1} = [\ell_2]_1[\ell_3]_2[\ell_4]_3\{R_{PO}\}_{4/4}$$

which, on differentiation with respect to time gives

$$\{V_{PO}\}_{1/1} = \left[[\omega_2]_{1/1}[\ell_4]_1 + [\ell_2]_1[\omega_3]_{2/2}[\ell_4]_3\right]\{R_{PO}\}_{4/4}$$

$$+ [\ell_4]_1\{V_{PO}\}_{4/4}$$

and therefore

$$\{V_{PO}\}_{1/4} = \left[[\omega_2]_{1/4} + [\omega_3]_{2/4}\right]\{R_{PO}\}_{4/4} + \{V_{PO}\}_{4/4}$$

$$= [\omega_3]_{1/4}\{R_{PO}\}_{4/4} + \{V_{PO}\}_{4/4}$$

$$= \omega\begin{bmatrix} 0 & -c\alpha & s\alpha \\ c\alpha & 0 & 0 \\ -s\alpha & 0 & 0 \end{bmatrix}\begin{bmatrix} x \\ y \\ z \end{bmatrix} + \begin{bmatrix} \dot{x} \\ \dot{y} \\ \dot{z} \end{bmatrix}$$

$$= \begin{bmatrix} \omega(z\sin\alpha - y\cos\alpha) + \dot{x} \\ \omega x\cos\alpha + \dot{y} \\ -\omega x\sin\alpha + \dot{z} \end{bmatrix}$$

where $\omega = \Omega + \omega_e$. Since

$$\{V_{PO}\}_{1/1} = [\ell_4]_1\{V_{PO}\}_{1/4},$$

$$\{A_{PO}\}_{1/1} = [\omega_4]_{1/1}[\ell_4]_1\{V_{PO}\}_{1/4} + [\ell_4]_1\frac{d}{dt}\{V_{PO}\}_{1/4}$$

and therefore

$$\{A_{PO}\}_{1/4} = [\omega_4]_{1/4}\{V_{PO}\}_{1/4} + \frac{d}{dt}\{V_{PO}\}_{1/4}$$

$$= \omega\begin{bmatrix} 0 & -c\alpha & s\alpha \\ c\alpha & 0 & 0 \\ -s\alpha & 0 & 0 \end{bmatrix}\begin{bmatrix} \omega(zs\alpha - yc\alpha) + \dot{x} \\ \omega xc\alpha + \dot{y} \\ -\omega xs\alpha + \dot{z} \end{bmatrix}$$

$$+ \begin{bmatrix} \omega(\dot{z}s\alpha - \dot{y}c\alpha) + \ddot{x} \\ \omega\dot{x}c\alpha + \ddot{y} \\ -\omega\dot{x}s\alpha + \ddot{z} \end{bmatrix}$$

$$= \begin{bmatrix} \ddot{x} - \omega^2 x - 2\omega(\dot{y}c\alpha - \dot{z}s\alpha) \\ \ddot{y} + \omega^2(zs\alpha c\alpha - yc^2\alpha) + 2\omega\dot{x}c\alpha \\ \ddot{z} + \omega^2(ys\alpha c\alpha - zs^2\alpha) - 2\omega\dot{x}s\alpha \end{bmatrix}.$$

Now $\Omega = 200 \times 10^{-9}$ rad/s and $\omega_e = 72.7 \times 10^{-6}$ rad/s therefore terms involving ω^2 and Ω can be neglected. Thus

$$\{A_{PO}\}_{1/4} \simeq \begin{bmatrix} \ddot{x} - 2\omega_e(\dot{y}\cos\alpha - \dot{z}\sin\alpha) \\ \ddot{y} + 2\omega_e \dot{x}\cos\alpha \\ \ddot{z} - 2\omega_e \dot{x}\sin\alpha \end{bmatrix}.$$

Accelerations measured in frame 4, $\ddot{x}$, $\ddot{y}$ and $\ddot{z}$, will thus be close to those measured in frame 1, provided that the accelerations $\omega_e \dot{x}$, $\omega_e \dot{y}$ and $\omega_e \dot{z}$ are small compared with $\ddot{x}$, $\ddot{y}$ and $\ddot{z}$.

Problem 3.11. Figure 3.11 shows part of an epicyclic gear train in which the epicyclic arm, body 2, and the annular wheel, body 4, are driven at a constant rate relative to the gear case, body 1. The planet wheel, body 3, is carried on the arm and meshes with the annular wheel.

At a given instant of time, frames 1, 2, 3 and 4 are aligned and B on the planet wheel is coincident with C on the annular wheel. In the subsequent motion frames 2 and 4 rotate about the y_1 axis through the angles α and β respectively relative to frame 1, while frame 3 rotates relative to frame 2 about the y_3 axis through the angle γ. Obtain a general expression for

$$\{A_{BO}\}_{1/2}.$$

Solution. Now

$$\{R_{BO}\}_{1/1} = [\ell_2]_1 \{R_{AO}\}_{2/2} + [\ell_2]_1 [\ell_3]_2 \{R_{BA}\}_{3/3}$$

and therefore

$$\{V_{BO}\}_{1/1} = [\omega_2]_{1/1}[\ell_2]_1 \{R_{AO}\}_{2/2} + [\omega_2]_{1/1}[\ell_2]_1[\ell_3]_2\{R_{BA}\}_{3/3}$$
$$+ [\ell_2]_1[\omega_3]_{2/2}[\ell_3]_2\{R_{BA}\}_{3/3}$$

or

$$\{V_{BO}\}_{1/2} = [\omega_2]_{1/2}\{R_{AO}\}_{2/2}$$
$$+ \bigl[[\omega_2]_{1/2} + [\omega_3]_{2/2}\bigr][\ell_3]_2\{R_{BA}\}_{3/3}$$

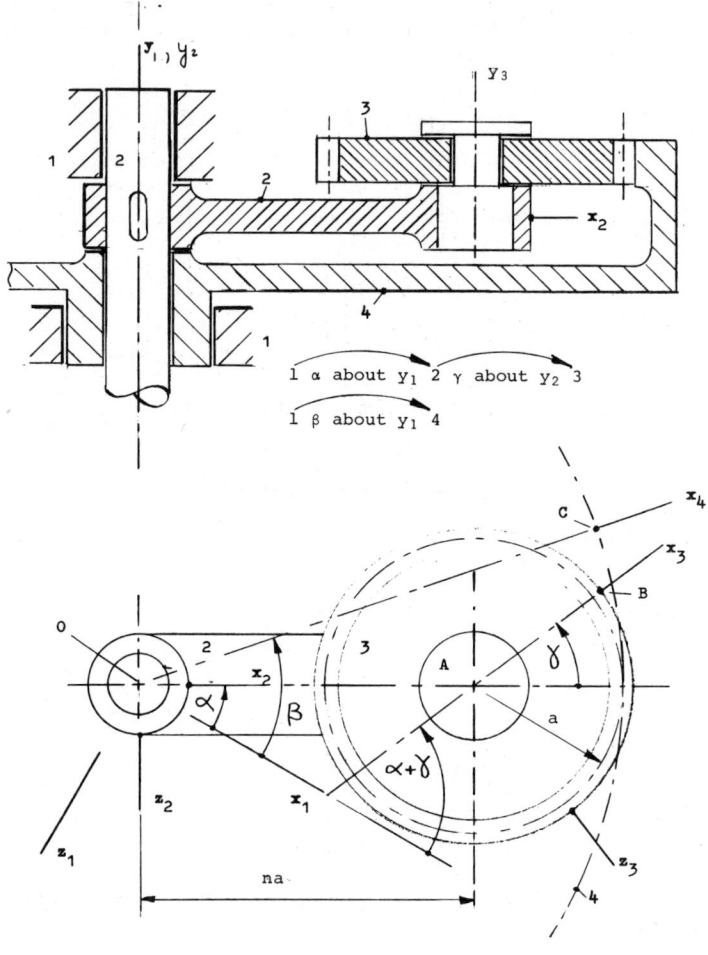

Fig. 3.11.

$$= na\dot{\alpha} \begin{bmatrix} 0 & 0 & 1 \\ 0 & 0 & 0 \\ -1 & 0 & 0 \end{bmatrix} \begin{bmatrix} 1 \\ 0 \\ 0 \end{bmatrix}$$

$$+ (\dot{\alpha} + \dot{\gamma})a \begin{bmatrix} 0 & 0 & 1 \\ 0 & 0 & 0 \\ -1 & 0 & 0 \end{bmatrix} \begin{bmatrix} c\gamma & 0 & s\gamma \\ 0 & 1 & 0 \\ -s\gamma & 0 & c\gamma \end{bmatrix} \begin{bmatrix} 1 \\ 0 \\ 0 \end{bmatrix}$$

$$= \begin{bmatrix} -a(\dot{\alpha} + \dot{\gamma})\sin\gamma \\ 0 \\ -na\dot{\alpha} - a(\dot{\alpha} + \dot{\gamma})\cos\gamma \end{bmatrix}.$$

Also
$$\{R_{CO}\}_{1/1} = [\ell_4]_1 \{R_{CO}\}_{4/4}$$
and therefore
$$\{V_{CO}\}_{1/1} = [\omega_4]_{1/1}[\ell_4]_1 \{R_{CO}\}_{4/4}$$
or
$$\{V_{CO}\}_{1/2} = [\omega_4]_{1/2}[\ell_1]_2[\ell_4]_1 \{R_{CO}\}_{4/4}$$

$$= ma\dot\beta \begin{bmatrix} 0 & 0 & 1 \\ 0 & 0 & 0 \\ -1 & 0 & 0 \end{bmatrix} \begin{bmatrix} c\alpha & 0 & -s\alpha \\ 0 & 1 & 0 \\ s\alpha & 0 & c\alpha \end{bmatrix} \begin{bmatrix} c\beta & 0 & s\beta \\ 0 & 1 & 0 \\ -s\beta & 0 & c\beta \end{bmatrix} \begin{bmatrix} 1 \\ 0 \\ 0 \end{bmatrix}$$

$$= ma\dot\beta \begin{bmatrix} \cos\beta\sin\alpha - \sin\beta\cos\alpha \\ 0 \\ -(\cos\beta\cos\alpha + \sin\beta\sin\alpha) \end{bmatrix}$$

where $m = (n+1)$. Equating

$$\{V_{BO}\}_{1/2}\bigg|_{\gamma=0} \quad \text{to} \quad \{V_{CO}\}_{1/2}\bigg|_{\substack{\alpha=0 \\ \beta=0}}$$

gives $\dot\gamma = m(\dot\beta - \dot\alpha)$ and $\dot\alpha + \dot\gamma = m\dot\beta - n\dot\alpha = p$. Therefore

$$\{V_{BO}\}_{1/2} = -a \begin{bmatrix} p\sin\gamma \\ 0 \\ n\dot\alpha + p\cos\gamma \end{bmatrix}$$

where $\gamma = m(\dot\beta - \dot\alpha)t$, t being the time which has elapsed since B and C were coincident. Hence

$$\{A_{BO}\}_{1/2} = [\omega_2]_{1/2}\{V_{BO}\}_{1/2} + \frac{d}{dt}\{V_{BO}\}_{1/2}$$

$$= -a\dot\alpha \begin{bmatrix} 0 & 0 & 1 \\ 0 & 0 & 0 \\ -1 & 0 & 0 \end{bmatrix} \begin{bmatrix} ps\gamma \\ 0 \\ n\dot\alpha + pc\gamma \end{bmatrix} + a p\dot\gamma \begin{bmatrix} -c\gamma \\ 0 \\ s\gamma \end{bmatrix}$$

$$= \begin{bmatrix} -(ap^2\cos\gamma + n\dot\alpha^2) \\ 0 \\ ap^2\sin\gamma \end{bmatrix}.$$

106   Matrix Methods in Engineering Mechanics

Problem 3.12. Body 2 rotates at a constant rate of relative to body 1 about the z_1 axis as shown in Fig. 3.12a. Body 2 carries a rotor, body 3, which rotates at a constant rate ω relative to body 2 about an axis which is parallel to the x_2 axis.

Find
$$\{\omega_3\}_{1/2} , \{\omega_3\}_{1/1} , \{\dot{\omega}_3\}_{1/2} \text{ and } \{\dot{\omega}_3\}_{1/1}.$$
Illustrate the results by drawing appropriate vectors.

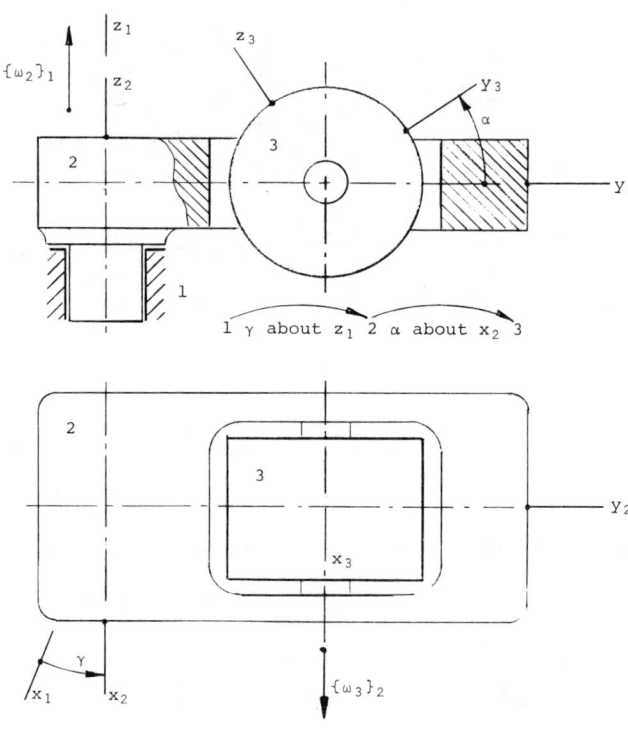

Fig. 3.12a.

Solution. Now
$$\{\omega_3\}_1 = \{\omega_2\}_1 + \{\omega_3\}_1$$
as shown in Fig. 3.12b. Hence
$$\{\omega_3\}_{1/1} = \{\omega_2\}_{1/1} + \{\omega_3\}_{2/1}$$
$$= \{\omega_2\}_{1/1} + [\ell_2]_1 \{\omega_3\}_{2/2} \qquad (1)$$

Solution of Kinematics Problems

$$= \begin{bmatrix} 0 \\ 0 \\ \Omega \end{bmatrix} + \begin{bmatrix} c\gamma & -s\gamma & 0 \\ s\gamma & c\gamma & 0 \\ 0 & 0 & 1 \end{bmatrix} \begin{bmatrix} \omega \\ 0 \\ 0 \end{bmatrix} = \begin{bmatrix} \omega\cos\gamma \\ \omega\sin\gamma \\ \Omega \end{bmatrix} \quad (2)$$

as show in Fig. 3.12c.

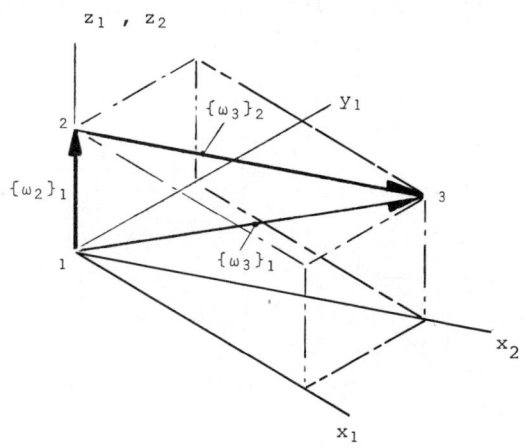

Fig. 3.12b.

Also

$$\{\omega_3\}_{1/2} = [\ell_1]_2 \{\omega_2\}_{1/1} + \{\omega_3\}_{2/2}$$

$$= \begin{bmatrix} c\gamma & s\gamma & 0 \\ -s\gamma & c\gamma & 0 \\ 0 & 0 & 1 \end{bmatrix} \begin{bmatrix} 0 \\ 0 \\ \Omega \end{bmatrix} + \begin{bmatrix} \omega \\ 0 \\ 0 \end{bmatrix} = \begin{bmatrix} \omega \\ 0 \\ \Omega \end{bmatrix}$$

or alternatively,

$$\{\omega_3\}_{1/2} = [\ell_1]_2 \{\omega_3\}_{1/1} .$$

Differentiation of Eq. 1 with respect to time gives

$$\{\dot{\omega}_3\}_{1/1} = \{\dot{\omega}_2\}_{1/1} + [\omega_2]_{1/1}[\ell_2]_1\{\omega_3\}_{2/2} + [\ell_2]_1\{\dot{\omega}_3\}_{2/2},$$

but in this case $\{\dot{\omega}_2\}_1$ and $\{\dot{\omega}_3\}_2$ are null vectors since Ω and ω are constants. Therefore

$$\{\dot{\omega}_3\}_{1/1} = [\omega_2]_{1/1}[\ell_2]_1\{\omega_3\}_{2/2}$$

and since, from Eq. 1

$$[\ell_2]_1\{\omega_3\}_{2/2} = \{\omega_3\}_{1/1} - \{\omega_2\}_{1/1} ,$$

$$\{\dot{\omega}_3\}_{1/1} = [\omega_2]_{1/1}\{\{\omega_3\}_{1/1} - \{\omega_2\}_{1/1}\} = [\omega_2]_{1/1}\{\omega_3\}_{1/1}$$

because

108   Matrix Methods in Engineering Mechanics

$[\omega_2]_{1/1}\{\omega_2\}_{1/1}$

is a null matrix. Hence

$$\{\dot{\omega}_3\}_{1/1} = \begin{bmatrix} 0 & -\Omega & 0 \\ \Omega & 0 & 0 \\ 0 & 0 & 0 \end{bmatrix} \begin{bmatrix} \omega c\gamma \\ \omega s\gamma \\ \Omega \end{bmatrix} = \begin{bmatrix} -\omega\Omega\sin\gamma \\ \omega\Omega\cos\gamma \\ 0 \end{bmatrix}$$

as shown in Fig. 3.12d.

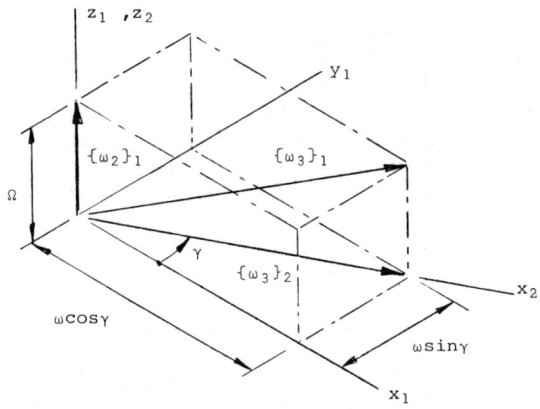

Fig. 3.12c

Alternatively, $\{\dot{\omega}_3\}_{1/1}$ can be obtained directly by differentiation of Eq. 2. Also

$$\{\dot{\omega}_3\}_{1/2} = [\ell_1]_2 \{\dot{\omega}_3\}_{1/1} = \begin{bmatrix} c\gamma & s\gamma & 0 \\ -s\gamma & c\gamma & 0 \\ 0 & 0 & 1 \end{bmatrix} \begin{bmatrix} -\omega\Omega s\gamma \\ \omega\Omega c\gamma \\ 0 \end{bmatrix} = \begin{bmatrix} 0 \\ \omega\Omega \\ 0 \end{bmatrix}.$$

Alternatively

$$\{\dot{\omega}_3\}_{1/2} = [\omega_2]_{1/2}\{\omega_3\}_{1/2} + \frac{d}{dt}\{\omega_3\}_{1/2}$$

$$= \begin{bmatrix} 0 & -\Omega & 0 \\ \Omega & 0 & 0 \\ 0 & 0 & 0 \end{bmatrix} \begin{bmatrix} \omega \\ 0 \\ \Omega \end{bmatrix} + \begin{bmatrix} 0 \\ 0 \\ 0 \end{bmatrix}$$

$$= \begin{bmatrix} 0 \\ \omega\Omega \\ 0 \end{bmatrix}.$$

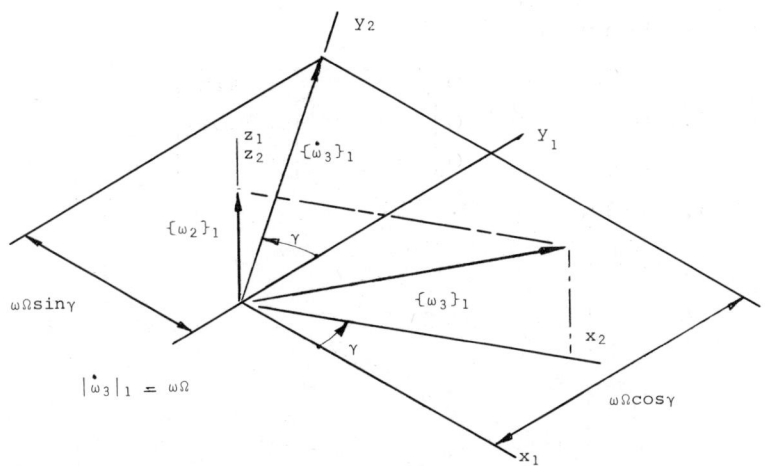

Fig. 3.12d.

Problem 3.13. Figure 3.13a shows a part section of a ball thrust race. Determine, for the case in which the upper track, body 2, runs at a constant angular velocity relative to the lower track, body 1, and the balls roll without slip, the angular acceleration of a typical ball, body 4. Also find the linear acceleration of the centre of mass C of the ball and the velocity with which the cage rubs on the ball.

Solution. Consider the motion of the ball and in particular its angular velocity. There will be a component of angular velocity parallel to the y_3 axis of a frame fixed in the cage, body 3, as a result of the velocity which A has relative to B. The component angular velocity parallel to the x_3 axis must be zero since there is no slip at the points of contact A and B. If the ball has a component angular velocity parallel to the z_3 axis, then an instant later when the line BCA has moved to the position indicated in Fig. 3.13b a component of this angular velocity will exist parallel to the x_3 axis. Since this component parallel to the x_3 axis must be zero, so alos must the component parallel to the z_3 axis. The relative positions of the frames indicated in Fig. 3.13a are thus adequate in the description of the motion. While the angle θ can be made positive, the angles γ and β will not necessarily be positive.

Consider the point O as a point fixed in body 2 and also as a point fixed in an imaginary extension of body 4. Such points have no relative motion and are therefore points on the instantaneous axis for the relative motion of bodies 2 and 4. Further consider a point A in body 2 and as a coincident point in body 4. Such points have, instantaneously, no relative motion and are similarly points on the instan-

110 Matrix Methods in Engineering Mechanics

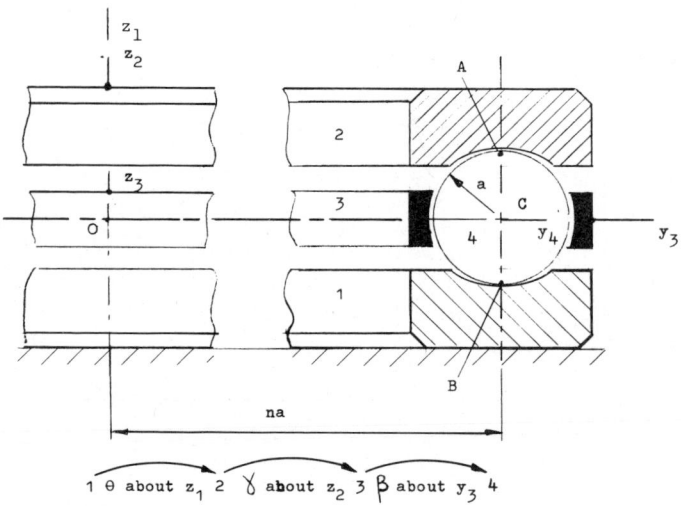

Fig. 3.13a.

taneous axis for the motion of body 2 relative to body 4. Hence OA is the instantaneous axis for the relative motion of bodies 2 and 4 and $\{\omega_4\}_2$ is parallel to OA. By a similar argument, the vector $\{\omega_4\}_1$ is parallel to OB. The angular velocity vector diagram can thus be constructed as follows.

$\{\omega_2\}_1$ 1 ⟶ 2 ω long and parallel to the z_1 axis.

$\{\omega_4\}_1$ 1 ⟶ 4 parallel to OB. The position of point 4 is not defined at this stage.

$\{\omega_4\}_2$ 2 ⟶ 4 parallel to OA. The intersection of lines 1 4 and 1 2 fixes the position of point 4.

$\{\omega_3\}_1$ 1 ⟶ 3 parallel to the z_1 axis. The position of point 3 on 1 2 is not defined at this stage.

$\{\omega_3\}_4$ 4 ⟶ 3. If the ball had a spin about the axis AC, then an instant later the existence of such an angular velocity would require a component angular velocity parallel to the x_3 axis, and this is not possible as explained above. Thus 4 ⟶ 3 is parallel to the y_3 axis and this vector fixes the position of 3 on the line 1 2.

The angular velocity vector diagram is thus as shown in Fig. 3.13c. The vector $\{\omega_3\}_2$ is negatively directed along the z_3 axis and therefore γ is negative. Similarly, since $\{\omega_4\}_3$ is negatively directed along the y_3 axis β is negative.

Now
$$\{\omega_4\}_1 = \{\omega_2\}_1 + \{\omega_3\}_2 + \{\omega_4\}_3 \tag{1}$$

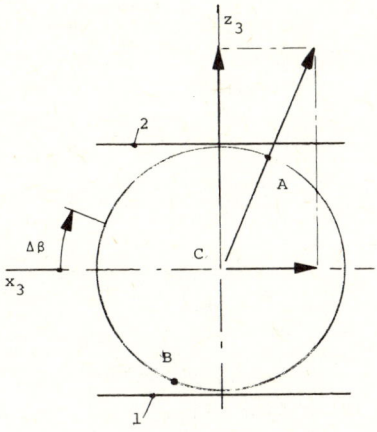

Fig. 3.13b.

and therefore, by reference to the angular velocity vector diagram,

$$\{\omega_4\}_{1/3} = \{\omega_2\}_{1/3} + \{\omega_3\}_{2/3} + \{\omega_4\}_{3/3}$$

$$= \begin{bmatrix} 0 \\ 0 \\ \omega \end{bmatrix} + \begin{bmatrix} 0 \\ 0 \\ -\omega/2 \end{bmatrix} + \begin{bmatrix} 0 \\ -n\omega/2 \\ 0 \end{bmatrix} = \begin{bmatrix} 0 \\ -n\omega/2 \\ \omega/2 \end{bmatrix} . \quad (2)$$

The angular acceleration of body 4, measured in frame 1 and referred to frame 3, is given by

$$\{\dot{\omega}_4\}_{1/3} = [\omega_3]_{1/3}\{\omega_4\}_{1/3} + \frac{d}{dt}\{\omega_4\}_{1/3} \quad (3)$$

From Eq. 2

$$\{\omega_3\}_{1/3} = \{\omega_4\}_{1/3} - \{\omega_4\}_{3/3}$$

and this result in Eq. 3, noting the the last term in this equation is a null matrix, gives

$$\{\dot{\omega}_4\}_{1/3} = -[\omega_4]_{3/3}\{\omega_4\}_{1/3} = -\frac{n\omega^2}{4}\begin{bmatrix} 0 & 0 & -1 \\ 0 & 0 & 0 \\ 1 & 0 & 0 \end{bmatrix}\begin{bmatrix} 0 \\ -n \\ 1 \end{bmatrix}$$

$$= \frac{n\omega^2}{4}\begin{bmatrix} 1 \\ 0 \\ 0 \end{bmatrix} .$$

The velocity of the centre of the ball is found by noting that C is a

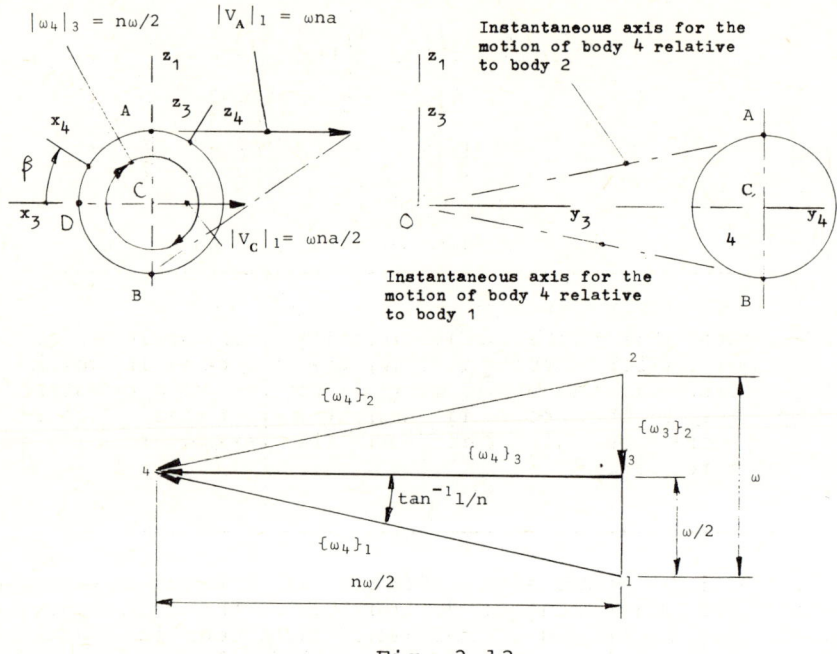

Fig. 3.13c.

fixed point in frame 3, is given by

$$\{V_C\}_{1/3} = \{V_4\}_{1/3} = [\omega_3]_{1/3}\{R_C\}_{3/3} = \frac{na\omega}{4}\begin{bmatrix} 0 & 0 & -1 \\ 0 & 0 & 0 \\ 1 & 0 & 0 \end{bmatrix}\begin{bmatrix} 0 \\ 1 \\ 0 \end{bmatrix}$$

$$= \frac{na\omega}{2}\begin{bmatrix} -1 \\ 0 \\ 0 \end{bmatrix}.$$

The acceleration of the centre of the ball is thus

$$\{\dot{V}_4\}_{1/3} = [\omega_3]_{1/3}\{V_4\}_{1/3} + \frac{d}{dt}\{V_4\}_{1/3}$$

and since the last term in this equation is a null matrix,

$$\{A_4\}_{1/3} = \frac{na\omega^2}{4}\begin{bmatrix} 0 & -1 & 0 \\ 1 & 0 & 0 \\ 0 & 0 & 0 \end{bmatrix}\begin{bmatrix} -1 \\ 0 \\ 0 \end{bmatrix} = \frac{na\omega^2}{4}\begin{bmatrix} 0 \\ -1 \\ 0 \end{bmatrix}.$$

The rubbing velocity between D a point on the ball (Fig. 3.13c) and E, a corresponding point on the cage, will be given by

Solution of Kinematics Problems

$$\{V_{DC}\}_{1/3} - \{V_{EC}\}_{1/3} = [\omega_4]_{1/3}\{R_{DC}\}_{3/3} - [\omega_3]_{1/3}\{R_{EC}\}_{3/3}$$

$$= [\omega_4]_{3/3}\{R_{DC}\}_{3/3}$$

$$= \frac{n\omega}{2}\begin{bmatrix} 0 & 0 & -1 \\ 0 & 0 & 0 \\ 1 & 0 & 0 \end{bmatrix}\begin{bmatrix} a \\ 0 \\ 0 \end{bmatrix} = \frac{na\omega}{2}\begin{bmatrix} 0 \\ 0 \\ 1 \end{bmatrix}.$$

Problem 3.14. Determine the angular velocity and acceleration of a roller in a taper roller thrust bearing for the case in which the inner ring is fixed and the outer ring is driven at a constant angular velocity. Assume that rolling without slip takes place at the roller and ring contacts. Also determine the maximum rubbing velocity of the cage on a roller.

Solution. Figure 3.14 shows a half section of a taper roller thrust bearing. The fixed inner ring is designated 1, the cage 2, a roller 4 and the outer ring 5. Frames of reference have been introduced with the relative positions indicated. The angular velocity vector diagram is constructed as follows.

$\{\omega_5\}_1$ 1 ⟶ 5   long and parallel to the z_1 axis.

$\{\omega_4\}_1$ 1 ⟶ 4   parallel to the generator of the cone on the inner ring in the given section. The position of point 4 is not defined at this stage.

$\{\omega_5\}_4$ 5 ⟶ 4   parallel to the generator of the cone on the outer ring given in the section. The intersection of lineds 1 4 and 5 4 fixes the position of the point 4.

$\{\omega_4\}_3$ 3 ⟶ 4   parallel to the z_3 or roller axis to fix the position of point 3 on the line 1 5. Points 3 and 2 are coincident because frames 2 and 3 have no relative motion. Hence $\{\omega_4\}_3 = \{\omega_4\}_2$.

By the application of the sine rule to triangles 145 and 124

$$|\omega_4|_1 = \frac{\omega\sin(\theta + \beta)}{\sin 2\beta}, \quad |\omega_2|_1 = \frac{\omega\sin(\theta + \beta)\sin\beta}{\sin 2\beta \sin\theta}$$

and

$$|\omega_4|_2 = \frac{\omega\sin(\theta + \beta)\sin(\theta - \beta)}{\sin 2\beta \sin\theta}.$$

Hence

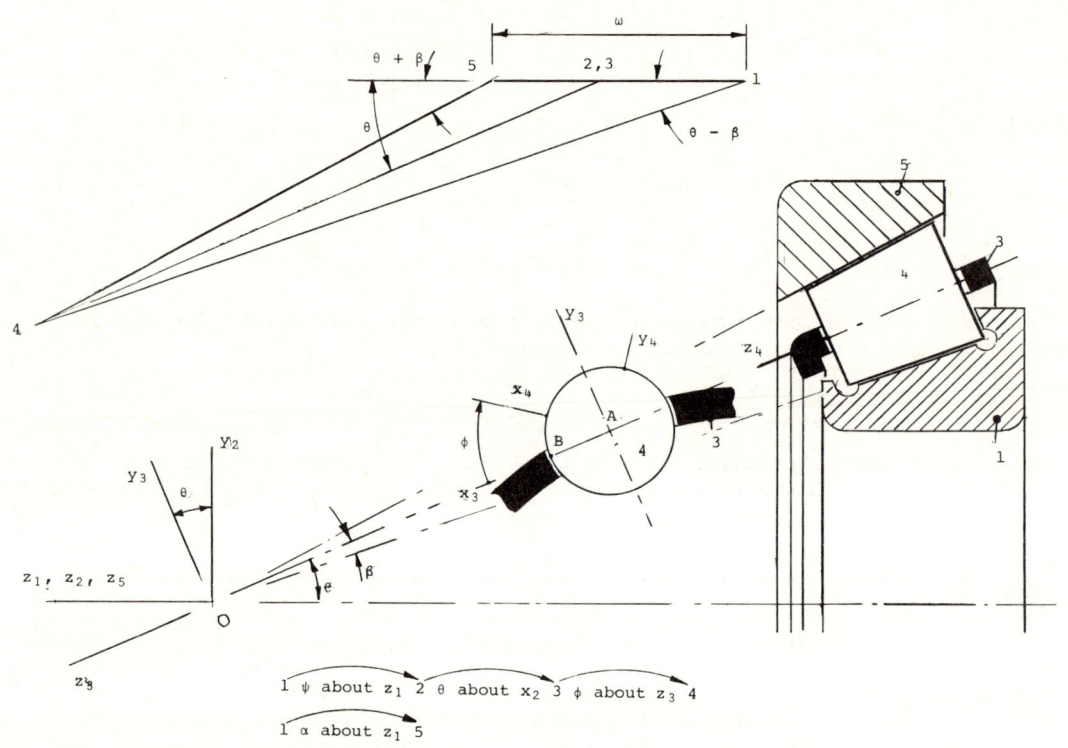

Fig. 3.14.

$$\{\omega_4\}_{1/3} = |\omega_4|_1 \begin{bmatrix} 0 \\ \sin\beta \\ \cos\beta \end{bmatrix} \quad , \quad \{\omega_2\}_{1/3} = |\omega_2|_1 \begin{bmatrix} 0 \\ \sin\theta \\ \cos\theta \end{bmatrix} \quad \text{and}$$

$$\{\omega_4\}_{3/3} = \{\omega_4\}_{2/3} = |\omega_4|_2 \begin{bmatrix} 0 \\ 0 \\ 1 \end{bmatrix}.$$

Now
$$\{\dot{\omega}_4\}_{1/3} = [\omega_2]_{1/3} \{\omega_4\}_{1/3} + \frac{d}{dt} \{\omega_4\}_{1/3}$$

and
$$\{\omega_4\}_{1/3} = \{\omega_2\}_{1/3} + \{\omega_3\}_{2/3} + \{\omega_4\}_{3/3}$$

giving

$$\{\dot{\omega}_4\}_{1/3} = [\omega_2]_{1/3}\{\omega_4\}_{3/3}$$

$$= |\omega_2|_1|\omega_4|_2 \begin{bmatrix} 0 & -c\theta & s\theta \\ c\theta & 0 & 0 \\ -s\theta & 0 & 0 \end{bmatrix} \begin{bmatrix} 0 \\ 0 \\ 1 \end{bmatrix}$$

$$= |\omega_2|_1|\omega_4|_2 \cos\theta \begin{bmatrix} -1 \\ 0 \\ 0 \end{bmatrix}$$

since $\{\omega_3\}_{2/3}$ and $d\{\omega_4\}_{1/3}/dt$ are null matrices.

The magnitude of the maximum rubbing velocity between cage and roller is given by

$$a|\omega_4|_2$$

where a is the magnitude of the position of B relative to A or the maximum roller diameter.

Problem 3.15. A rotor, body 4, turns at a constant rate ω_s relative to a bearing, body 3, about the y_3 axis as shown in Fig.3.15a. The bearing is free to turn with respect to body 2 about the x_2 axis and body 2 rotates at a constant rate Ω about the z_1 axis in body 1. Find $\{\dot{\omega}_4\}_{1/2}$ for the case in which $\dot{\alpha}$ and $\ddot{\alpha}$ are not zero.

Solution. Now

$$\{\omega_4\}_1 = \{\omega_2\}_1 + \{\omega_3\}_2 + \{\omega_4\}_3$$

and therefore

$$\{\omega_4\}_{1/1} = \{\omega_2\}_{1/1} + [\ell_2]_1\{\omega_3\}_{2/2} + [\ell_2]_1[\ell_3]_2\{\omega_4\}_{3/3}.$$

Since $\{\dot{\omega}_2\}_1$ and $\{\dot{\omega}_4\}_3$ are null vectors

$$\{\dot{\omega}_4\}_{1/1} = [\omega_2]_{1/1}[\ell_2]_1\{\omega_3\}_{2/2} + [\ell_2]_1\{\dot{\omega}_3\}_{2/2}$$
$$+ [\omega_2]_{1/1}[\ell_2]_1[\ell_3]_2\{\omega_4\}_{3/3}$$
$$+ [\ell_2]_1[\omega_3]_{2/2}[\ell_3]_2\{\omega_4\}_{3/3} .$$

On premultiplication by $[\ell_1]_3$ and on the introduction of appropriate $[\ell_3]_1[\ell_1]_3$ products, this equation becomes

$$\{\dot{\omega}_4\}_{1/3} = [\ell_1]_3[\omega_2]_{1/1}[\ell_3]_1[\ell_1]_3[\ell_2]_1\{\omega_3\}_{2/2}$$
$$+ [\ell_1]_3[\ell_2]_1\{\dot{\omega}_3\}_{2/2}$$

$$+ [\ell_1]_3[\omega_2]_{1/1}[\ell_3]_1[\ell_1]_3[\ell_2]_1[\ell_3]_2\{\omega_4\}_{3/3}$$
$$+ [\ell_1]_3[\ell_2]_1[\omega_3]_{2/2}[\ell_3]_2\{\omega_4\}_{3/3}$$
$$= [\omega_2]_{1/3}\{\omega_3\}_{2/3} + \{\omega_3\}_{2/3} + [\omega_2]_{1/3}\{\omega_4\}_{3/3}$$
$$+ [\omega_3]_{2/3}\{\omega_4\}_{3/3}$$
$$= [\omega_2]_{1/3}\{\{\omega_3\}_{2/3} + \{\omega_4\}_{3/3}\} + [\omega_3]_{2\ 3}\{\omega_4\}_{3/3}$$
$$+ \{\dot{\omega}_3\}_{2/3} \ .$$

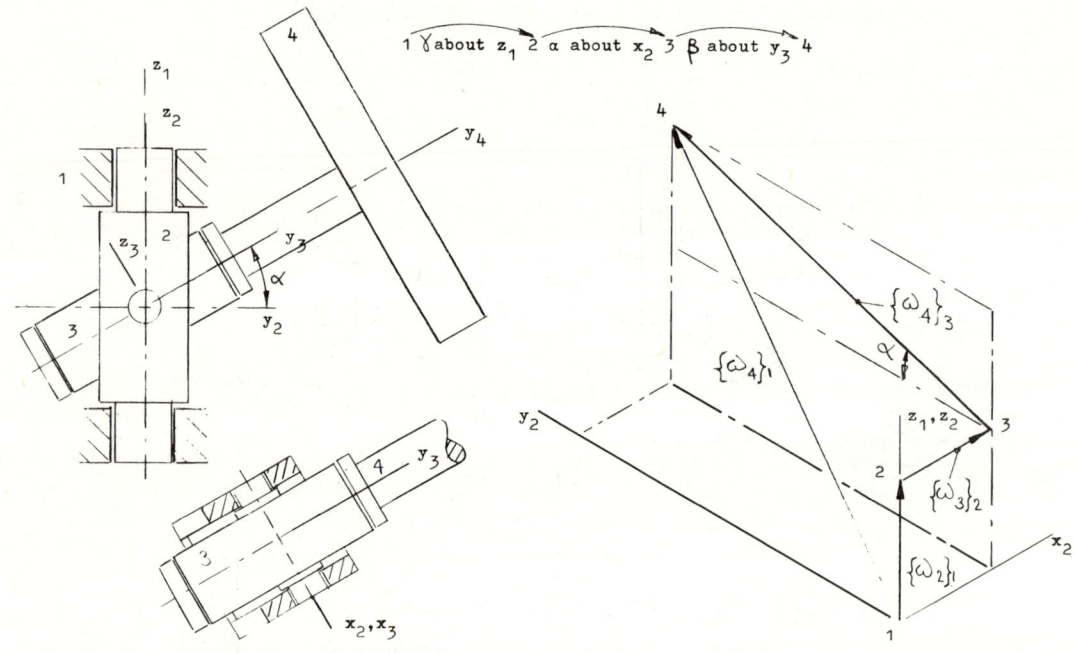

Fig. 3.15a.

Writing $\dot{\alpha} = \omega$ and evaluating individual terms in the above equation

$$\{\omega_2\}_{1/3} = [\ell_2]_3[\ell_1]_2\{\omega_2\}_{1/1} = \Omega\begin{bmatrix}0\\s\alpha\\c\alpha\end{bmatrix} ,$$

$$\{\omega_3\}_{2/3} = [\ell_2]_3\{\omega_3\}_{2/2} = \begin{bmatrix}\omega\\0\\0\end{bmatrix} ,$$

Solution of Kinematics Problems

Components of $\{\dot{\omega}_4\}_1$ along the axes of frame 2

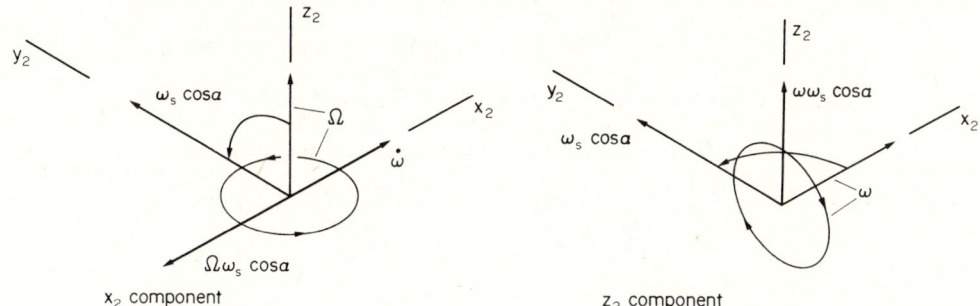

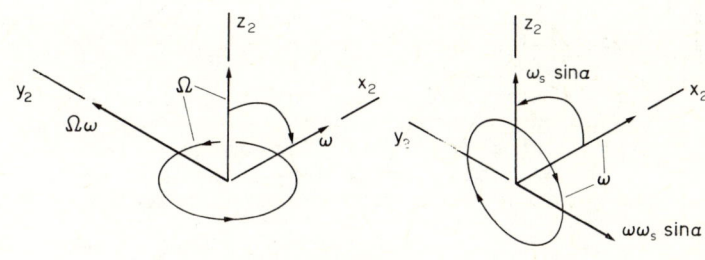

Fig. 3.15b.

$$[\omega_2]_{1/3}\{\{\omega_3\}_{2/3} + \{\omega_4\}_{3/3}\} = \Omega \begin{bmatrix} 0 & -c\alpha & s\alpha \\ c\alpha & 0 & 0 \\ -s\alpha & 0 & 0 \end{bmatrix} \begin{bmatrix} \omega \\ \omega_s \\ 0 \end{bmatrix}$$

$$= \Omega \begin{bmatrix} -\omega_s c\alpha \\ \omega c\alpha \\ -\omega s\alpha \end{bmatrix}$$

and

$$[\omega_3]_{2/3}\{\omega_4\}_{3/3} = \omega\omega_s \begin{bmatrix} 0 & 0 & 0 \\ 0 & 0 & -1 \\ 0 & 1 & 0 \end{bmatrix} \begin{bmatrix} 0 \\ 1 \\ 0 \end{bmatrix} = \omega\omega_s \begin{bmatrix} 0 \\ 0 \\ 1 \end{bmatrix}.$$

Therefore

$$\{\dot{\omega}_4\}_{1/3} = \begin{bmatrix} \dot{\omega} - \Omega\omega_s \cos\alpha \\ \Omega\omega\cos\alpha \\ \omega\omega_s - \Omega\omega\sin\alpha \end{bmatrix}$$

and

$$\{\dot\omega_4\}_{1/2} = [\ell_3]_2 \{\omega_4\}_{1/3} = \begin{bmatrix} \dot\omega - \Omega\omega_s \cos\alpha \\ \Omega\omega - \omega\omega_s \sin\alpha \\ \omega\omega_s \cos\alpha \end{bmatrix}.$$

The terms in this last statement for the angular acceleration are shown in Fig.3.15b. The reader is invited to determine $\{\omega_4\}_{1/2}$ and then determine $\{\dot\omega_4\}_{1/2}$ from

$$\{\dot\omega_4\}_{1/2} = \lfloor\omega_2\rfloor_{1/2}\{\omega_4\}_{1/2} + \frac{d}{dt}\{\omega_4\}_{1/2}.$$

Problem 3.16. In the system of Fig. 3.16 body 2 turns at a constant rate relative to body 1 about the z_1 axis. Body 3 is free to turn on body 2 about the y_2 axis and rolls on body 1. Conditions are such that on the line of contact between bodies 1 and 3 the velocity of slip is zero at the point A. Determine the rubbing velocity at B and D. Locate a point Q on the central axis for the for the motion of body 3 relative to body 1 and show that it is on the line OA. Also show that instantaneously the velocity of Q measured in frame 1 is zero, i.e. OQA is the instantaneous axis for the motion of body 3 relative to body 1.

Solution. The angular velocity vector diagram is constructed as follows.

$\{\omega_2\}_1$ 1⟶2 ω long parallel to the z_1 axis.

$\{\omega_3\}_2$ 2⟶3   parallel to the y_2 axis. The position of point 3 is not defined.

$\{\omega_3\}_1$ 1⟶3   parallel to OA, the instantaeous axis for the motion of body 3 relative to body 1. The intersection of lines 2 3 and 1 3 defines the position of point 3.

By reference to the angular velocity vector diagram

$$\{\omega_3\}_{1/2} = \{\omega_2\}_{1/2} + \{\omega_3\}_{2/2} = \begin{bmatrix} 0 \\ 0 \\ \omega \end{bmatrix} + \begin{bmatrix} 0 \\ -\omega a/r \\ 0 \end{bmatrix} = \begin{bmatrix} 0 \\ -\omega a/r \\ \omega \end{bmatrix}.$$

For the point B, fixed in body 3

$$\{V_{BO}\}_{1/2} = \lfloor\omega_3\rfloor_{1/2}[\ell_3]_2\{R_{BO}\}_{3/3}$$

Solution of Kinematics Problems

$$= \omega \begin{bmatrix} 0 & -1 & -a/r \\ 1 & 0 & 0 \\ a/r & 0 & 0 \end{bmatrix} \begin{bmatrix} c\beta & 0 & -s\beta \\ 0 & 1 & 0 \\ s\beta & 0 & c\beta \end{bmatrix} \begin{bmatrix} 0 \\ b \\ -r \end{bmatrix}$$

$$= \omega \begin{bmatrix} a\cos\beta - b \\ r\sin\beta \\ a\sin\beta \end{bmatrix} .$$

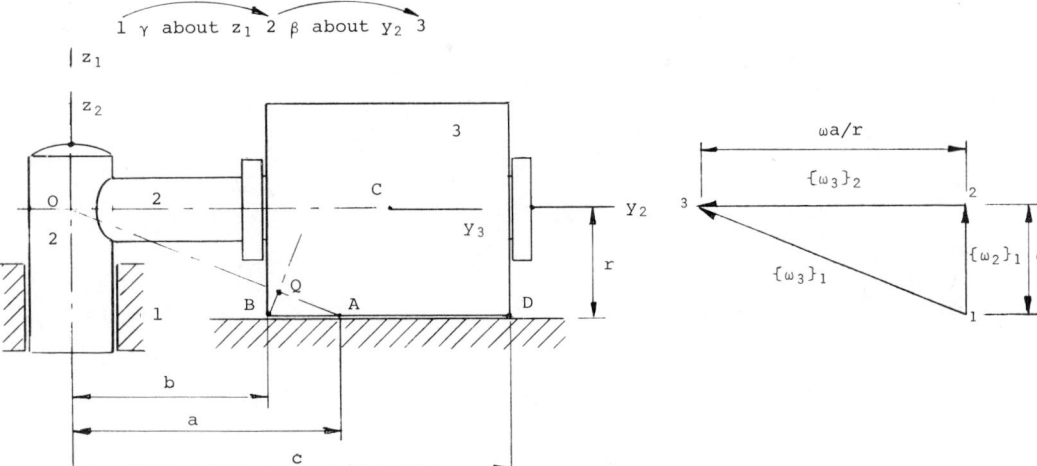

Fig. 3.16.

The rubbing velocity at B is thus given by

$$\{V_{BO}\}_{1/2}\big|_{\beta=0} = \omega(a-b) \begin{bmatrix} 1 \\ 0 \\ 0 \end{bmatrix} .$$

Similarly, the velocity of rubbing at D is given by

$$\{V_{BO}\}_{1/2}\big|_{\beta=0} = \omega(a-c) \begin{bmatrix} 1 \\ 0 \\ 0 \end{bmatrix} .$$

If Q is a point on the central axis, then by Eq. 1.48

$$\{R_{QB}\}_{1/2}\big|_{\beta=0} = \frac{[\omega_3]_{1/2} \{V_{BO}\}_{1/2}\big|_{\beta=0}}{|\omega_3|_1^2}$$

$$= \frac{(a-b)}{a^2/r^2 + 1} \begin{bmatrix} 0 & -1 & -a/r \\ 1 & 0 & 0 \\ a/r & 0 & 0 \end{bmatrix} \begin{bmatrix} 1 \\ 0 \\ 0 \end{bmatrix}$$

$$= \frac{r(a-b)}{a^2 + r^2} \begin{bmatrix} 0 \\ r \\ a \end{bmatrix}$$

and for $\beta = 0$

$$\{R_{QO}\}_{1/2} = \{R_{BO}\}_{1/2} + \{R_{QB}\}_{1/2}$$

$$= \begin{bmatrix} 0 \\ b \\ -r \end{bmatrix} + \frac{r(a-b)}{a^2 + r^2} \begin{bmatrix} 0 \\ r \\ a \end{bmatrix} = \frac{ab + r^2}{a^2 + r^2} \begin{bmatrix} 0 \\ a \\ -r \end{bmatrix} = \lambda \{R_{AO}\}_{1/2}$$

showing that Q is on the line OA. Also, since

$$\{V_{QO}\}_{1/2}\big|_{\beta=0} = [\omega_3]_{1/2} \{R_{QO}\}_{1/2}\big|_{\beta=0}$$

$$= \lambda \omega \begin{bmatrix} 0 & -1 & -a/r \\ 1 & 0 & 0 \\ a/r & 0 & 0 \end{bmatrix} \begin{bmatrix} 0 \\ a \\ -r \end{bmatrix} = \begin{bmatrix} 0 \\ 0 \\ 0 \end{bmatrix}$$

OQA is the instantaneous axis for the motion of the roller relative to body 1.

Problem 3.17. In the system of Fig. 3.17 bodies 2 and 3 are connected by a simple pin joint, the axis of which is parallel to the y_3 axis. Body 2 is free to turn relative to body 1 about the z_1 axis and body 4 is free to turn relative to body 3 about the z_3 axis. Body 5 is to rotate relative to body 1 about the z_1 axis. Find, for the case in which body 4 rolls without slip on body 5,

$$\{\omega_4\}_{1/3} \text{ and } \{\dot{\omega}_4\}_{1/3}$$

when bodies 2 and 3 are driven at constant rates ω_2 and ω_5 respectively relative to body 1.

Solution. Now

$$\{R_{AO}\}_{1/1} = \{R_{BO}\}_{1/1} + \{R_{AB}\}_{1/1}$$

$$= [\ell_2]_1 [\ell_3]_2 \{R_{BO}\}_{3/3} + [\ell_2]_1 [\ell_3]_2 [\ell_4]_3 \{R_{AB}\}_{4/4} ,$$

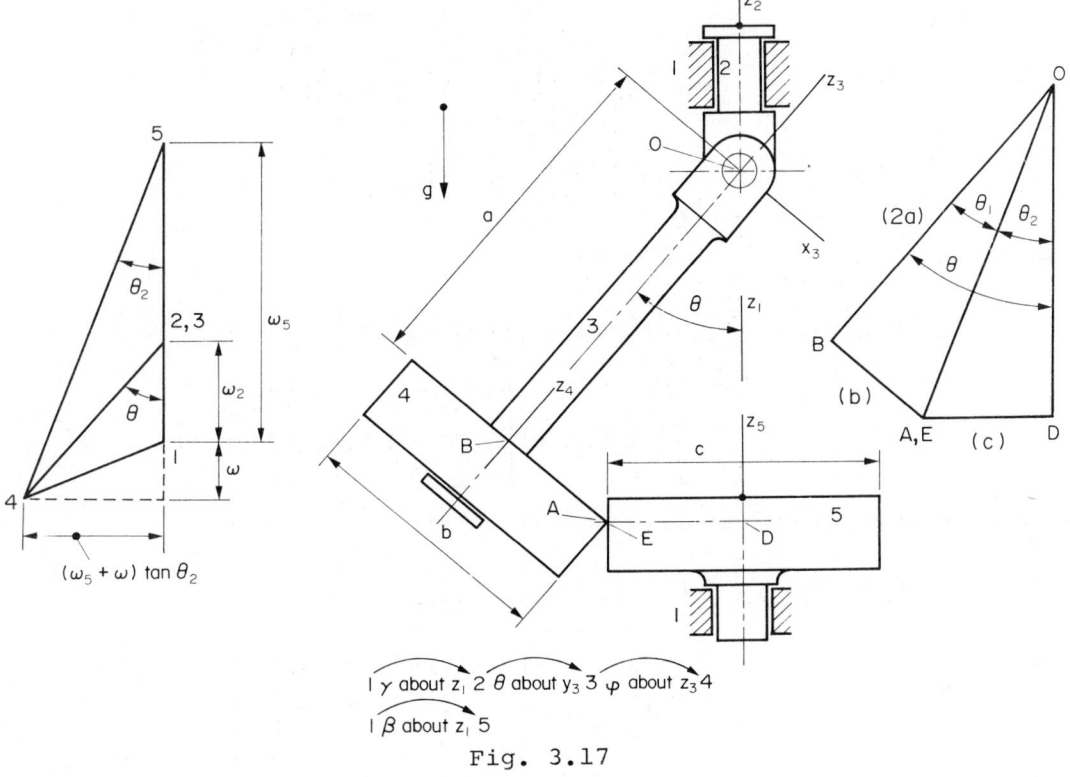

Fig. 3.17

giving

$$\{V_{AO}\}_{1/1} = [\omega_2]_{1/1}[\ell_2]_1[\ell_3]_2\{R_{BO}\}_{3/3}$$
$$+ [\omega_2]_{1/1}[\ell_2]_1[\ell_3]_2\{R_{AB}\}_{4/4}$$
$$+ [\ell_2]_1[\ell_3]_2[\omega_4]_{3/3}[\ell_4]_3\{R_{AB}\}_{4/4} ,$$

since $\{\omega_3\}_2$ is a null vector and therefore

$$\{V_{AO}\}_{1/3} = [\omega_2]_{1/3}\{R_{BO}\}_{3/3}$$
$$+ \Big[[\omega_2]_{1/3} + [\omega_4]_{3/3}\Big][\ell_4]_3\{R_{AB}\}_{4/4} .$$

Hence

$$\{V_{AO}\}_{1/3}\Big|_{\phi=0} = \omega_2 \begin{bmatrix} 0 & -c\theta & 0 \\ c\theta & 0 & s\theta \\ 0 & -s\theta & 0 \end{bmatrix} \begin{bmatrix} 0 \\ 0 \\ -a \end{bmatrix}$$

$$+ \omega_2 \begin{bmatrix} 0 & -c\theta & 0 \\ c\theta & 0 & s\theta \\ 0 & -s\theta & 0 \end{bmatrix} + \dot{\phi} \begin{bmatrix} 0 & -1 & 0 \\ 1 & 0 & 0 \\ 0 & 0 & 0 \end{bmatrix} \begin{bmatrix} b/2 \\ 0 \\ 0 \end{bmatrix}$$

$$= \omega_2 a \sin\theta \begin{bmatrix} 0 \\ -1 \\ 0 \end{bmatrix} + \frac{\omega_2 b \cos\theta}{2} \begin{bmatrix} 0 \\ 1 \\ 0 \end{bmatrix} + \frac{b\dot{\phi}}{2} \begin{bmatrix} 0 \\ 1 \\ 0 \end{bmatrix} .$$

Also

$$\{R_{ED}\}_{1/1} = [\ell_5]_1 \{R_{ED}\}_{5/5} \quad \text{and} \quad \{V_{ED}\}_{1/5} = [\omega_5]_{1/5} \{R_{ED}\}_{5/5}$$

giving

$$\{V_{EB}\}_{1/5}\Big|_{\beta=0} = \omega_5 \begin{bmatrix} 0 & -1 & 0 \\ 1 & 0 & 0 \\ 0 & 0 & 0 \end{bmatrix} \begin{bmatrix} -c/2 \\ 0 \\ 0 \end{bmatrix} = \frac{c\omega_5}{2} \begin{bmatrix} 0 \\ -1 \\ 0 \end{bmatrix} .$$

Since

$$\{V_{AO}\}_{1/3}\Big|_{\dot{\phi}=0} = \{V_{EB}\}_{1/5}\Big|_{\beta=0} ,$$

$$\dot{\phi} = \omega_2 \{(2a/b)\sin\theta - \cos\theta\} - (c/b)\omega_5 .$$

By reference to Fig. 3.17

$$\tan\theta = \tan(\theta_1 + \theta_2) = \frac{b/2a + c/r}{1 - bc/2ar} = \frac{br + 2ac}{2ar - bc}$$

where $r^2 = 4a^2 + b^2 - c^2$, giving

$$\sin\theta = (br + 2ac)/R \quad \text{and} \quad \cos\theta = (2ar - bc)/R$$

where $R = \sqrt{\{(br + 2ac)^2 + (2ar - bc)^2\}} = (r^2 + c^2)$.

With these values for $\sin\theta$ and $\cos\theta$ the expression for $\dot{\phi}$ reduces to

$$\dot{\phi} = c(\omega_2 - \omega_5)/b.$$

Therefore

$$\{\omega_4\}_{1/3} = \{\omega_2\}_{1/3} + \{\omega_4\}_{3/3}$$

$$= \omega_2 \begin{bmatrix} -\sin\theta \\ 0 \\ \cos\theta \end{bmatrix} + \frac{c(\omega_2 - \omega_5)}{b} \begin{bmatrix} 0 \\ 0 \\ 1 \end{bmatrix} .$$

This expression for $\{\omega_4\}_{1/3}$ can also be determined from the angular velocity vector diagram of Fig. 3.17. $\{\omega_2\}_1$ and $\{\omega_5\}_1$ are drawn parallel to the z_1 axis. $\{\omega_4\}_3$ drawn parallel to OB and $\{\omega_5\}_4$ drawn parallel to OA, the instantaneous axis for the motion of body 4 relative to body 3, intersect to define the position of point 4. From the diagram

Solution of Kinematics Problems

$$(\omega_2 + \omega)\tan\theta = (\omega_5 + \omega)\tan\theta_2$$

giving

$$\omega = \frac{\omega_5\tan\theta_2 - \omega_2\tan\theta}{\tan\theta - \tan\theta_2}$$

and therefore

$$|\dot\phi| = \frac{\omega + \omega_2}{\cos\theta} = \frac{(\omega_5 - \omega_2)\tan\theta_2}{\cos\theta(\tan\theta - \tan\theta_2)} = c(\omega_5 - \omega_2)/b .$$

which confirms the previous result.

The angular acceleration $\{\dot\omega_4\}_{1/3}$, since $d\{\omega_4\}_{1/3}/dt$ is null, is given by

$$\{\dot\omega_4\}_{1/3} = [\omega_3]_{1/3}\{\omega_4\}_{1/3} = [\omega_2]_{1/3}\{\omega_4\}_{3/3}$$

$$= \frac{c\omega_2(\omega_2 - \omega_5)}{b}\begin{bmatrix} 0 & -c\theta & 0 \\ c\theta & 0 & s\theta \\ 0 & -s\theta & 0 \end{bmatrix}\begin{bmatrix} 0 \\ 0 \\ 1 \end{bmatrix}$$

$$= \frac{\omega_2 c\sin\theta(\omega_2 - \omega_5)}{b}\begin{bmatrix} 0 \\ 1 \\ 0 \end{bmatrix} .$$

Problem 3.18. In the system of Fig. 3.18 bodies 2 and 3 rotate about axes fixed in body 4. The relative motion of bodies 2 and 3 is governed by mating bevel gear wheels cut in them. Body 2 rotates about an axis fixed in body 1, while body 3 rolls without slip on body 1. For the case in which body 2 is driven at a constant rate relative to body 1, draw the angular velocity vector diagram for the system and hence determine

$$\{\omega_3\}_{1/4} \text{ and } \{\dot\omega_3\}_{1/4} .$$

Also determine

$$\{V_{CO}\}_{1/4}, \{V_{AO}\}_{1/4} \text{ and } \{A_{AO}\}_{1/4}$$

where A is a point fixed in body 3 which is shown at the point of contact with body .

Solution. The angular velocity vector diagram is constructed as follows.

$\{\omega_2\}_1$ 1⟶2 ω long and parallel to the y_1 axis.

$\{\omega_3\}_2$ 2⟶3   parallel to the common generator of the pitch cones of the mating bevel gear wheels, the instantaneous axis for the relative motion of bodies 2 and 3. The position of point 3 is not defined at this stage.

$\{\omega_3\}_1$ 1⟶3   parallel to the instantaneous axis for the relative motion of bodies 1 and 3, OA. Lines 2 3 and 1 3 intersect to define the position of point 3.

$\{\omega_4\}_3$ 3⟶4   parallel to OC.

$\{\omega_4\}_1$ 1⟶4   parallel to the y_1 axis.

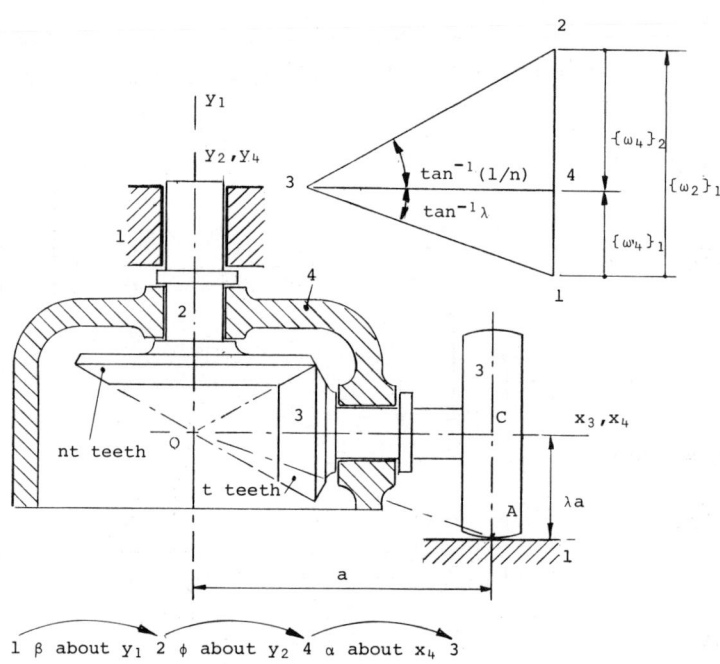

Fig. 3.18.

From the angular velocity vector diagram, writing $|\omega_4|_2 = \omega_1$,

$$|\omega_3|_4 = n\omega_1, |\omega_2|_1 = \omega = (1 + \lambda n)\omega_1, |\omega_4|_1 = n\lambda\omega_1 = r\lambda\omega$$

and
$$|\omega_3|_4 = r\omega$$

where
$$r = n/(1 + \lambda n).$$

Now

Solution of Kinematics Problems

$$\{\omega_3\}_1 = \{\omega_2\}_1 + \{\omega_4\}_2 + \{\omega_3\}_4$$
$$= \{\omega_4\}_1 + \{\omega_3\}_4$$

and therefore

$$\{\omega_3\}_{1/4} = \{\omega_4\}_{1/4} + \{\omega_3\}_{4/4}$$

$$= \omega r \begin{bmatrix} 0 \\ \lambda \\ 0 \end{bmatrix} + \omega r \begin{bmatrix} -1 \\ 0 \\ 0 \end{bmatrix} = \frac{\omega n}{1 + \lambda n} \begin{bmatrix} -1 \\ \lambda \\ 0 \end{bmatrix}$$

The angular acceleration $\{\dot{\omega}_3\}_{1/4}$, since $d\{\omega_3\}_{1/4}/dt$ is null, is given by

$$\{\dot{\omega}_3\}_{1/4} = [\omega_4]_{1/4}\{\omega_3\}_{1/4} = [\omega_4]_{1/4}\{\omega_3\}_{4/4}$$

$$= \omega^2 r^2 \lambda \begin{bmatrix} 0 & 0 & 1 \\ 0 & 0 & 0 \\ -1 & 0 & 0 \end{bmatrix} \begin{bmatrix} -1 \\ 0 \\ 0 \end{bmatrix} = \frac{\omega^2 n^2 \lambda}{(1 + \lambda n)^2} \begin{bmatrix} 0 \\ 0 \\ 1 \end{bmatrix}.$$

Since C is a point fixed in frame 4

$$\{V_{CO}\}_{1/4} = [\omega_4]_{1/4}\{R_{CO}\}_{4/4} = \omega r \lambda \begin{bmatrix} 0 & 0 & 1 \\ 0 & 0 & 0 \\ -1 & 0 & 0 \end{bmatrix} \begin{bmatrix} a \\ 0 \\ 0 \end{bmatrix}$$

$$= \frac{\omega a \lambda n}{1 + \lambda n} \begin{bmatrix} 0 \\ 0 \\ -1 \end{bmatrix}.$$

Also, for A a point fixed in body 3,

$$\{R_{AO}\}_{1/1} = \{R_{CO}\}_{1/1} + \{R_{AC}\}_{1/1}$$
$$= [\ell_4]_1\{R_{CO}\}_{4/4} + [\ell_3]_1\{R_{AC}\}_{3/3}$$

and therefore

$$\{V_{AO}\}_{1/1} = [\omega_4]_{1/1}[\ell_4]_1\{R_{CO}\}_{4/4} + [\omega_3]_{1/1}[\ell_3]_1\{R_{AC}\}_{3/3},$$

or, noting that α is negative,

$$\{V_{AO}\}_{1/4} = [\omega_4]_{1/4}\{R_{CO}\}_{4/4} + [\omega_3]_{1/4}[\ell_3]_4\{R_{AC}\}_{3/3}$$

$$= \omega a r \begin{bmatrix} 0 \\ 0 \\ -\lambda \end{bmatrix} + \omega r \begin{bmatrix} 0 & 0 & \lambda \\ 0 & 0 & 1 \\ -\lambda & -1 & 0 \end{bmatrix} \begin{bmatrix} 1 & 0 & 0 \\ 0 & c\alpha & s\alpha \\ 0 & -s\alpha & c\alpha \end{bmatrix} \begin{bmatrix} 0 \\ -\lambda a \\ 0 \end{bmatrix}$$

$$= \frac{\omega a \lambda n}{1 + \lambda n} \begin{bmatrix} \lambda \sin\alpha \\ \sin\alpha \\ \cos\alpha - 1 \end{bmatrix}.$$

Hence

$$\{A_{AO}\}_{1/4} = [\omega_4]_{1/4} \{V_{AO}\}_{1/4} + \frac{d}{dt}\{V_{AO}\}_{1/4}$$

$$= \omega^2 ar^2 \lambda \begin{bmatrix} 0 & 0 & \lambda \\ 0 & 0 & 0 \\ -\lambda & 0 & 0 \end{bmatrix} \begin{bmatrix} \lambda s\alpha \\ s\alpha \\ c\alpha - 1 \end{bmatrix} + \omega^2 ar^2 \lambda \begin{bmatrix} \lambda c\alpha \\ c\alpha \\ -s\alpha \end{bmatrix}$$

$$= \frac{\omega^2 a^2 n^2 \lambda}{(1 + \lambda n)^2} \begin{bmatrix} \lambda(2\cos\alpha - 1) \\ \cos\alpha \\ -(1 + \lambda^2)\sin\alpha \end{bmatrix}.$$

Problem 3.19. Figure 3.19 shows a bevel wheel epicyclic gear train in which body 2 is driven at a constant rate ω relative to body 1. Determine

$\{\dot{\omega}_4\}_{1/3}$ and $\{\omega_5\}_{1/1}$.

Solution. The angular velocity vector diagram is constructed as follows.

$\{\omega_2\}_1$ 1 ⟶ 2 ω long and parallel to the x_1 axis.

$\{\omega_4\}_1$ 1 --⟶ 4   parallel to OA, the common generator of the pitch cones on bodies 1 and 4.

$\{\omega_4\}_2$ 2 ⟶ 4   parallel to the z axis. Lines 1 4 and 2 4 intersect to define the position of point 4.

$\{\omega_5\}_1$ 1 ⟶ 5   parallel to the x_1 axis.

$\{\omega_5\}_4$ 4 ⟶ 5   parallel to OB, the common generator of the pitch cones on bodies 4 and 5. Lines 1 5 and 4 5 intersect to define the position of point 5.

The angles specified in Fig. 3.19 are related to the gear teeth numbers by the following expressions

$$\sin\beta = (t_1 - t_5)/2t_4 , \quad \cos\beta = CD/OC ,$$

$$\tan\alpha = AC/OC = 2t_4\cos\beta/(t_1 + t_5) \text{ and } \sin\phi = t_1\sin\alpha/t_4.$$

Solution of Kinematics Problems

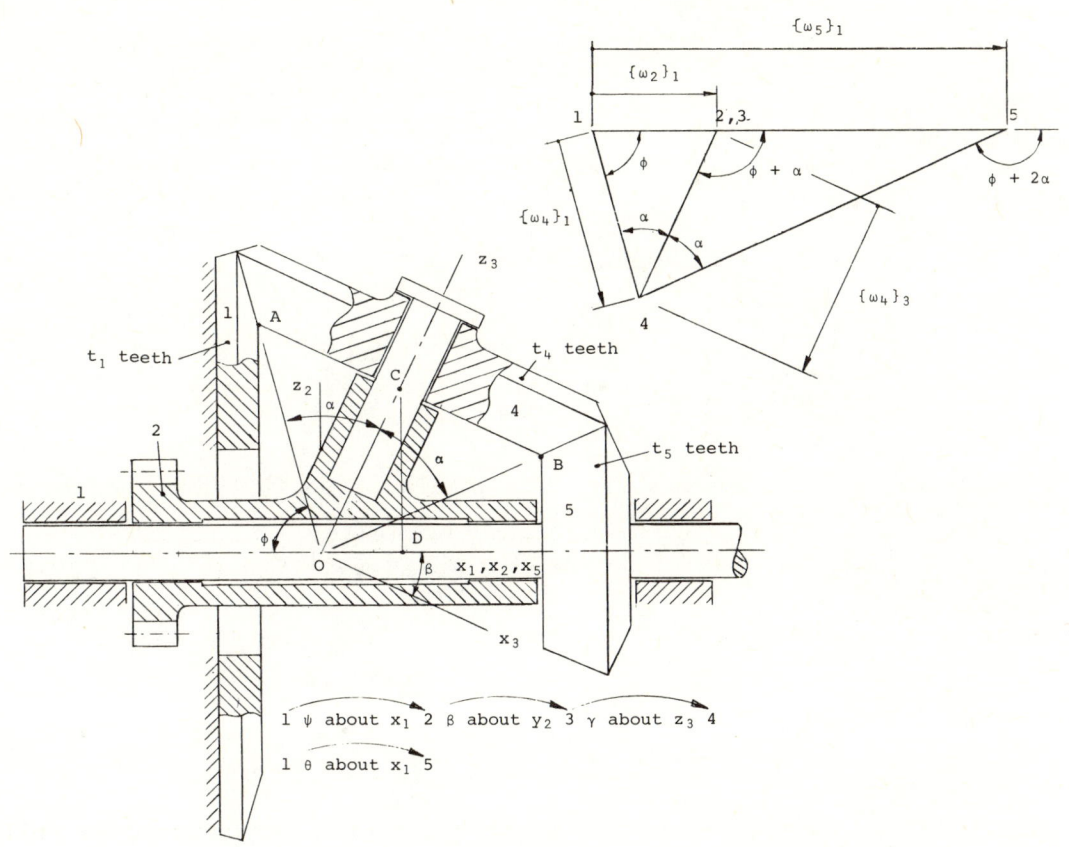

Fig. 3.19

Application of the sine rule to the angular velocity vector diagram triangle 124 gives

$$|\omega_2|_1/\sin\alpha = |\omega_4|_1/\sin(\phi + \alpha) = |\omega_4|_3/\sin\phi .$$

The relationship $\phi + \alpha = \beta + \pi/2$ can be used to eliminate ϕ to give

$$|\omega_2|_1/\sin\alpha = |\omega_4|_1/\cos\beta = |\omega_4|_3/\cos(\beta - \alpha) .$$

Hence

$$\{\omega_4\}_{1/3} = \frac{\omega\cos\beta}{\sin\alpha}\begin{bmatrix}\sin\alpha \\ 0 \\ -\cos\alpha\end{bmatrix} = \omega\cos\beta\begin{bmatrix}1 \\ 0 \\ -\cot\alpha\end{bmatrix}$$

and

$$\{\omega_4\}_{3/3} = \frac{\omega\cos(\beta - \alpha)}{\sin\alpha}\begin{bmatrix} 0 \\ 0 \\ -1 \end{bmatrix}.$$

In this case the required angular acceleration of body 4 is given by

$$\{\dot{\omega}_4\}_{1/3} = [\omega_2]_{1/3}\{\omega_4\}_{3/3}$$

$$= \frac{\omega^2\cos(\beta - \alpha)}{\sin\alpha}\begin{bmatrix} 0 & -s\beta & 0 \\ s\beta & 0 & -c\beta \\ 0 & c\beta & 0 \end{bmatrix}\begin{bmatrix} 0 \\ 0 \\ -1 \end{bmatrix}$$

$$= \frac{\omega^2 t_1 \sqrt{\{4t_4^2 - (t_1 - t_5)^2\}}}{2t_4^2}\begin{bmatrix} 0 \\ 1 \\ 0 \end{bmatrix}.$$

Application of the sine rule to the angular velocity vector diagram triangle 145 gives

$$|\omega_5|_1/\sin 2\alpha = |\omega_4|_1/\sin(\phi + 2\alpha) = |\omega_4|_1/\cos(\alpha + \beta)$$

and therefore

$$|\omega_5|_1 = 2\omega\cos\beta\cos\alpha/\cos(\alpha + \beta)$$

giving

$$\{\omega_5\}_{1/1} = \omega(1 + t_1/t_5)\begin{bmatrix} 1 \\ 0 \\ 0 \end{bmatrix}$$

Problem 3.20. Hooke's joint, shown diagrammatically in Fig. 3.20, is a device for coupling shafts which have intersecting non colinear axes. With this arrangement the output shaft, body 4, has a variable angular velocity when the input shaft, body 2, is driven at a constant rate relative to the bearings, body 1, in which the shafts are constrained to turn. Find

$$\{\omega_4\}_{1/5}, \{\omega_3\}_{2/2} \text{ and } \{\omega_4\}_{3/3}.$$

Solution. Frame 4, fixed in the output shaft, can be reached from frame 1, fixed in the bearings, by either of the sequences of rotations indicated in Fig. 3.20. Hence

$$\{\omega_4\}_{1/1} = \{\omega_2\}_{1/1} + [\ell_2]_1\{\omega_3\}_{2/2} + [\ell_2]_1[\ell_3]_2\{\omega_4\}_{3/3} \qquad (1)$$

$$= \{\omega_5\}_{1/1} + [\ell_5]_1\{\omega_4\}_{5/5} \qquad (2)$$

where $\{\omega_5\}_1$ is a null vector.

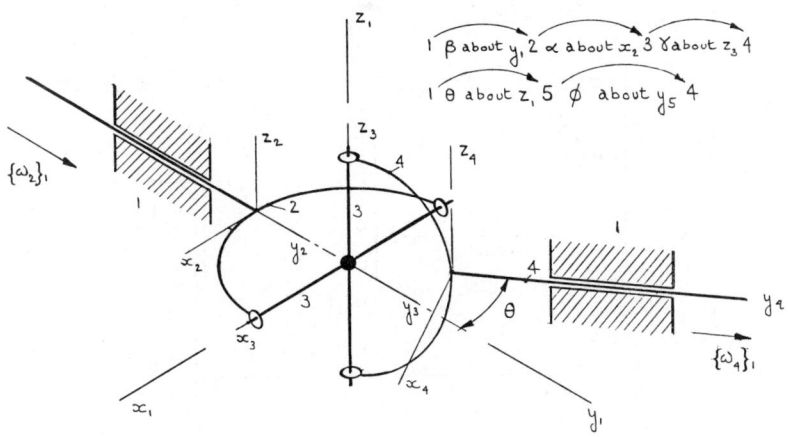

Fig. 3.20.

By Eq. 1

$$\omega_4 \begin{bmatrix} -\sin\theta \\ \cos\theta \\ 0 \end{bmatrix} = \omega_2 \begin{bmatrix} 0 \\ 1 \\ 0 \end{bmatrix} + \begin{bmatrix} c\beta & 0 & s\beta \\ 0 & 1 & 0 \\ -s\beta & 0 & c\beta \end{bmatrix} \begin{bmatrix} \dot{\alpha} \\ 0 \\ 0 \end{bmatrix}$$

$$+ \begin{bmatrix} c\beta & 0 & s\beta \\ 0 & 1 & 0 \\ -s\beta & 0 & c\beta \end{bmatrix} \begin{bmatrix} 1 & 0 & 0 \\ 0 & c\alpha & -s\alpha \\ 0 & s\alpha & c\alpha \end{bmatrix} \begin{bmatrix} 0 \\ 0 \\ \dot{\gamma} \end{bmatrix}$$

$$= \begin{bmatrix} \dot{\alpha}c\beta + \dot{\gamma}s\beta c\alpha \\ \omega_2 - \dot{\gamma}s\alpha \\ -\dot{\alpha}s\beta + \dot{\gamma}c\beta c\gamma \end{bmatrix} \quad (3)$$

By Eq. 2

$$\omega_4 \begin{bmatrix} -\sin\theta \\ \cos\theta \\ 0 \end{bmatrix} = \begin{bmatrix} c\theta & -s\theta & 0 \\ s\theta & c\theta & 0 \\ 0 & 0 & 1 \end{bmatrix} \begin{bmatrix} 0 \\ \dot{\phi} \\ 0 \end{bmatrix} \quad (4)$$

From the z component of Eqs. 3

$$\dot{\alpha} = \dot{\gamma}\cos\alpha/\tan\beta \quad (5)$$

and from the y component of Eqs. 4 $\omega_4 = \dot{\phi}$. By Eq. 5 and the x component of Eqs. 3

$$\dot{\gamma} = -\omega_4 \sin\beta\cos\theta/\cos\alpha \quad (6)$$

By Eq. 6 and the y component of Eqs. 3

$$\omega_4 = \omega_2/(\cos\theta - \sin\beta\sin\theta\tan\alpha) \,. \tag{7}$$

A relationship between α, β and θ can be found by comparing corresponding terms in the two expressions for $[\ell_4]_1$.

$$[\ell_4]_1 = \begin{bmatrix} c\theta c\phi & -s\theta & c\theta s\phi \\ s\theta c\phi & c\theta & s\theta s\phi \\ -s\phi & 0 & c\phi \end{bmatrix}$$

$$= \begin{bmatrix} c\beta c\gamma + s\beta s\gamma c\alpha & -c\beta s\gamma + s\beta c\gamma s\alpha & s\beta c\alpha \\ s\gamma c\alpha & c\gamma c\alpha & -s\alpha \\ -s\beta c\gamma + c\beta s\gamma s\alpha & s\beta s\gamma + c\beta c\gamma s\alpha & c\beta c\alpha \end{bmatrix} \tag{8}$$

and therefore

$$a_{23}/a_{13} = \tan\theta = -\tan\alpha/\sin\beta. \tag{9}$$

Hence, by Eqs. 7 and 9

$$\omega_4 = \omega_2\cos\theta/(1 - \sin^2\theta\cos^2\beta)$$

and

$$\{\omega_4\}_{1/5} = \frac{\omega_2\cos\theta}{1 - \sin^2\theta\cos^2\beta} \begin{bmatrix} 0 \\ 1 \\ 0 \end{bmatrix} \tag{10}$$

Also, by Eqs. 5, 6, 9 and 10

$$|\omega_4|_3 = \dot{\gamma} = \frac{-\omega_2\sin\beta\sin\theta\cos\theta\sqrt{(1 + \tan^2\theta\sin^2\beta)}}{1 - \sin^2\theta\cos^2\beta}$$

and

$$|\omega_3|_2 = \dot{\alpha} = \frac{-\omega_2\sin\theta\cos\theta}{\cos\beta(1 - \sin^2\theta\cos^2\beta)}$$

giving

$$\{\omega_4\}_{3/3} = |\omega_4|_3 \begin{bmatrix} 0 \\ 0 \\ 1 \end{bmatrix} \tag{11}$$

and

$$\{\omega_3\}_{2/2} = |\omega_3|_2 \begin{bmatrix} 1 \\ 0 \\ 0 \end{bmatrix}. \tag{12}$$

Problem 3.21. Figure 3.21 shows a shaft, body 2, and and axle and disc, body 3, coupled by a constant velocity joint. The shaft turns at a constant rate ω_2 about the y_1 axis relative to body 1 and the disc rolls without slip on body 1. An imaginary body 4 is introduced to assist in a description of the kinematic properties of the joint. The relative angular velocities

$$\{\omega_2\}_4 \text{ and } \{\omega_3\}_4$$

are such that their magnitudes are equal. Construct the angular velocity diagram and use it to determine

$$\{\dot{\omega}_3\}_{1/5} .$$

The reader is referred to Morrison,J.L.M. and Crossland,B. (1964). *An Introduction to the Mechanics of Machines*,Longmans,London. Chap.2,p.123. for a description of the Birfield universal joint.

Solution. The angular velocity vector diagram is constructed as follows.

$\{\omega_2\}_1$ 1⟶2 ω_2 long and parallel to the y_1 axis.

$\{\omega_3\}_1$ 1⟶3 parallel to OP, the instantaneous axis for the relative motion of bodies 3 and 1. The position of point 3 is not defined.

$\{\omega_2\}_4$ 4⟶2 parallel to the y_1 axis. The position of point 4 is not defined.

$\{\omega_4\}_5$ 5⟶4 a null vector, points 4 and 5 are coincident.

$\{\omega_3\}_5$ 5⟶3 parallel to the x_5 axis.

Since

$$|\omega_2|_4 = |\omega_3|_4$$

the angles 432 and 423 are equal so that the angle 132, the inclination of $\{\omega_3\}_2$ to OP, is $(90° - \gamma)/2$. Now the bisector of the angle AOC is inclined at

$$(90° + \gamma)/2 - \gamma = (90° - \gamma)/2$$

to OP and $\{\omega_3\}_2$ is therefore parallel to OP. This allows the vectors 1⟶3 and 2⟶3 to be drawn on the vector diagram to locate the point 3, and consequently the coincident points 4 and 5. Let

$$|\omega_2|_4 = |\omega_3|_4 = \omega .$$

Then

$$|\omega_3|_1 = \omega\cos\gamma \text{ and } \tan\dot\gamma = (\omega - \omega_2)/\omega\cos\gamma ,$$

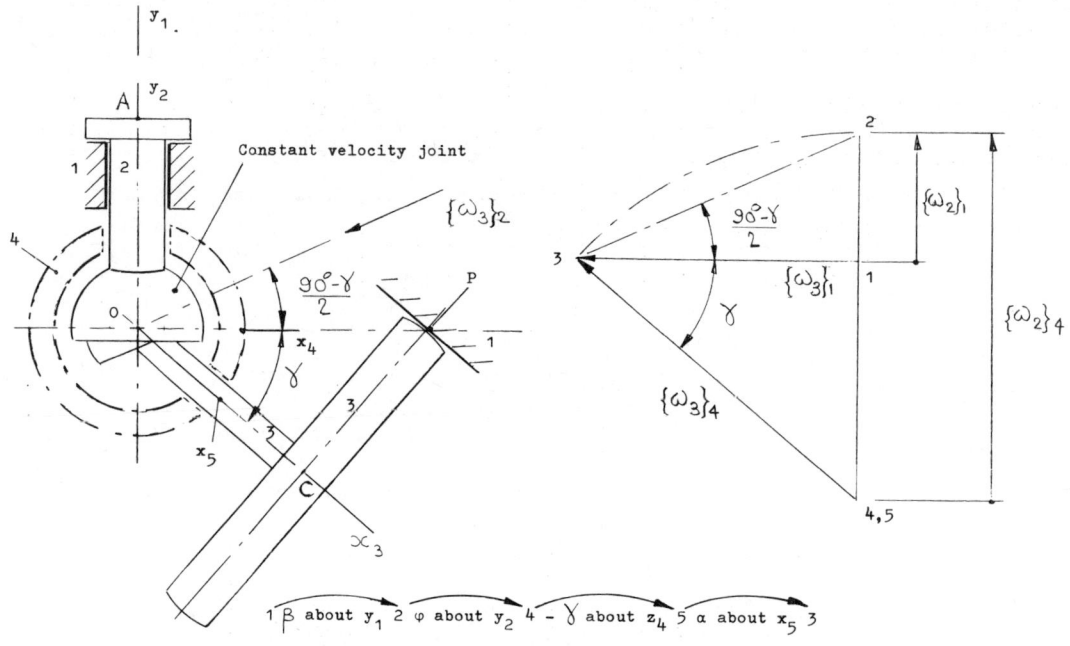

Fig. 3.21

giving

$$\omega = \omega_2/(1 - \sin\gamma) \quad \text{and} \quad |\omega_5|_1 = \omega_2 \sin\gamma/(1 - \sin\gamma).$$

Therefore

$$\{\omega_3\}_{1/5} = \frac{\omega_2 \cos\gamma}{1 - \sin\gamma} \begin{bmatrix} -\cos\gamma \\ -\sin\gamma \\ 0 \end{bmatrix}$$

and

$$\{\omega_5\}_{1/5} = \frac{\omega_2 \sin\gamma}{1 - \sin\gamma} \begin{bmatrix} \sin\gamma \\ -\cos\gamma \\ 0 \end{bmatrix}$$

giving

$$\{\dot{\omega}_3\}_{1/5} = [\omega_5]_{1/5} \{\omega_3\}_{1/5} = \frac{\omega_2^2 s\gamma c\gamma}{(1 - s\gamma)^2} \begin{bmatrix} 0 & 0 & -c\gamma \\ 0 & 0 & -s\gamma \\ c\gamma & s\gamma & 0 \end{bmatrix} \begin{bmatrix} -c\gamma \\ -s\gamma \\ 0 \end{bmatrix}$$

$$= \frac{\omega_2^2 \sin\gamma \cos\gamma}{(1 - \sin\gamma)^2} \begin{bmatrix} 0 \\ 0 \\ -1 \end{bmatrix}.$$

Solution of Kinematics Problems

Problem 3.22. Body 2 moves from a position in which frame 2 fixed in it is aligned with frame 1, to a position in which frame 2 is aligned with frame 3. This motion can be reduced to a linear displacement of a point in body 2 combined with a rotation about some axis. Find the direction of the axis and the angle through which the body rotates.

The plate in Fig. 3.22a is to be moved from the position ABC to the position A'B'C' by a simple rotation about some axis. Determine the direction of the axis and the magnitude of the rotation about the axis.

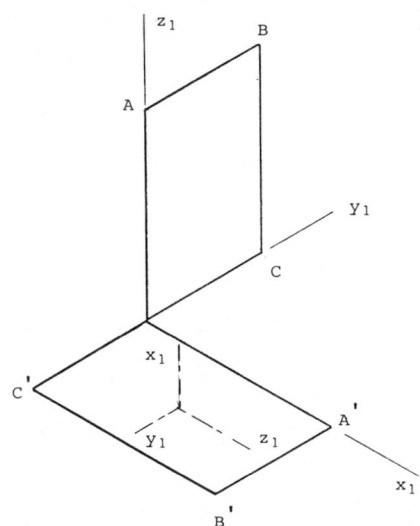

Fig. 3.22a.

Solution. One approach to the solution of this problem exploits two of the properties of the similarity transformation. A similarity transformation of the square matrix $[A]$ is given by

$$[B] = [T][A][T]^{-1} \qquad (1)$$

or

$$= [T]^{-1}[A][T] \qquad (2)$$

where $[T]$ is a nonsingular matrix of the same order as $[A]$. $[B]$ is is said to be similar to $[A]$.

The particular similarity transformation to be employed here is

$$[B] = [\ell][A][\ell]^{-1} \qquad (3)$$

where $[B]$ is a given transformation matrix and $[\ell]$ is a transformation matrix which is to be determined.

The particular properties of the similarity transformation to be exploited here are:

(a) that $[B]$ and $[A]$ have equal eigenvalues and
(b) that $[B]$ and $[A]$ have equal traces.

These properties can be shown to exist as follows. Consider the equation

$$[A]\{X\} = \lambda\{X\} \tag{4}$$

The values of λ which satisfy this equation are the eigenvalues of $[A]$ and the $\{X\}$'s are its eigenvectors. Premultiplication of Eq. 4 by $[\ell]$ and the introduction of $[\ell]^{-1}[\ell]$ gives

$$[\ell][A][\ell]^{-1}[\ell]\{X\} = \lambda[\ell]\{X\}$$

and by Eq. 3

$$[B][\ell]\{X\} = \lambda[\ell]\{X\}$$

or

$$[B]\{Y\} = \lambda\{Y\} \tag{5}$$

where

$$[\ell]\{X\} = \{Y\}. \tag{6}$$

Hence the eigenvalues of $[B]$ are equal to the eigenvalues of $[A]$ and their eigenvectors are related by Eq. 6.

The equation

$$[A]\{X\} = \lambda\{X\}$$

can be written

$$\big[[A] - \lambda[1]\big]\{X\} = \{0\}$$

and if $\{X\}$ is not null, then

$$\det\big[[A] - \lambda[1]\big] = 0. \tag{7}$$

Writing

$$[A] = \begin{bmatrix} a_1 & b_1 & c_1 \\ a_2 & b_2 & c_2 \\ a_3 & b_3 & c_3 \end{bmatrix}$$

gives the requirement of Eq. 7 as

$$\begin{vmatrix} a_1 - \lambda & b_1 & c_1 \\ a_2 & b_2 - \lambda & c_2 \\ a_3 & b_3 & c_3 - \lambda \end{vmatrix} = 0$$

or

Solution of Kinematics Problems

$$(a_1 - \lambda)\{(b_2 - \lambda)(c_3 - \lambda) - b_3 c_2\}$$

$$- b_1\{a_2(c_3 - \lambda) - a_3 c_2\}$$

$$+ c_1\{a_2 b_3 - a_3(b_2 - \lambda)\} = 0$$

and this expands to

$$\lambda^3 - (a_1 + b_2 + c_3)\lambda^2 + \ldots = 0. \tag{8}$$

If λ_1, λ_2 and λ_3 are the eigenvalues, then

$$(\lambda - \lambda_1)(\lambda - \lambda_2)(\lambda - \lambda_3) = 0$$

or

$$\lambda^3 - (\lambda_1 + \lambda_2 + \lambda_3)\lambda^2 + \ldots = 0. \tag{9}$$

Thus, the trace of $[A]$, $a_1 + b_2 + c_3$, is the sum of the eigenvalues of either $[A]$ or $[B]$.

Let $\{C\}$ be a vector which is parallel to the axis about which body 2 rotates when frame 2 moves from alignment with frame 1 to alignment with frame 3. The components of such a vector $\{C\}$ will be the same when referred to either of the frames 1 or 3. Therefore

$$\{C\}_3 = \{C\}_1 = [\ell_3]_1 \{C\}_1$$

and since the length of the vector $\{C\}$ is unimportant,

$$[\ell_3]_1 \{C\}_1 = \lambda \{C\}_1$$

also satisfies the requirement for $\{C\}$. Hence, for a given $[\ell_3]_1$, the direction of $\{C\}$ can be determined.

The angle through which frame 2 turns about an axis parallel to C while it moves from alignment with frame 1 to alignment with frame 3 can be found by considering the following sequence of rotations

1 γ about z_1 4 β about y_4 5 θ about x_5 6

6 $-\beta$ about y_6 7 $-\gamma$ about z_7 3

so that

$$[\ell_3]_1 = [\ell_4]_1 [\ell_5]_4 [\ell_6]_5 [\ell_7]_6 [\ell_3]_7$$

$$= [\ell_4]_1 [\ell_5]_4 [\ell_6]_5 [\ell_5]_4^{-1} [\ell_4]_1^{-1}$$

$$= [\ell_5]_1 [\ell_6]_5 [\ell_5]_1^{-1}.$$

The net rotation is thus θ, while

$$[\ell_3]_1 = [\ell_5]_1 [\ell_6]_5 [\ell_5]_1^{-1}$$

is a similarity transformation. The traces of $[\ell_3]_1$ and $[\ell_6]_5$ are thus equal, allowing θ to be determined.

Consider the equation

$$[\ell_3]_1 \{C\}_1 = \{C\}_1$$

or

$$\left[[\ell_3]_1 - [1]\right]\{C\}_1 = \{0\}$$

$$\begin{bmatrix} a_1 - 1 & B & C \\ D & b_2 - 1 & F \\ G & H & c_3 - 1 \end{bmatrix} \begin{bmatrix} x \\ y \\ z \end{bmatrix} = \begin{bmatrix} 0 \\ 0 \\ 0 \end{bmatrix}$$

$$\begin{bmatrix} A & B & C \\ D & E & F \\ G & H & J \end{bmatrix} \begin{bmatrix} x \\ y \\ z \end{bmatrix} = \begin{bmatrix} 0 \\ 0 \\ 0 \end{bmatrix}$$

$$Ax + By + Cz = 0$$
$$Dx + Ey + Fz = 0 \;.$$
$$Gx + Hy + Jz = 0$$

Putting $z = 1$, since the length of $\{C\}$ is unimportant, these equations reduce to

$$Ax + By + C = 0$$
$$Dx + Ey + F = 0$$

giving

$$x = \frac{\begin{vmatrix} -C & B \\ -F & E \end{vmatrix}}{\begin{vmatrix} A & B \\ D & E \end{vmatrix}} = (FB - CE)/(AE - CE) \qquad (10)$$

and

$$y = \frac{\begin{vmatrix} A & -C \\ D & -F \end{vmatrix}}{\begin{vmatrix} A & B \\ D & E \end{vmatrix}} = (DC - AF)/(AE - CE) \qquad (11)$$

The direction of the axis about which rotation takes place is thus determined. Now

$$[\ell_6]_5 = \begin{bmatrix} 1 & 0 & 0 \\ 0 & c\theta & -s\theta \\ 0 & s\theta & c\theta \end{bmatrix}$$

and since

$$\text{trace}[\ell_3]_1 = \text{trace}[\ell_6]_5 \;,$$

$$\cos\theta = (a_1 + b_2 + c_3 - 1)/2 \ . \qquad (12)$$

The magnitude of the angle through which frame 2 turns about an axis parallel to the vector {C} in moving from alignment with frame 1 to alignment with frame 3 is thus determined. The direction of the rotation remains to be determined. Recalling that $[\ell_3]_1$ can be regarded as being made up of three column vectors

$$[\ell_3]_1 = \begin{bmatrix} \{\ell_{x\,3}\}_1 & \{\ell_{y\,3}\}_1 & \{\ell_{z\,3}\}_1 \end{bmatrix} = \begin{bmatrix} a_1 & B & C \\ D & b_2 & F \\ G & H & c_3 \end{bmatrix}$$

where $\{\ell_{x\,3}\}_1$, $\{\ell_{y\,3}\}_1$ and $\{\ell_{z\,3}\}_1$ are the direction cosines of the x_3, y_3 and z_3 axes respectively, with respect to frame 1. Thus, a well proportioned sketch of frame 1, the vector {C} and, say, the z_1 axis will allow the direction of the rotation to be determined as shown in Fig. 3.22b.

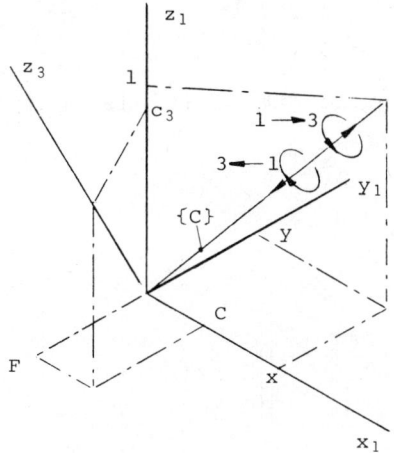

Fig. 3.22b.

The plate of Fig. 3.22a can be moved from the position ABC to the position A'B'C' by a number of alternative rotation sequences. Consider four of the possible alternatives.

1. Refer to Fig. 3.22c.

$$[\ell_3]_1 = \begin{bmatrix} c\beta & 0 & s\beta \\ 0 & 1 & 0 \\ -s\beta & 0 & c\beta \end{bmatrix} \begin{bmatrix} c\gamma & -s\gamma & 0 \\ s\gamma & c\gamma & 0 \\ 0 & 0 & 1 \end{bmatrix}$$

$$= \begin{bmatrix} 0 & 0 & 1 \\ 0 & 1 & 0 \\ -1 & 0 & 0 \end{bmatrix} \begin{bmatrix} -1 & 0 & 0 \\ 0 & -1 & 0 \\ 0 & 0 & 1 \end{bmatrix} = \begin{bmatrix} 0 & 0 & 1 \\ 0 & -1 & 0 \\ 1 & 0 & 0 \end{bmatrix} \ .$$

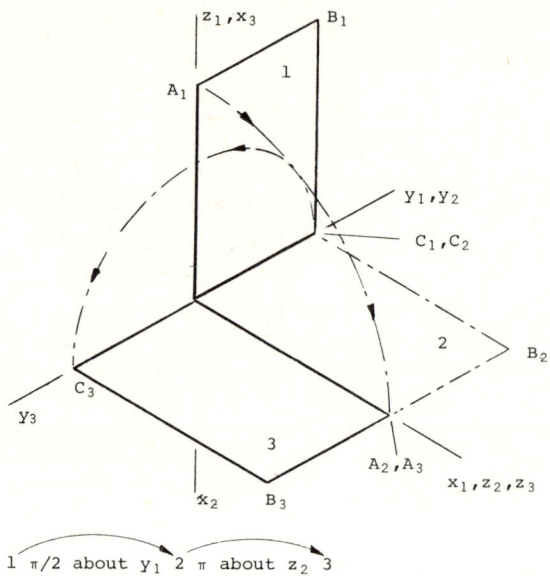

Fig. 3.22c

2. Refer to Fig. 3.22d.

$$[\ell_3]_1 = \begin{bmatrix} c\gamma & -s\gamma & 0 \\ s\gamma & c\gamma & 0 \\ 0 & 0 & 1 \end{bmatrix} \begin{bmatrix} c\beta & 0 & s\beta \\ 0 & 1 & 0 \\ -s\beta & 0 & c\beta \end{bmatrix}$$

$$= \begin{bmatrix} -1 & 0 & 0 \\ 0 & -1 & 0 \\ 0 & 0 & 1 \end{bmatrix} \begin{bmatrix} 0 & 0 & -1 \\ 0 & 1 & 0 \\ 1 & 0 & 0 \end{bmatrix} = \begin{bmatrix} 0 & 0 & 1 \\ 0 & -1 & 0 \\ 1 & 0 & 0 \end{bmatrix}.$$

3. Refer to Fig. 3.22e.

$$[\ell_3]_1 = \begin{bmatrix} c\beta & 0 & s\beta \\ 0 & 1 & 0 \\ -s\beta & 0 & c\beta \end{bmatrix} \begin{bmatrix} 1 & 0 & 0 \\ 0 & c\alpha & -s\alpha \\ 0 & s\alpha & c\alpha \end{bmatrix}$$

$$= \begin{bmatrix} 0 & 0 & -1 \\ 0 & 1 & 0 \\ 1 & 0 & 0 \end{bmatrix} \begin{bmatrix} 1 & 0 & 0 \\ 0 & -1 & 0 \\ 0 & 0 & -1 \end{bmatrix} = \begin{bmatrix} 0 & 0 & 1 \\ 0 & -1 & 0 \\ 1 & 0 & 0 \end{bmatrix}.$$

$[\ell_3]_1$ is thus independent of the sequence of rotations. Of course $[\ell_3]_1$ can be formed immediately by reference to Fig. 3.22a.

Solution of Kinematics Problems

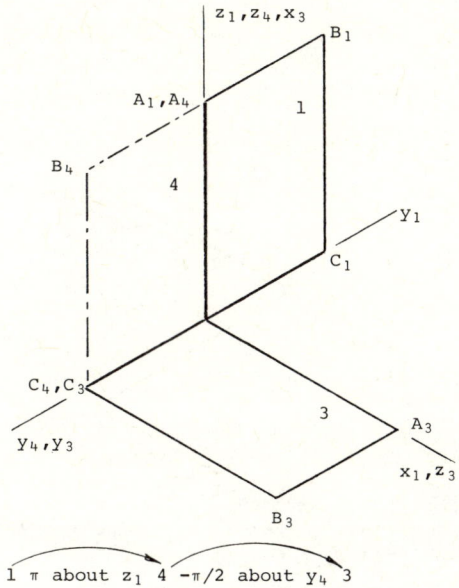

1 π about z_1 4 -π/2 about y_4 3

Fig. 3.22d.

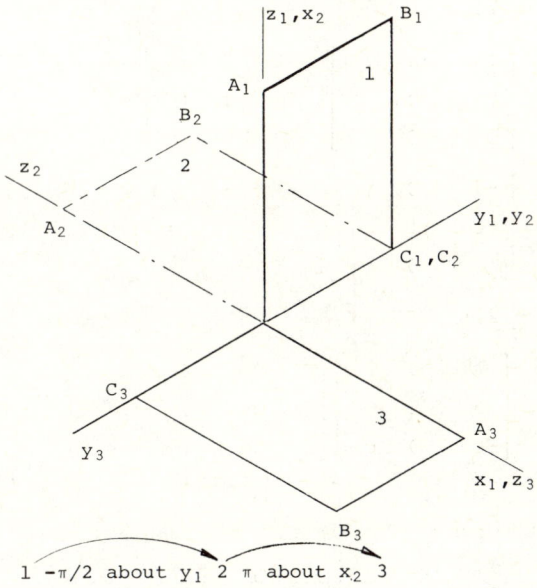

1 -π/2 about y_1 2 π about x_2 3

Fig. 3.22e.

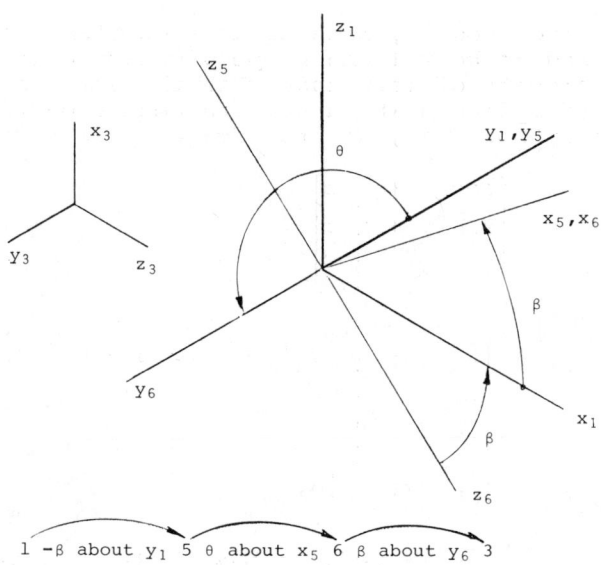

1 $-\beta$ about y_1 5 θ about x_5 6 β about y_6 3

Fig. 3.22f.

4. Refer to Fig. 3.22f.

$$[\ell_3]_1 = \begin{bmatrix} c\beta & 0 & -s\beta \\ 0 & 1 & 0 \\ s\beta & 0 & c\beta \end{bmatrix} \begin{bmatrix} 1 & 0 & 0 \\ 0 & c\theta & -s\theta \\ 0 & s\theta & c\theta \end{bmatrix} \begin{bmatrix} c\beta & 0 & s\beta \\ 0 & 1 & 0 \\ -s\beta & 0 & c\beta \end{bmatrix}$$

$$= \begin{bmatrix} c^2\beta + s^2\beta c\theta & -s\theta c\beta & s\beta c\beta - s\beta c\beta c\theta \\ s\theta s\beta & c\theta & -s\theta c\beta \\ s\beta c\beta + s\beta c\beta c\theta & s\theta c\beta & s^2\beta + c^2\beta c\theta \end{bmatrix}$$

$$= \begin{bmatrix} 0 & 0 & 1 \\ 0 & -1 & 0 \\ 1 & 0 & 1 \end{bmatrix}.$$

Hence
$\cos\theta = -1$, giving $\theta = \pi$.

Also
$\cos^2\beta + \sin^2\beta(-1) = 0$ or $\cos 2\beta = 0$, giving $\beta = \pi/4$.

Or, by reference to Eqs. 12, 10 and 11 respectively,

$\cos\theta = (0 - 1 + 0 - 1)/2 = -1$, giving $\theta = \pi$,

and
$x = \{0 - 1(-2)\}/\{(-1)(-2) - 0\} = 1$

$y = \{(0)(1) - (-1)(0)\}/\{(-1)(-2) - 0\} = 0$

thus confirming the previous result.

3.2. Problems For Solution

Problem 3.23. A disc, body 2, rotates at a constant rate ω about an axis which is fixed in body 1 and perpendicular to its plane. A point P moves with a velocity of magnitude v on the surface of the disc in a circular path of radius a which has its centre on the axis of rotation of the disc. Show that, for the frames specified in Fig. 3.23,

$$\{R_P\}_{1/1} = a \begin{bmatrix} \cos(\omega + \Omega)t \\ \sin(\omega + \Omega)t \\ 0 \end{bmatrix},$$

$$\frac{d}{dt}\{R_P\}_{1/1} = \{\dot{R}_P\}_{1/1} = \{V_P\}_{1/1} = a(\omega + \Omega) \begin{bmatrix} -\sin(\omega + \Omega)t \\ \cos(\omega + \Omega)t \\ 0 \end{bmatrix},$$

$$\frac{d}{dt}\{V_P\}_{1/1} = \{\dot{V}_P\}_{1/1} = \{A_P\}_{1/1}$$
$$= -(\omega^2 + v^2/a^2 + 2\omega v/a)\{R_P\}_{1/1},$$

$$\{V_P\}_{1/2} = a(\omega + \Omega) \begin{bmatrix} -\sin\Omega t \\ \cos\Omega t \\ 0 \end{bmatrix},$$

$$\{V_P\}_{1/3} = a(\omega + \Omega) \begin{bmatrix} 0 \\ 1 \\ 0 \end{bmatrix},$$

$$\{A_P\}_{1/2} = -(\omega^2 a + v^2/a + 2\omega v) \begin{bmatrix} \cos\Omega t \\ \sin\Omega t \\ 0 \end{bmatrix}$$

and

$$\{A_P\}_{1/3} = -(\omega^2 a + v^2/a + 2\omega v) \begin{bmatrix} 1 \\ 0 \\ 0 \end{bmatrix}$$

where $\Omega = v/a$.

Problem 3.24. A vector is specified by

$$\{B\}_{1/4} = \begin{bmatrix} -0.56813 \\ 0.76761 \\ 1.00000 \end{bmatrix}.$$

Find

$$\{B\}_{1/1}$$

for the case in which

$$[\ell_4]_1 = \begin{bmatrix} 0.741516 & 0.45315 & -0.494731 \\ -0.595012 & 0.784856 & -0.172904 \\ 0.30995 & 0.4226 & 0.85165 \end{bmatrix}.$$

(Comment. It will be found that $\{B\}_{1/1} = \{B\}_{1/4}$. It is clear that the vector $\{B\}_1$ has special properties in relation to the given $[\ell_4]_1$. The components of this vector are the same along the axes of both frames. Frame 1 can be aligned with frame 4 by rotating it about an axis parallel to $\{B\}_1$.

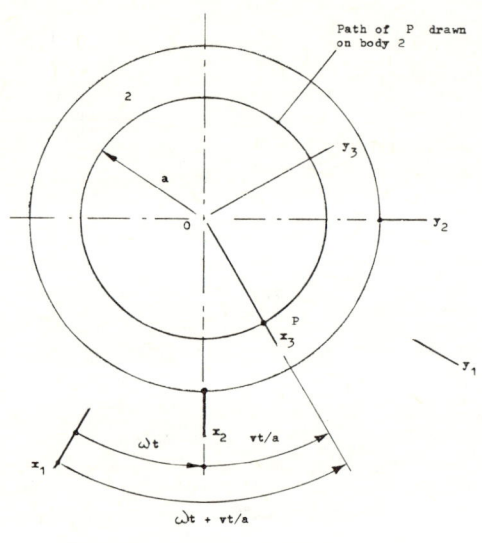

Fig. 3.23.

Problem 3.25. Frame 1 can be aligned with frame 2 by a simple positive rotation α about the x_1 axis. Thus

$$[\ell_2]_1 = \begin{bmatrix} 1 & 0 & 0 \\ 0 & \cos\alpha & -\sin\alpha \\ 0 & \sin\alpha & \cos\alpha \end{bmatrix}.$$

Show that $d[\ell_2]_1/dt$ can be written $[\dot{\alpha}][\ell_2]_1$ where

$$[\dot{\alpha}] = \begin{bmatrix} 0 & 0 & 0 \\ 0 & 0 & -\dot{\alpha} \\ 0 & \dot{\alpha} & 0 \end{bmatrix}.$$

Repeat the problem for the case in which alignment is achieved by a

Solution of Kinematics Problems

simple positive rotation about either y_1 or z_1.

Problem 3.26. Aircraft A and B fly in the x_1y_1 plane as shown in Fig. 3.26. Aircraft A flies in a circular path, centred at O and of radius a, at a constant speed v_A, while aircraft B flies at a constant speed v_B on a straight path parallel to the y_1 axis and at a distance b from it.

Find the velocity of B relative to A which is measured in and referred to a frame 2 fixed in aircraft A, when aircraft A is at A_o and aircraft B is at B_o. Also find the corresponding acceleration.

(Assistance.

$$\{R_{BA}\}_{2/2} = [\ell_1]_2 \{R_{BA}\}_{1/1} = [\ell_1]_2 \{R_{BA}\}_{2/1}$$

$$\{V_{BA}\}_{2/2} = d\{R_{BA}\}_{2/2}/dt \quad \text{and} \quad \{A_{BA}\}_{2/2} = d\{V_{BA}\}_{2/2}/dt.$$

Also

$$\{R_{BA}\}_{1/1} = \begin{bmatrix} b - a\cos\theta \\ v_B t - a\sin\theta \\ 0 \end{bmatrix} \quad .\ )$$

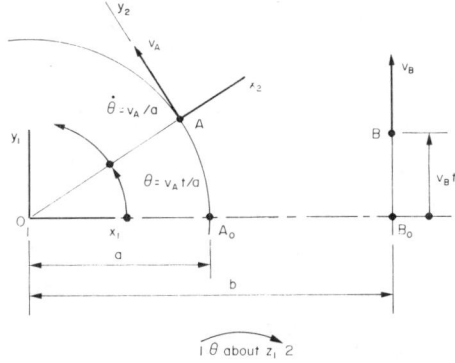

Fig. 3.26.

Problem 3.27. Aircraft A flies at a constant speed v_A in a circular path of radius a drawn on the x_1y_1 plane. Aircraft B flies at a constant speed v_B in a circular path drawn on a plane parallel to the y_1z_1 plane as shown in Fig. 3.27.

Find the velocity and acceleration of B relative to A which are measured in and referred to frame 2 fixed in aircraft A, when aircraft A is at A_o and aircraft B is at B_o.

Also, find the velocity and acceleration of A relative to B which are measured in and referred to frame 3 fixed in aircraft B, when aircraft B is at B_o and aircraft A is at A_o.

(Assistance.
$$\{R_{BA}\}_{1/1} = \begin{bmatrix} c - a\cos\theta \\ b\sin\alpha - a\sin\theta \\ b(1 - \cos\alpha) \end{bmatrix}$$

where $\theta = v_A t/a$ and $\alpha = v_B t/b$.)

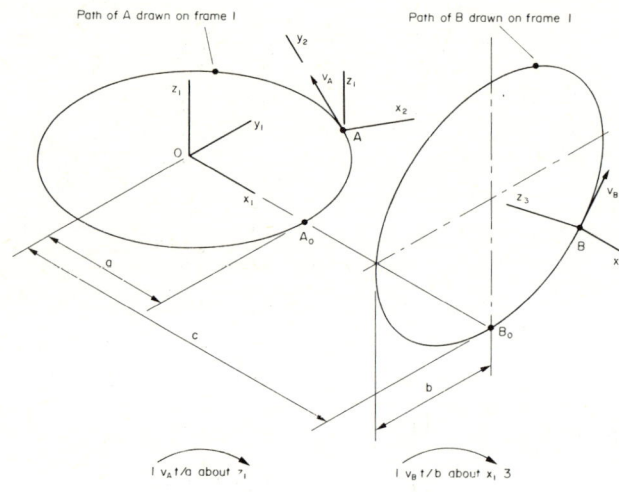

Fig. 3.27.

Problem 3.28. Figure 3.28 shows frame 3 positioned in relation to frame 1. A vector referred to frame 3 has the components

$$\begin{bmatrix} x_R \\ y_\theta \\ z_\phi \end{bmatrix}$$

Show that the vector is referred to frame 1 by the transformation

$$\begin{bmatrix} x_1 \\ y_1 \\ z_1 \end{bmatrix} = \begin{bmatrix} \cos\theta\cos\phi & -\sin\theta & -\cos\theta\sin\phi \\ \sin\theta\cos\phi & \cos\theta & -\sin\theta\sin\phi \\ \sin\phi & 0 & \cos\phi \end{bmatrix} \begin{bmatrix} x_R \\ y_\theta \\ z_\phi \end{bmatrix}.$$

Problem 3.29. Figure 3.29 is drawn to show the relative positions of frames 1 and 2. The z_1 axis is perpendicular to the plane ABE, the y_1 axis perpendicular to AB and in the plane ABE and the x_2 axis is along AB. Find $[\ell_2]_1$. Frame 1 can be aligned with frame 2 by either of the sequences of rotations given in Fig. 3.29. Show that $\alpha = -35°$, $\beta = 23.83°$, $\gamma = 29.96°$, $\phi = -37.57°$ and $\psi = 41.45°$.

(Assistance.

Solution of Kinematics Problems

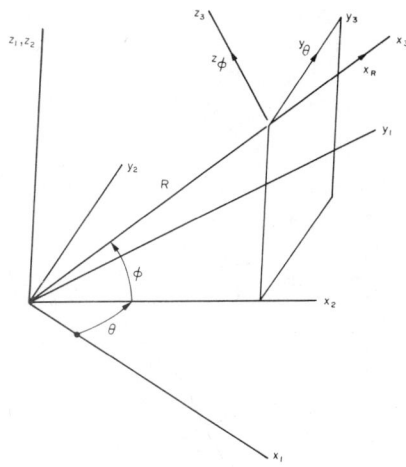

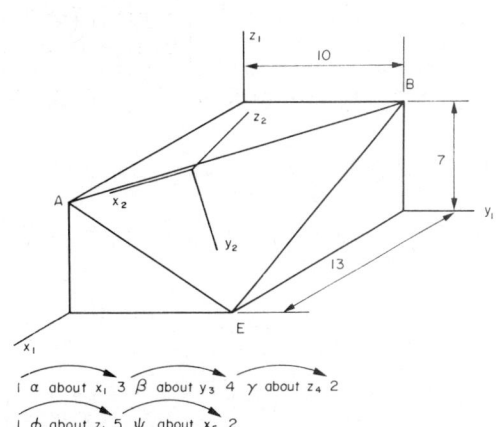

Fig. 3.28.   Fig. 3.29.

$$\{N_{Z_2}\}_1 = [R_{AB}]_{1/1}\{R_{AB}\}_{1/1}/|[R_{AB}]_{1/1}\{R_{AB}\}_{1/1}|$$

$$= [R_{EA}]_{1/1}\{R_{EA}\}_{1/1}/|[R_{EA}]_{1/1}\{R_{EA}\}_{1/1}|,$$

$$\{N_{Y_2}\}_1 = [N_{Z_2}]_1\{N_{X_2}\}_1/|[N_{Z_2}]_1\{N_{X_2}\}_1|$$

and

$$\{N_{X_2}\}_1 = \{R_{AB}\}_{1/1}/|R_{AB}|$$

where $\{N\}_1$ is unit vector, in the direction specified by the suffix, referred to frame 1. Hence

$$[\ell_2]_1 = \left[\{N_{X_2}\}_1 \quad \{N_{Y_2}\}_1 \quad \{N_{Z_2}\}_1 \right].$$

Problem 3.30. An aircraft flies in a horizontal circular path, of radius a, drawn on an earth fixed frame 1, which has its $x_1 y_1$ plane tangential to the surface of the earth as shown in Fig. 3.30a. The velocity of the aircraft has a constant magnitude v and it is tracked by a radar antenna, body 3, at O. Show that the acceleration of the aircraft, measured in the earth fixed frame 1 and referred to frame 3 fixed in the antenna as shown in Fig. 3.30b, is given by

$$\{A_{AO}\}_{1/3} = (v^2/a) \begin{bmatrix} -\cos\beta\cos(\gamma - \theta) \\ \sin(\gamma - \theta) \\ \cos(\gamma - \theta) \end{bmatrix}$$

where

$$\tan\gamma = (b + a\sin\theta)/a\cos\theta$$

and

$$\tan\beta = h/\sqrt{(a^2 + 2ab\sin\theta + b^2)}.$$

Also show that the acceleration of the aircraft, measured in frame 3 and referred to frame 1, is given by

$$\{A_{AO}\}_{3/1} = \frac{d}{dt}\{V_{AO}\}_{3/1} - [\omega_3]_{1/1}\{V_{AO}\}_{3/1}$$

where

$$\{V_{AO}\}_{3/1} = \{V_{AO}\}_{1/1} - [\omega_3]_{1/1}\{R_{AO}\}_{1/1}.$$

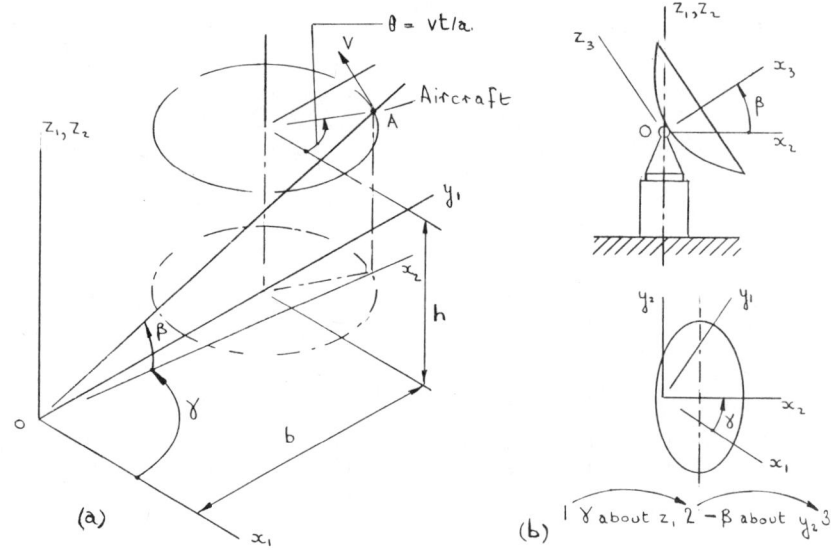

Fig. 3.30.

Problem 3.31. Points A and B fixed in body 2 have velocities given by

$$\{V_{AO}\}_{1/1} = [\omega_2]_{1/1}\{R_{AO}\}_{1/1}$$

and

$$\{V_{BO}\}_{1/1} = [\omega_2]_{1/1}\{R_{BO}\}_{1/1}$$

respectively relative to point O which is also fixed in body 2. Show that, given information about the relative positions of points O, A and B, and the linear velocities of points A and B relative to O, the angular velocity of body 2 can be determined from

$$\{\omega_2\}_{1/1} = [V_{AO}]_{1/1}\{V_{BO}\}_{1/1}/(\{V_{AO}\}^T_{1/1}\{R_{BO}\}_{1/1})$$

$$= [V_{BO}]_{1/1}\{V_{AO}\}_{1/1}/(\{V_{BO}\}^T_{1/1}\{R_{AO}\}_{1/1})$$

Solution of Kinematics Problems

provided that the points O, A and B are not collinear.

(Assistance. Multiply the first two equations by

$$[V_{BO}]_{1/1} \quad \text{and} \quad [V_{AO}]_{1/1}$$

respectively and expand their right hand sides using the vector triple product expansion

$$[A][B]\{C\} = (\{A\}^T\{C\})\{B\} - (\{A\}^T\{B\})\{C\} \quad).$$

Problem 3.32. At a certain instant of time body 3 (the connecting rod of Problem 3.5) has a motion specified as follows

$$\{R_A\}_{1/1} = \begin{bmatrix} 43.3 \\ 25 \\ 0 \end{bmatrix} \text{mm}, \quad \{R_B\}_{1/1} = \begin{bmatrix} 286.9 \\ 0 \\ 50 \end{bmatrix} \text{mm},$$

$$\{V_A\}_{1/1} = \begin{bmatrix} -249 \\ 433 \\ 0 \end{bmatrix} \text{mm/s}, \quad \{V_B\}_{1/1} = \begin{bmatrix} -294 \\ 0 \\ 0 \end{bmatrix} \text{mm/s}$$

and

$$\{\omega_3\}_{1/1} = \begin{bmatrix} 0.346 \\ -0.0355 \\ -1.706 \end{bmatrix} \text{rad/s}$$

where A and B are points fixed in body 3.
Obtain the equation to the central axis for the motion of body 3 relative to body 1 and find the point at which the axis meets the $x_1 y_1$ plane. Also find the velocity of the axis.

Problem 3.33. In the system shown in Fig. 3.33, body 3 turns relative to body 2 about the x_1 axis and body 2 turns relative to body 1 about the z_1 axis. Show that

$$\{V_{AO}\}_{1/2} = \begin{bmatrix} -\omega b \sin\alpha \\ \omega a - \dot{\alpha} b \cos\alpha \\ \dot{\alpha} b \sin\alpha \end{bmatrix}$$

where $\omega = \dot{\gamma}$ and A is fixed in body 3 as shown. Hence show that

$$\{R_{QA}\}_{1/2} = [\omega_3]_{1/2}\{V_{AO}\}_{1/2}/|\omega_3|_1^2$$

$$= 1/(\omega^2 + \dot{\alpha}^2) \begin{bmatrix} -\omega^2 a - \omega\dot{\alpha} b \cos\alpha \\ -\omega^2 b \sin\alpha - \dot{\alpha}^2 b \sin\alpha \\ \omega\dot{\alpha} a + \dot{\alpha}^2 b \cos\alpha \end{bmatrix}$$

where Q is a point on the central axis corresponding to A for the motion of body 3 relative to body 1, and

148   Matrix Methods in Engineering Mechanics

$$\{R_{SC}\}_{1/2} = \lfloor \omega_3 \rfloor_{1/2} \{V_{CO}\}_{1/2} / |\omega_3|_1^2$$

$$= 1/(\omega^2 + \dot{\alpha}^2) \begin{bmatrix} -\omega^2 a \\ 0 \\ \dot{\alpha}\omega a \end{bmatrix}$$

where S is the point on the central axis corresponding to C for the motion of body 3 relative to body 1.

Find

$$\{R_{QA}\}_{1/2}\big|_{\alpha=0}$$

for the case in which body 3 rolls without slip on body 1 at B ($\dot{\alpha} = -\omega a/b$). Also show that

$$\{R_{SC}\}_{1/2} = ab/(a^2 + b^2) \begin{bmatrix} -b \\ 0 \\ -a \end{bmatrix}$$

for this case and hence show that S is on the line OB.

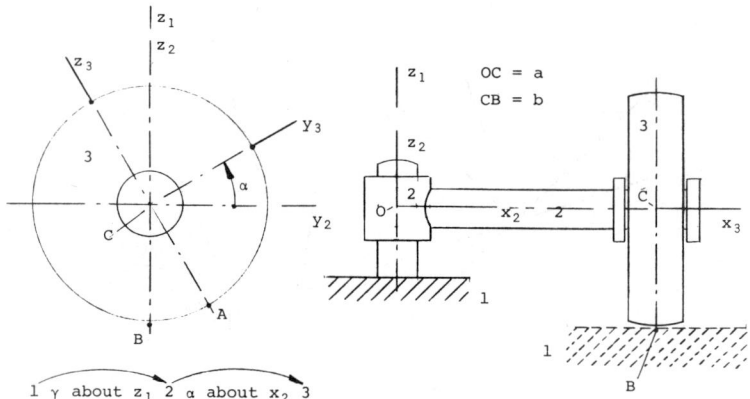

Fig. 3.33

Find

$$\{\omega_3\}_{1/2}^T \{V_{AO}\}_{1/2} \, , \, \{\omega_3\}_{1/2}^T \{V_{CO}\}_{1/2}$$

and

$$\{V_{SO}\}_{1/2} = \{V_{CO}\}_{1/2} + \lfloor \omega_3 \rfloor_{1/2} \{R_{SC}\}_{1/2} \, .$$

(Assistance. Obtain the expression for the velocity of A by differentiating the equation

$$\{R_{AO}\}_{1/1} = \lfloor \ell_2 \rfloor_1 \{R_{CO}\}_{2/2} + \lfloor \ell_2 \rfloor_1 \lfloor \ell_3 \rfloor_2 \{R_{AC}\}_{3/3} \,).$$

Problem 3.34. One set of rotations for positioning body 4 relative to body 1 is shown in Fig. 3.34. This particular sequence of angles are known as Euler angles. Show that

$$\{\omega_4\}_{1/4} = \begin{bmatrix} \omega_x \\ \omega_y \\ \omega_z \end{bmatrix} = \dot{\psi} \begin{bmatrix} s\theta s\phi \\ s\theta c\phi \\ c\theta \end{bmatrix} + \dot{\theta} \begin{bmatrix} c\phi \\ -s\phi \\ 0 \end{bmatrix} + \dot{\phi} \begin{bmatrix} 0 \\ 0 \\ 1 \end{bmatrix}$$

$$= \begin{bmatrix} \sin\theta\sin\phi & \cos\phi & 0 \\ \sin\theta\cos\phi & -\sin\phi & 0 \\ \cos\theta & 0 & 1 \end{bmatrix} \begin{bmatrix} \dot{\psi} \\ \dot{\theta} \\ \dot{\phi} \end{bmatrix}$$

and hence, by inversion, that

$$\begin{bmatrix} \dot{\psi} \\ \dot{\theta} \\ \dot{\phi} \end{bmatrix} = 1/\sin\theta \begin{bmatrix} \sin\phi & \cos\phi & 0 \\ \sin\theta\cos\phi & -\sin\theta\sin\phi & 0 \\ -\cos\theta\sin\phi & -\cos\theta\cos\phi & \sin\theta \end{bmatrix} \begin{bmatrix} \omega_x \\ \omega_y \\ \omega_z \end{bmatrix}$$

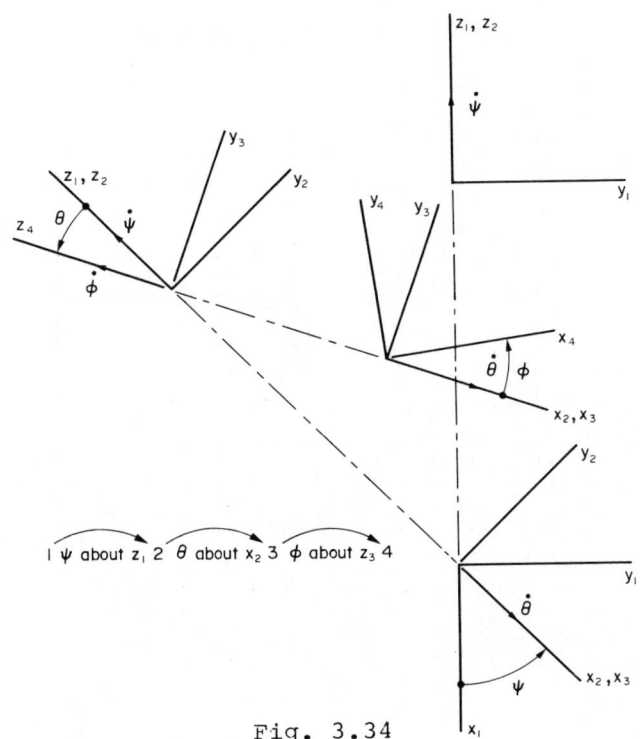

Fig. 3.34

Note that with this set of angles difficulties will arise when θ is close to $n\pi$ ($n = 0, 1 \ldots$).

Problem 3.35. One set of rotations for positioning body 4 relative to body 1 is shown in Fig. 3.35. This particular sequence of angles are known as Bryant angles. Show that

$$\{\omega_4\}_{1/4} = \begin{bmatrix} \omega_x \\ \omega_y \\ \omega_z \end{bmatrix} = \dot{\psi} \begin{bmatrix} c\theta c\phi \\ -c\theta s\phi \\ s\theta \end{bmatrix} + \dot{\theta} \begin{bmatrix} s\phi \\ c\phi \\ 0 \end{bmatrix} + \dot{\phi} \begin{bmatrix} 0 \\ 0 \\ 1 \end{bmatrix}$$

$$= \begin{bmatrix} \cos\theta\cos\phi & \sin\phi & 0 \\ -\cos\theta\sin\phi & \cos\phi & 0 \\ \sin\theta & 0 & 1 \end{bmatrix} \begin{bmatrix} \dot{\psi} \\ \dot{\theta} \\ \dot{\phi} \end{bmatrix}$$

and hence, by inversion, that

$$\begin{bmatrix} \dot{\psi} \\ \dot{\theta} \\ \dot{\phi} \end{bmatrix} = 1/\cos\theta \begin{bmatrix} \cos\phi & -\sin\theta & 0 \\ \sin\phi\cos\theta & \cos\phi\cos\theta & 0 \\ -\cos\phi\sin\theta & \sin\phi\sin\theta & \cos\theta \end{bmatrix} \begin{bmatrix} \omega_x \\ \omega_y \\ \omega_z \end{bmatrix}$$

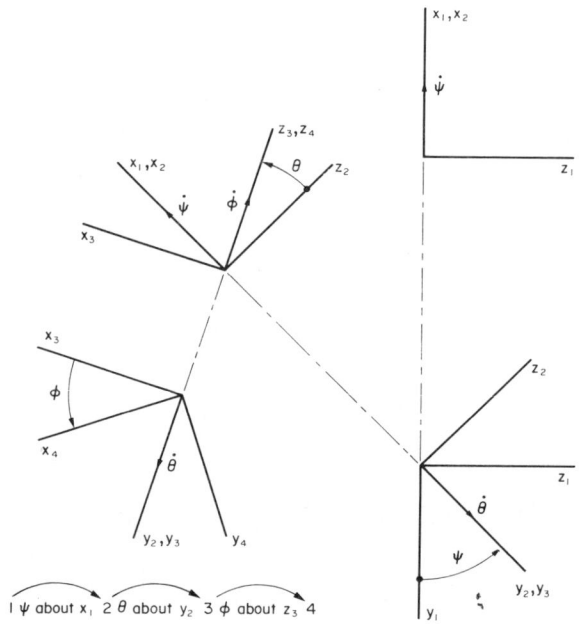

Fig. 3.35.

To what do these relationships reduce when ψ, θ and ϕ are small angles? Note that with this set of angles difficulties will arise when θ is close to $(\pi/2) + n$ ($n = 0, 1 \ldots$).

Problem 3.36. In Fig. 3.36 body 2 turns about the z_1 axis at a constant rate relative to body 1 and body 3 turns about the z_3 axis at a constant rate relative to body 2.

Find, for the given axis system,

$$\{A_{BO}\}_{1/2}$$

where B is a point fixed in body 3. Point B should first be treated as a moving point in frame 3, where frame 3 is fixed relative to frame 2 and aligned with it. Point then should then be treated as a point fixed in the rotating frame 3 as shown.

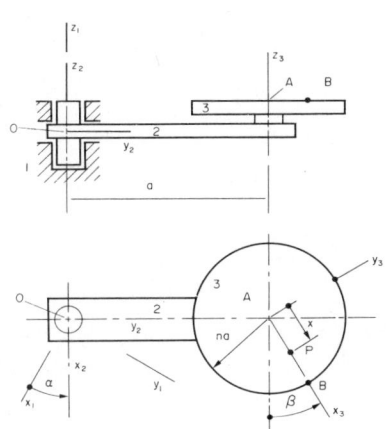

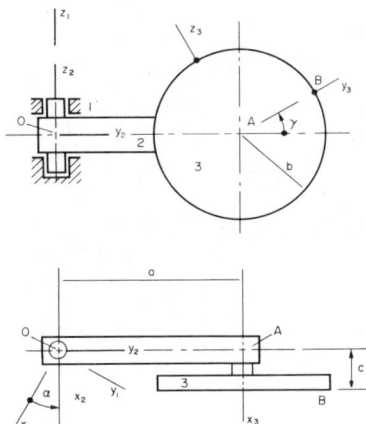

Fig. 3.36.   Fig. 3.38.

Problem 3.37. Find

$$\{A_{PO}\}_{1/2}$$

in the system of Fig. 3.36, where the point P moves along the radial line AB fixed in body 3 at the constant rate $\dot{x} = v$.

Problem 3.38. In Fig. 3.38 body 2 turns about the z_1 axis at a constant rate relative to body 1 and body 3 turns about the x_3 axis at a constant rate relative to body 2. Find, for the given axis system,

$$\{A_{BO}\}_{1/2}.$$

Problem 3.39. Figure 3.39 shows a wheel and axle system. The axle, body 2, turns at a constant rate $\dot{\gamma} = \omega$ relative to body 1 about the z_1 axis. The wheel, body 3, rolls without slip on body 1. Obtain expressions for

$$\{V_{QO}\}_{1/2} \;,\; \{A_{QO}\}_{1/2} \;\text{and}\; \{\dot{\omega}_3\}_{1/2}$$

where Q is a point fixed in body 3 as shown. Sart with the relationship

$$\{R_{QO}\}_{1/1} = [\ell_2]_1\{R_{CO}\}_{2/2} + [\ell_2]_1[\ell_3]_2\{R_{QC}\}_{3/3}$$

152   Matrix Methods in Engineering Mechanics

and apply the condition that

$$\{V_{PO}\}_{1/2}$$

is a null vector when $\gamma = \beta = 0$ to show that $\dot{\beta} = -\omega a/r$ and hence that β is a negative angle, P being a point on the periphery of the wheel then at the point of contact between the wheel any body 1.

Alternatively, after justifying the given angular velocity vector diagram, use its properties to obtain

$$\{V_{QO}\}_{1/2}.$$

Find

$$\{V_{PO}\}_{1/2} \text{ and } \{A_{PO}\}_{1/2}$$

when P is at A, B, and D.

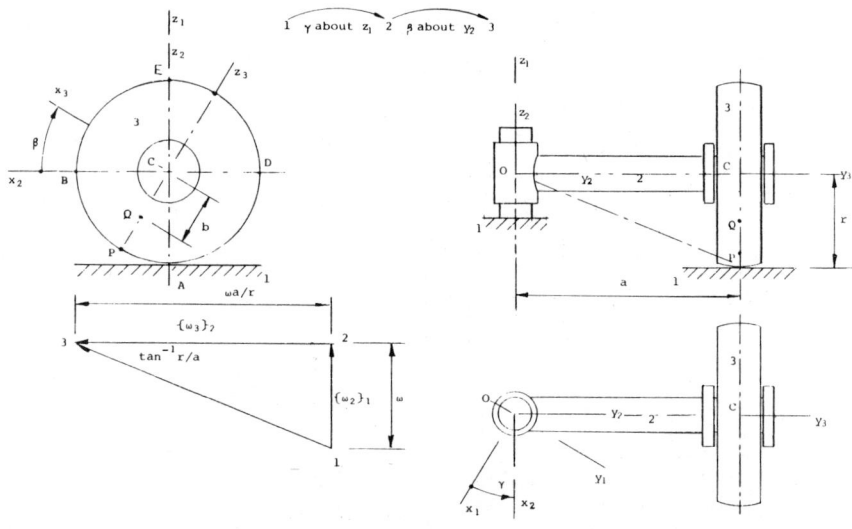

Fig. 3.39.

Problem 3.40. Figure 3.40 shows a wheel and axle system. The axle, body 2, turns at a constant rate $\dot{\gamma} = \omega_2$ relative to body 1 about the z_1 axis. The wheel, body 3, rolls without slip on body 4 which is turning at a constant rate $\dot{\phi} = \omega_4$ relative to body 1 about the z_1 axis. Verify the given angular velocity vector diagram (drawn for $\dot{\gamma} > \dot{\phi}$) and use it to determine

$$\{\omega_3\}_{1/2} \text{ and } \{\dot{\omega}_3\}_{1/2}.$$

Obtain expressions for

$$\{V_{AO}\}_{1/2} \text{ and } \{A_{AO}\}_{1/2}$$

Solution of Kinematics Problems

where A is a point fixed on the periphery of the wheel as shown.

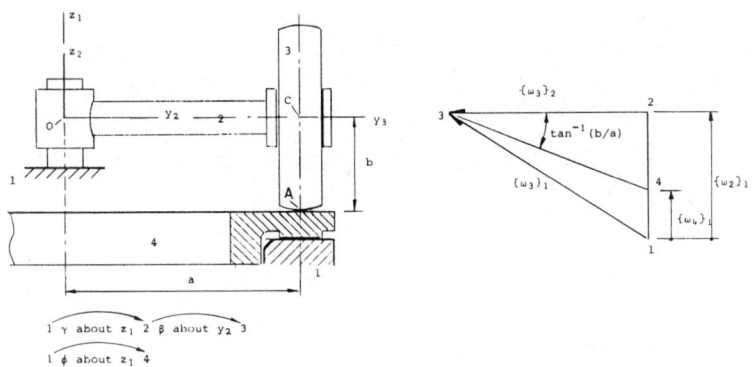

Fig. 3.40.

Problem 3.41. Figure 3.41 shows a wheel and axle system. A wheel and axle are rigidly connected and constitute body 4, while a wheel, body 5 is free to turn on the axle. The system moves on the inner surface of a conical track, body 1, so that the centre line of the axle remains parallel to a generator of the conical track and the centre of the axle C, traces out a circular path of radius R drawn on body 1. The velocity of C is given by

$$\{V_{CO}\}_{1/2} = V \begin{bmatrix} -1 \\ 0 \\ 0 \end{bmatrix}$$

Verify the given angular velocity vector diagram for the system, which is drawn for the case in which the wheels roll without slip, and use it to determine

$$\{\omega_4\}_{1/3}, \quad \{\omega_5\}_{1/3}, \quad \{\dot{\omega}_4\}_{1/3} \quad \text{and} \quad \{\dot{\omega}_5\}_{1/3}.$$

What is the velocity of slip at D if body 5 siezes on the axle and body 4 continues to roll without slip at E?

Problem 3.42. Figure 3.42 shows a rotating and telescoping antenna. Body 2 rotates at a constant rate $\dot{\beta}$ relative to body 1 about the z_1 axis. Body 3 rotates at a constant rate $\dot{\alpha}$ relative to body 2 about the x_2 axis and is extending at the constant rate $\dot{z} = v$. Obtain expressions for

$$\{\omega_3\}_{1/2}, \quad \{\dot{\omega}_3\}_{1/2}, \quad \{V_{AO}\}_{1/2} \quad \text{and} \quad \{A_{AO}\}_{1/2}.$$

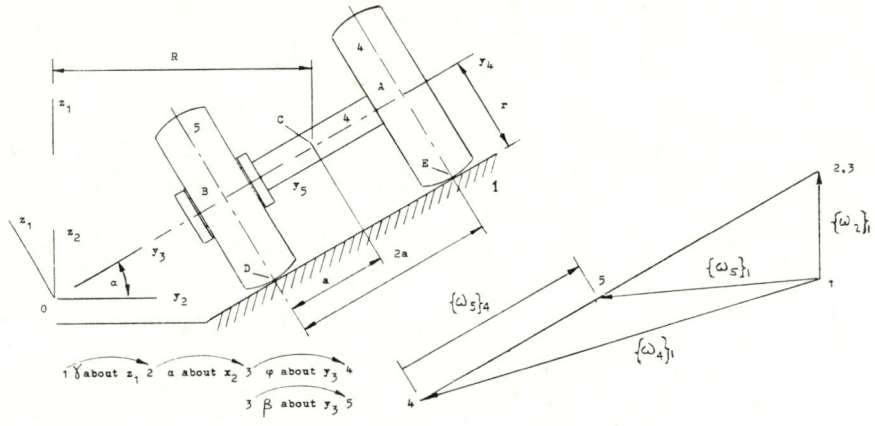

Fig. 3.41.

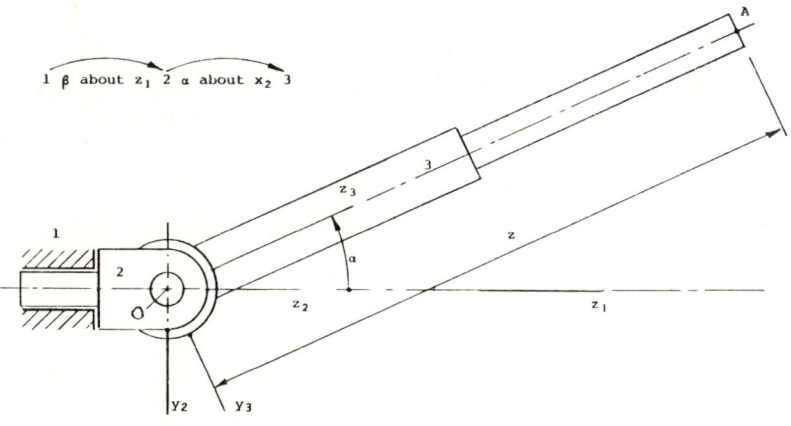

Fig. 3.42.

Problem 3.43. An articulated trailer is shown diagrammatically in Fig. 3.43. It consists of rigid members OA, body 3, ABCD, body 3, and two similar wheels, bodies 4 and 5, which are free to rotate relative to body 3 about the common axis CBD. The wheels roll without slip on a plane, body 1, which is parallel to the plane which contains OA and ABCD. OA and ABCD are joined by a smooth pivot at A which allows relative motion about an axis through A which is perpendicular to the plane on which the wheels roll.

Determine the attitude, θ, of AB relative to OA when OA rotates at a constant rate relative to body 1 about the z_1 axis. Also determine the angular velocity of each wheel under the above conditions and the accelerations of E and F when they are in contact with the plane.

Solution of Kinematics Problems

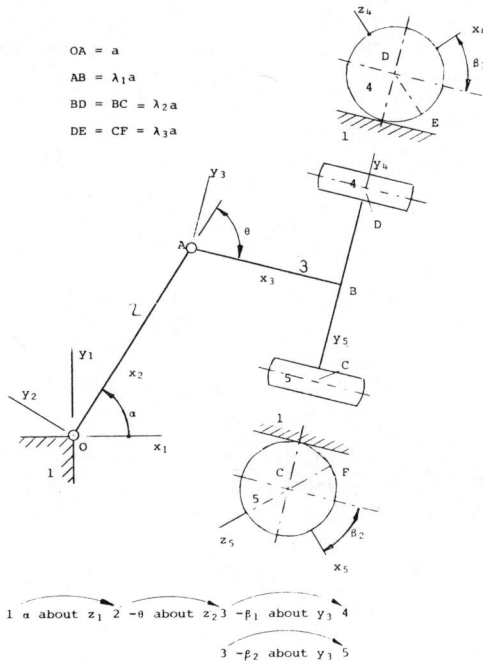

Fig. 3.43.

(Assistance. Find
$$\{V_{EO}\}_{1/3}$$
starting with the relationship
$$\{R_{EO}\}_{1/1} = [\ell_2]_1\{R_{AO}\}_{2/2} + [\ell_2]_1[\ell_3]_2\{R_{DA}\}_{3/3}$$
$$+ [\ell_2]_1[\ell_3]_2[\ell_3]_4\{R_{ED}\}_{4/4}$$
and apply the condition
$$\{V_{EO}\}_{1/3}\big|_{\beta_1=0}$$
is a null vector to obtain
$$\lambda_1\dot{\theta} = \dot{\alpha}(\cos\theta + \lambda_1).$$
When rolling without slip is taking place $\dot{\theta} = 0$, giving $\cos\theta = -\lambda_1$).

Problem 3.44. In the system shown in Fig. 3.44 body 2 turns at a constant rate ω relative to body 1 about the z_1 axis. Body 3 is constrained to turn about the y_2 axis in body 2 and roll without slip on body 1. Body 4 is contrained to rotate at a constant rate ω_s relative to body 3 about an axis parallel to the z_3 axis. Find $\{\dot{\omega}_4\}_{1/3}$.

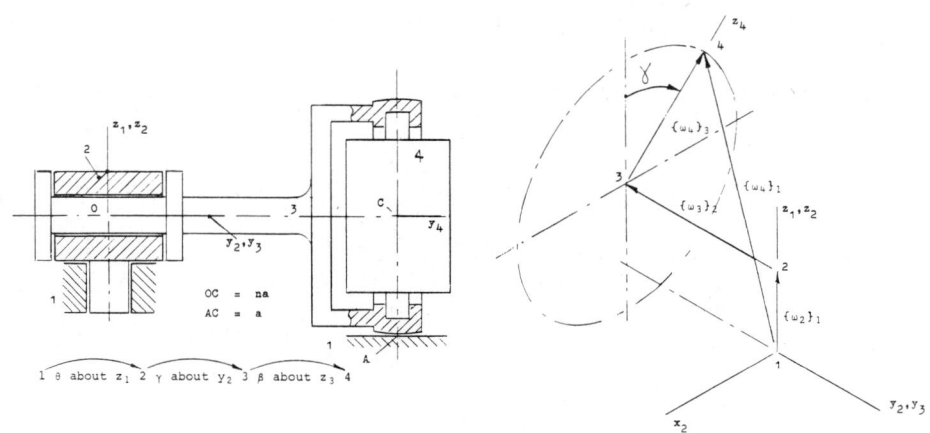

Fig. 3.44.

Problem 3.45. Figure 3.45 shows the arrangement of a conical thrust bearing which employs balls constrained by a track on which the balls roll without slip at the two points of contact, A and B, with the track. The balls also roll without slip on the shaft which carries an axial load. The track is designated body 1, the shaft body 5 and a typical ball body 4. Verify the given angular velocity vector diagram. For the case in which the shaft is driven at the constant rate $\dot{\phi} = \omega$ relative to the track, show that

$$\{\omega_4\}_{1/3} = \{\omega_2\}_{1/3} + \{\omega_4\}_{3/3}$$

$$= \begin{bmatrix} 0 \\ \dot{\gamma}\sin\alpha - \dot{\beta} \\ \dot{\gamma}\cos\alpha \end{bmatrix} = \dot{\phi}(\dot{\gamma}/\dot{\phi}) \begin{bmatrix} 0 \\ \sin\alpha - \dot{\beta}/\dot{\gamma} \\ \cos\alpha \end{bmatrix}$$

$$= \frac{\omega\tan\psi}{1 - \tan\psi} \begin{bmatrix} 0 \\ -(\sin\alpha + \cos\alpha) \\ n\cos\alpha \end{bmatrix}$$

where

$$\tan\alpha = n + 1, \quad \tan\psi = \frac{1 - n\cos\theta}{1 + n(1 + \sin\theta)},$$

Solution of Kinematics Problems

$$\frac{\dot{\gamma}}{\dot{\phi}} = \frac{n\tan\psi}{1 - \tan\psi} \quad \text{and} \quad \frac{\dot{\beta}}{\dot{\gamma}} = \frac{1}{n\cos\alpha} .$$

Also show that

$$\{\dot{\omega}_4\}_{1/3} = \frac{n\omega^2\tan^2\psi}{(1 - \tan\psi)^2}\begin{bmatrix}1\\0\\0\end{bmatrix} .$$

How should the angle θ be determined to ensure that rubbing between the balls and shaft will not occur at F?

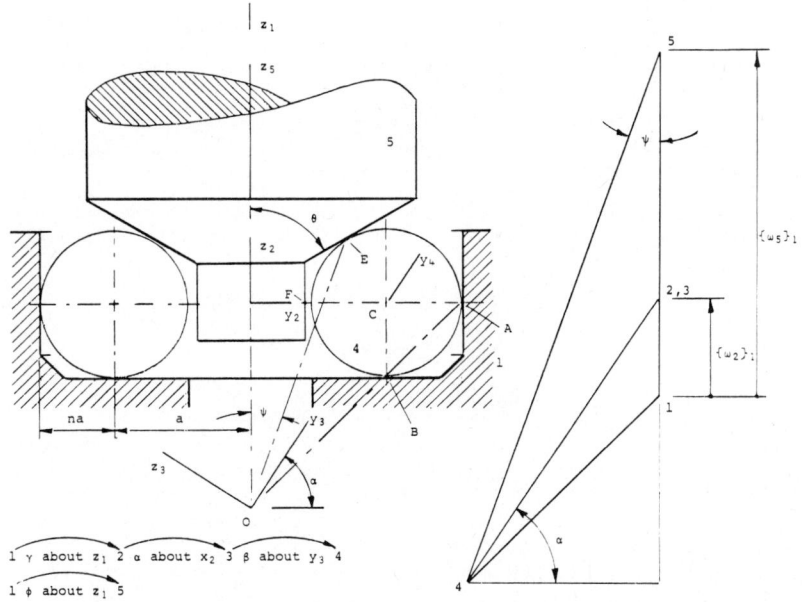

Fig. 3.45.

Problem 3.46. Figure 3.46 shows the arrangement of an automotive differential. The bevel pinion, body 2, is driven from the engine via a gear box and meshes with the crown wheel, body 3. The crown wheel carries planetary pinions of which body 4 is typical. These planetary pinions mesh with wheels 5 and 6 attached to the road wheels. Verify the given angular velocity vector diagram which is drawn for the case in which

$$\{\omega_2\}_{1/1} = \begin{bmatrix}0\\0\\\omega\end{bmatrix}, \quad \{\omega_5\}_{1/1} = \begin{bmatrix}\Omega\\0\\0\end{bmatrix}$$

and the road wheel speeds are unequal. Hence determine $\{\dot{\omega}_4\}_{1/3}$. Find $|\omega_5|_1$ to make $|\omega_6|_1$ zero. What will $|\omega_4|_3$ be under these conditions?

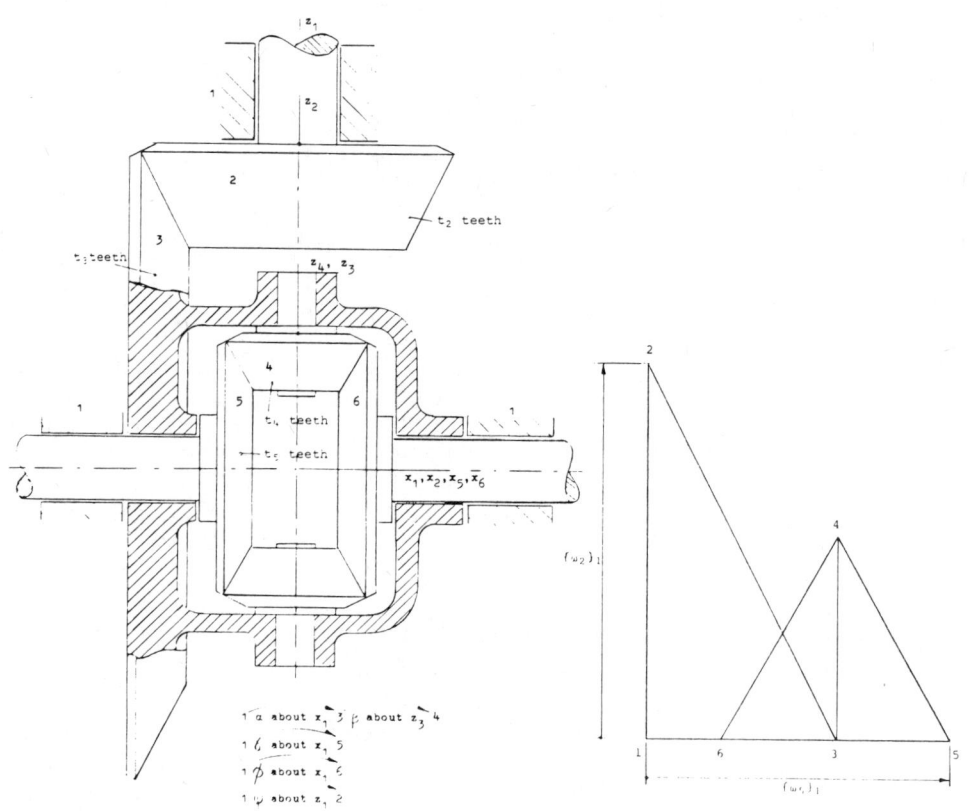

Fig. 3.46.

Problem 3.47. Figure 3.47 shows part of an epicyclic bevel wheel gear train. Verify the given angular velocity vector diagram which is drawn for the case in which

$$\{\omega_2\}_{1/1} = \begin{bmatrix} 0 \\ 0 \\ \omega \end{bmatrix} \text{ and } \{\omega_3\}_{1/1} = \begin{bmatrix} 0 \\ 0 \\ \Omega \end{bmatrix}$$

where ω and Ω are constants ($\omega > \Omega$). Find

$$\{\omega_4\}_{1/2} \text{ and } \{\dot{\omega}_4\}_{1/2} .$$

Also find

$$\{V_{AO}\}_{1/2} \text{ and } \{A_{AO}\}_{1/2}$$

where A is a point fixed in body 4 as shown.

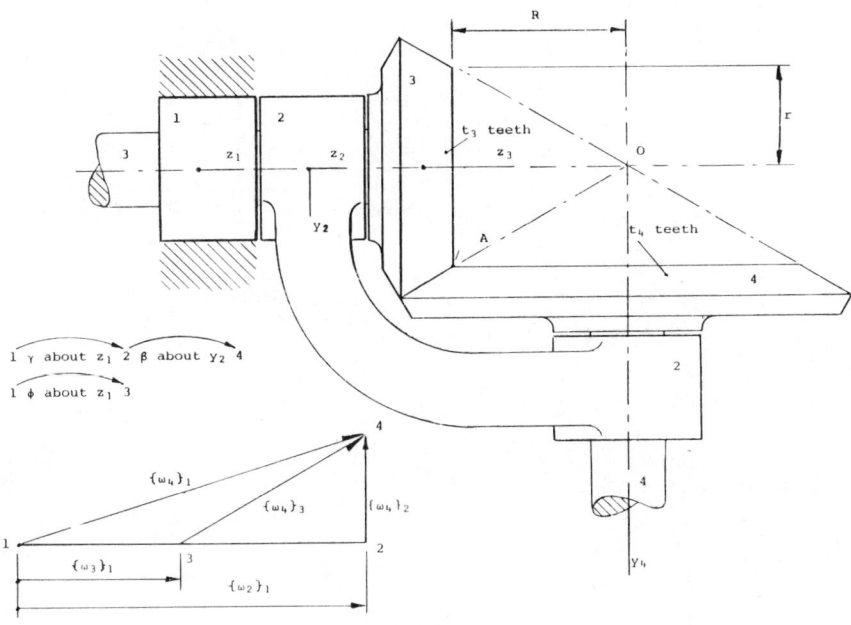

Fig. 3.47.

Problem 3.48. In the system of Fig. 3.48 a rotor, body 4, runs at a constant speed ω relative to an axle, body 3. The axle is pivoted to a rotating support, body 2. For the case in which body 2 is driven at the constant rate $\Omega = \dot{\alpha}$ relative to body 1, show that

$$\{\omega_4\}_{1/3} = \begin{bmatrix} \omega - \Omega\sin\beta \\ \dot{\beta} \\ \Omega\cos\beta \end{bmatrix},$$

$$\{\dot{\omega}_4\}_{1/3} = \begin{bmatrix} -\dot{\beta}\Omega\cos\beta \\ \omega\Omega\cos\beta + \ddot{\beta} \\ \dot{\beta}(\omega + \Omega\sin\beta) \end{bmatrix},$$

$$\{V_4\}_{1/3} = \begin{bmatrix} 0 \\ \Omega(a + b\cos\beta) \\ -b\dot{\beta} \end{bmatrix}$$

and

$$\{A_4\}_{1/3} = \begin{bmatrix} -\Omega^2(a + b\cos\beta)\cos\beta - b\dot{\beta}^2 \\ -2b\dot{\beta}\Omega\sin\beta \\ -\Omega^2(a + b\cos\beta)\sin\beta - b\ddot{\beta} \end{bmatrix}.$$

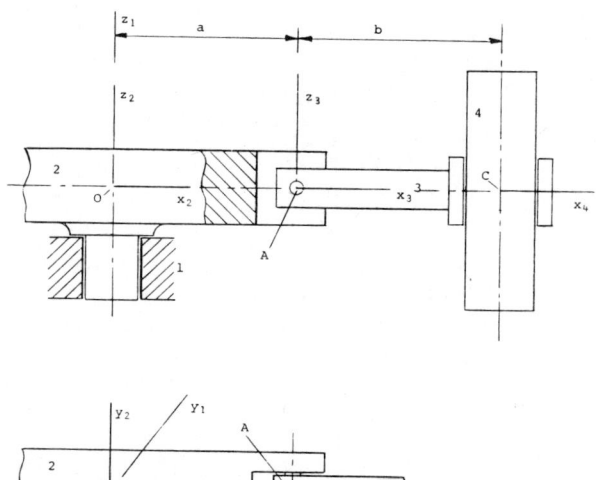

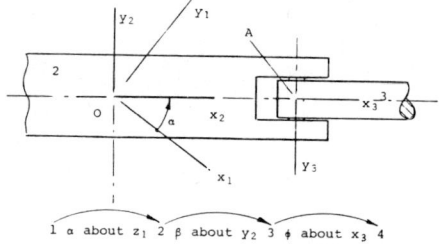

1 α about z_1 2 β about y_2 3 φ about x_3 4

Fig. 3.48.

Problem 3.49. In the system shown in Fig. 3.49, body 2 is free to rotate about the z_1 axis fixed in body 1 and the bevel wheel, body 4, which is mounted on body 3, meshes with a bevel wheel which is integral with body 2 and a bevel wheel, body 5, which is free to turn relative to body 2. Justify the given angular velocity vector diagram. For the case in which body 2 turns at the constant rate Ω relative to body 1 and body 3 rolls without slip on body 1, determine

$$\{\omega_4\}_{1/2}, \quad \{\omega_5\}_{1/2}, \quad \{\dot{\omega}_4\}_{1/2} \quad \text{and} \quad \{\dot{\omega}_5\}_{1/2}.$$

Problem 3.50. Figure 3.50 shows a mechanism. Body 2 rotates about the y_1 axis fixed in body 1. Body 3 slides in a radial groove cut in body 2. Body 4 has shperical ends which fit in hemishperical seatings cut in bodies 1 and 3. Determine, for the given position of the mechanism, the velocity and acceleration of sliding of body 3 relative to body 4 and the angular velocity and acceleration of body 4 relative to body 1 for the case in which

$$\{\omega_2\}_{1/1} = \begin{bmatrix} 0 \\ 10 \\ 0 \end{bmatrix} \text{ rad/s} \quad \text{and} \quad \{\dot{\omega}_2\}_{1/1} = \begin{bmatrix} 0 \\ -100 \\ 0 \end{bmatrix} \text{ rad/s}^2.$$

Solution of Kinematics Problems

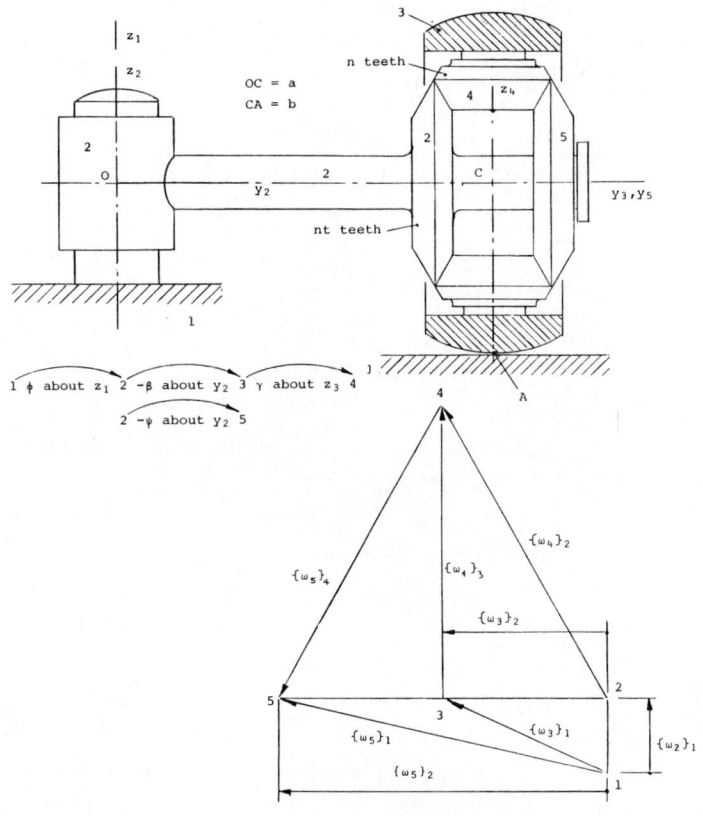

Fig. 3.49

Problem 3.51. A thin uniform disc, of radius b, rolls without slipping on an inertial horizontal plane. In preparation for analysing the motion of the disc it is necessary to obtain expressions for the velocity and acceleration of the centre of the disc and its angular velocity and acceleration. For the frames specified in Fig. 3.51, show that

$$\{\omega_3\}_{1/3} = \begin{bmatrix} \dot\psi\cos\theta \\ \dot\theta \\ \dot\psi\sin\theta \end{bmatrix}, \quad \{\omega_4\}_{1/3} = \begin{bmatrix} \dot\psi\cos\theta \\ \dot\theta \\ \dot\psi\sin\theta + \dot\phi \end{bmatrix},$$

$$\{\dot\omega_4\}_{1/3} = \begin{bmatrix} \dot\theta\dot\phi + \ddot\psi\cos\theta + \dot\psi\dot\theta\sin\theta \\ \ddot\theta - \dot\psi\dot\theta\cos\theta \\ \dot\psi\dot\theta\cos\theta + \ddot\psi\sin\theta + \ddot\phi \end{bmatrix},$$

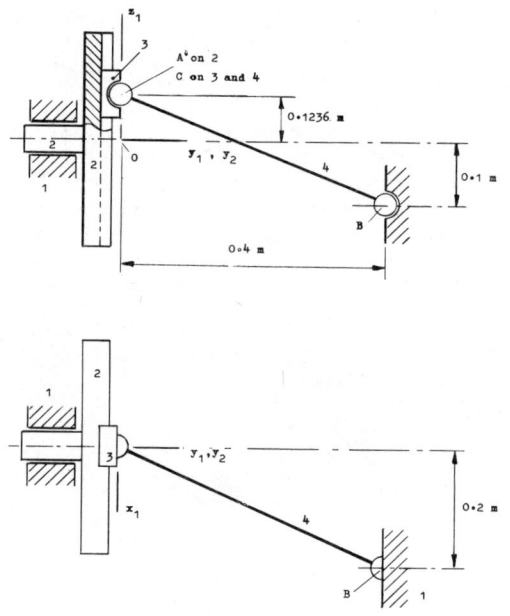

Fig. 3.50.

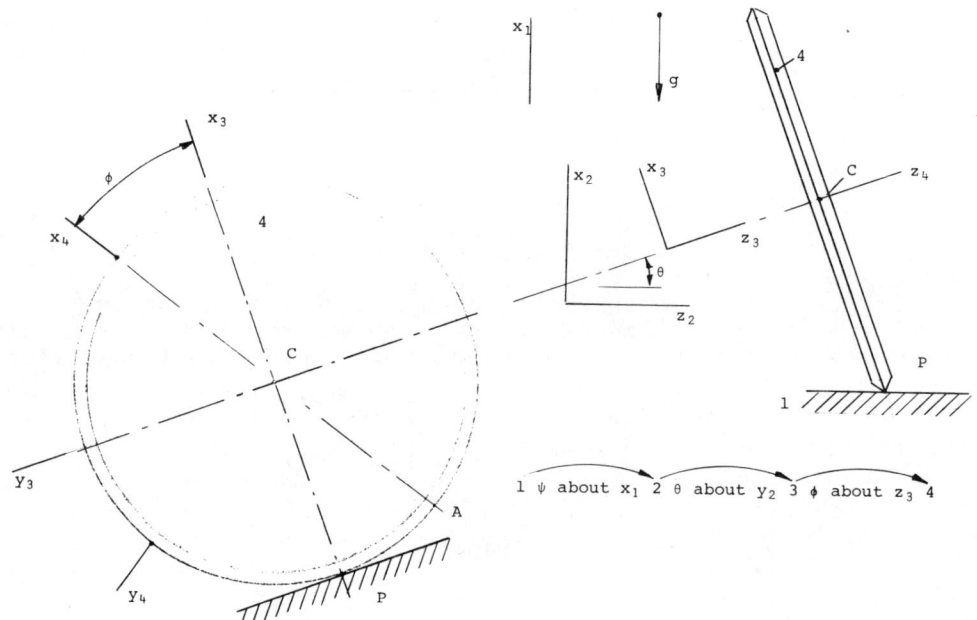

Fig. 3.51.

$$\{V_4\}_{1/3} = b \begin{bmatrix} 0 \\ \dot{\psi}\sin\theta + \dot{\phi} \\ -\dot{\theta} \end{bmatrix}$$

and

$$\{A_4\}_{1/3} = b \begin{bmatrix} -\dot{\psi}^2\sin^2\theta - \dot{\psi}\dot{\phi}\sin\theta - \dot{\theta}^2 \\ \ddot{\psi}\sin\theta + 2\dot{\theta}\dot{\psi}\cos\theta + \ddot{\phi} \\ \dot{\psi}^2\sin\theta\cos\theta + \dot{\psi}\dot{\phi}\cos\theta - \ddot{\theta} \end{bmatrix}.$$

Chapter 4

Solution of Dynamics Problems

4.1. Solved Problems

Problem 4.1. A particle of mass m is moving under the action of a force $\{F\}$. The acceleration of the particle is measured with respect to an inertial frame 1 and another frame, frame 2, which is moving relative to frame 1. Under what circumstances will the force predicted from accelerations measured relative to the moving frame be equal to the measured forces.

Solution. The measured force will be given by

$$\{F^m\}_2 = m[\ell_1]_2\{A_P\}_{1/1} = m[\ell_1]_2\{\{A_A\}_{1/1} + \{A_{PA}\}_{1/1}\} \quad (1)$$

where the position vectors are defined in Fig. 4.1. The force predicted from measurements relative to the moving frame 2 will be given by

$$\{F^P\}_2 = m\{A_{PA}\}_{2/2} \quad (2)$$

Now

$$\{R_{PA}\}_{2/2} = [\ell_1]_2\{R_{PA}\}_{1/1},$$

$$\{V_{PA}\}_{2/2} = [\omega_2]_{1/2}[\ell_1]_2\{R_{PA}\}_{1/1} + [\ell_1]_2\{V_{PA}\}_{1/1}$$

and

$$\{A_{PA}\}_{2/2} = [\dot{\omega}_2]_{1/2}[\ell_1]_2\{R_{PA}\}_{1/1} + [\omega_2]^2_{1/2}[\ell_1]_2\{R_{PA}\}_{1/1}$$

$$+ 2[\omega_2]_{1/2}[\ell_1]_2\{V_{PA}\}_{1/1} + [\ell_1]_2\{A_{PA}\}_{1/1}. \quad (3)$$

Thus the measured and predicted forces will be equal only if

    (i) $\{A_A\}_1$ is a null vector or $\{V_A\}_1$ is constant, that is the velocity of the origin of frame 2 must be zero or constant when measured with respect to an inertial reference, and

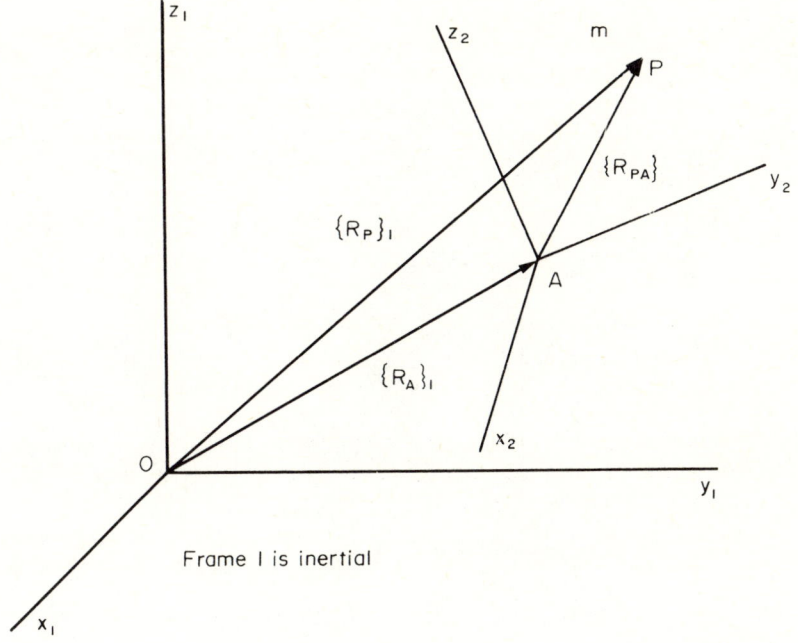

Fig. 4.1.

(ii) both $\{\omega_2\}_1$ and $\{\dot{\omega}_2\}_1$ are null vectors, that is the angular velocity and angular acceleration of frame 2 must be zero when measured with respect to an inertial reference.

Problem 4.2. A particle of mass m moves in the gravitational field of the earth. Find the work done by the force which the field exerts on the particle and the change of potential when it moves from point A to point B as shown in Fig. 4.2.

Solution.

$$W_{A \to B} = \int_A^B \{F\}^T d\{R\} = \int_A^B [0 \quad 0 \quad -mg] \begin{bmatrix} dx \\ dy \\ dz \end{bmatrix}$$

$$= -mgz \Big|_{h_A}^{h_B} = -mg(h_B - h_A).$$

166　　　　　　Matrix Methods in Engineering Mechanics

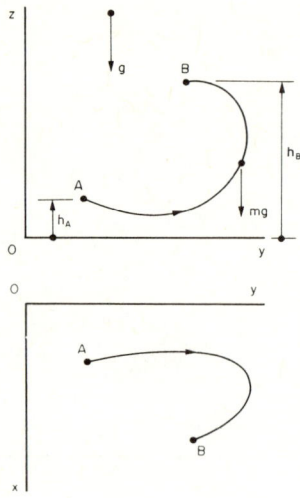

Fig. 4.2.

In this case

$$\frac{\partial F_z}{\partial y} = \frac{\partial F_z}{\partial x} = 0$$

and therefore by Eq. 2.11, the work done is independent of the path traced out by the particle in moving from A to B. Thus

$$dW = -dV = \{F\}^T d\{R\}$$

or

$$dV = mgdz$$

and the change of potential is given by

$$\int_{h_A}^{h_B} mgdz = mg(h_B - h_A) = -W_{A \to B}$$

Problem 4.3. A helical spring, of stiffness k, is stretched in the direction of its length. Obtain an expression for the work done by the force which the spring exerts and relate this to the change of potential.

Solution. If the natural length of the spring is a, then by reference

Solution of Dynamics Problems

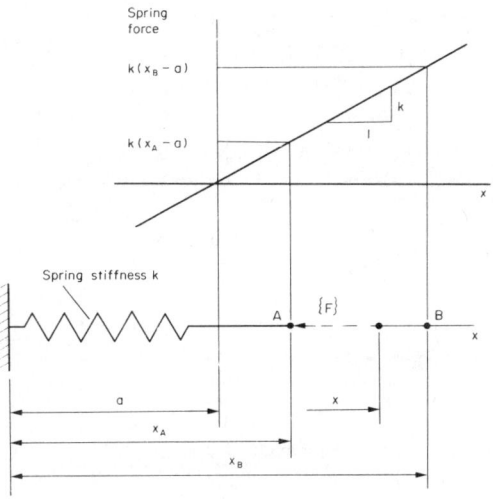

Fig. 4.3.

to Fig. 4.3 it can be seen that the magnitude of the force exerted by the spring is given by

$$k(x - a).$$

Hence

$$\{F\} = \begin{bmatrix} -k(x-a) \\ 0 \\ 0 \end{bmatrix}$$

and since

$$\{R_P\} = \begin{bmatrix} x \\ 0 \\ 0 \end{bmatrix}, \quad d\{R_P\} = \begin{bmatrix} dx \\ 0 \\ 0 \end{bmatrix}$$

giving

$$W_{A \to B} = \int_A^B \{F\}^T d\{R\} = \int_A^B [-k(x-a) \quad 0 \quad 0] \begin{bmatrix} dx \\ 0 \\ 0 \end{bmatrix}$$

$$= -k(x^2/2 - ax) \Big|_{x_A}^{x_B}$$

$$= -\frac{k}{2}(x_B^2 - x_A^2) + ka(x_B - x_A).$$

Since

$$\frac{\partial F_x}{\partial y} = \frac{\partial F_x}{\partial z} = 0$$

the spring force is conservative and the change of potential is

$$V_B - V_A = -W_{A \to B} \;.$$

The increment of elastic strain energy stored in the spring is found from the area under the spring force versus extension graph, which is given by

$$\frac{\{k(x_B - a) + k(x_A - a)\}(x_B - x_A)}{2}$$

$$= \frac{\{k(x_B + x_A) - 2ka\}(x_B - x_A)}{2}$$

$$= \frac{k}{2}(x_B^2 - x_A^2) - ka(x_B - x_A),$$

which is the change of potential.

Problem 4.4. A helical spring, of stiffness $k = 5$ kN/m, is fitted with shperical ends which fit into spherical sockets. The sockets are located in parts of a mechanism which have relative motion. In the unstretched condition the distance between the centres of the spherical ends is $a = 10$ cm and initially the co-ordinates of the socket centres A and A are

$$\{R_{A_1}\} = \begin{bmatrix} 2 \\ -2 \\ -1 \end{bmatrix} \text{cm} \quad \text{and} \quad \{R_{A_2}\} = \begin{bmatrix} 5 \\ 6 \\ 5 \end{bmatrix} \text{cm}.$$

Find the work done by the force which the spring exerts when the socket centres move the to points which have the co-ordinates

$$\{R_{B_1}\} = \begin{bmatrix} 3 \\ -3 \\ -2 \end{bmatrix} \text{cm} \quad \text{and} \quad \{R_{B_2}\} = \begin{bmatrix} 6 \\ 7 \\ 8 \end{bmatrix} \text{cm}$$

as shown in Fig. 4.4. Assume that the frictional effects due to the sockets can be neglected.

Solution. Let $\{R_{P_2 P_1}\}^T = [x \quad y \quad z]$

where P_1 and P_2 are points on the paths $A_1 B_1$ and $A_2 B_2$ respectively. The length of the spring at any extension is thus

Solution of Dynamics Problems

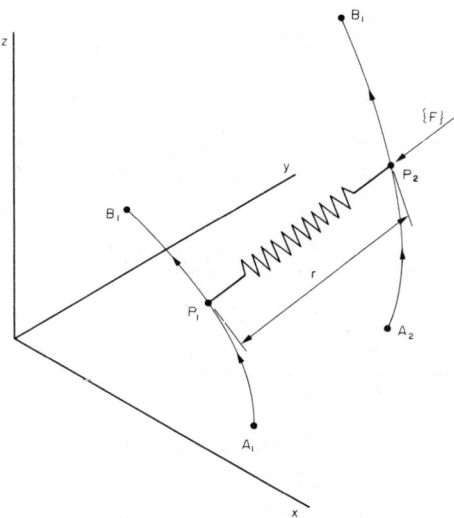

Fig. 4.4.

$$r = |R_{P_1 P_2}| = \sqrt{(x^2 + y^2 + z^2)} \ . \tag{1}$$

The magnitude of the spring force is thus given by

$$k(r - a)$$

and the force exerted by the spring is therefore

$$\{F\} = \begin{bmatrix} F_x \\ F_y \\ F_z \end{bmatrix} = -k(r - a) \begin{bmatrix} x/r \\ y/r \\ z/r \end{bmatrix}$$

The work done is thus given by

$$W_{\substack{A_1 \to B_1 \\ A_2 \to B_2}} = \int_A^B \{F\}^T d\{R_{P_2 P_1}\}$$

$$= \int_A^B -\frac{k(r - a)}{r} [x \quad y \quad z] \begin{bmatrix} dx \\ dy \\ dz \end{bmatrix} \tag{2}$$

If the spring force is conservative then $[\nabla]\{F\}$ will be a null vector, as it is in this case since

$$\frac{\partial F_x}{\partial y} = \frac{\partial F_x}{\partial z} = \frac{\partial F_y}{\partial x} = \frac{\partial F_y}{\partial z} = \frac{\partial F_z}{\partial x} = \frac{\partial F_z}{\partial y} = 0 \ .$$

Hence

$$dV = \frac{k(a-r)x}{r}dx + \frac{k(a-r)y}{r}dy + \frac{k(a-r)z}{r}dz$$

and therefore

$$V = \int \frac{k(a-r)x}{r} dx + f_1(y,z) \ ,$$

and

$$V = \int \frac{k(a-r)y}{r} dy + f_2(x,z)$$

$$V = \int \frac{k(a-r)z}{r} dz + f_3(x,y) \ .$$

Now

$$\int k(x - ax/r) dx = \frac{kx^2}{2} - kar$$

and

$$\int k(y - ay/r) dy = \frac{ky^2}{2} - kar$$

$$\int k(z - az/r) dz = \frac{kz^2}{2} - kar \ .$$

The functions

$$f_1(y,z), \ f_2(x,z) \text{ and } f_3(x,y)$$

are thus zero or constants and the potential is therefore given by

$$V = \frac{k}{2}(x^2 + y^2 + z^2) - ka\sqrt{(x^2 + y^2 + z^2)} + c \ .$$

The work done is therefore

$$W_{\substack{A_1 \to B_1 \\ A_2 \to B_2}} = -\frac{k}{2}(x^2 + y^2 + z^2) + ka\sqrt{(x^2 + y^2 + z^2)} \Big|_A^B$$

$$= -\frac{k}{2}(|R_{B_2B_1}|^2 - |R_{A_2A_1}|^2) + ka(|R_{B_2B_1}| - |R_{A_2A_1}|) \ .$$

Solution of Dynamics Problems

In this particular case

$$\{R_{B_2B_1}\} = \begin{bmatrix} 6-3 \\ 7+3 \\ 8+2 \end{bmatrix} = \begin{bmatrix} 3 \\ 10 \\ 10 \end{bmatrix} \text{ cm}, \quad |R_{B_2B_1}| = \sqrt{209} \text{ cm}$$

and

$$\{R_{A_2A_1}\} = \begin{bmatrix} 5-2 \\ 6+2 \\ 5+1 \end{bmatrix} = \begin{bmatrix} 3 \\ 8 \\ 6 \end{bmatrix} \text{ cm}, \quad |R_{A_2A_1}| = \sqrt{109} \text{ cm},$$

giving

$$\begin{aligned}
W_{\substack{A_1 \to B_1 \\ A_2 \to B_2}} &= -2.5(209 - 109) + 5 \times 10(\sqrt{209} - \sqrt{109}) \\
&= -49.2 \text{ kN m}^{-1} \text{ cm}^2 \\
&= -49.2 \text{ (kN m}^{-1} \text{ cm}^2)(10^3 \text{ N kN}^{-1})(10^{-4} \text{ m}^2 \text{ cm}^{-2}) \\
&= -4.92 \text{ J.}
\end{aligned}$$

Problem 4.5. Show that the force

$$\{F\} = \begin{bmatrix} 3x - 2y \\ y + 2z \\ -x^2 \end{bmatrix} \text{ N },$$

where x, y and z are measured in metres, is non-conservative.

Find the work done when the force moves its point of application from the point

$$\{R_A\} = \begin{bmatrix} 0 \\ 0 \\ 0 \end{bmatrix} \text{ m to the point } \{R_B\} = \begin{bmatrix} 2 \\ 3 \\ 5 \end{bmatrix} \text{ m}$$

when the paths of the point are

    (a) the curve $x = t$, $y = 3t^2/4$ and $z = 5t^3/8$

    (b) the straight lines

        A(0, 0, 0) m to C(2, 0, 0) m
        C(2, 0, 0) m to D(2, 3, 0) m
        D(2, 3, 0) m to B(2, 3, 5) m

    (c) the straight line

        A(0, 0, 0) m to B(2, 3, 5) m.

Solution. If V exists then

$$-\frac{\partial V}{\partial x} = 3x - 2y, \quad -\frac{\partial V}{\partial y} = y + 2z \quad \text{and} \quad -\frac{\partial V}{\partial z} = x^2,$$

which require that

$$-V = \frac{3x^2}{2} - 2xy + f_1(y, z),$$

$$-V = \frac{y^2}{2} + 2zy + f_2(x, z)$$

and

$$-V = -x^2 z + f_3(x, y).$$

From these results it is not possible to construct a function

$$V = f(x, y, z)$$

since, for example the term $-x^2 z$ in the third expression for V is not contained in the first and could not be accounted for in the unknown function $f_1(y, z)$ which does not contain x. The force must therefore be non-conservative.

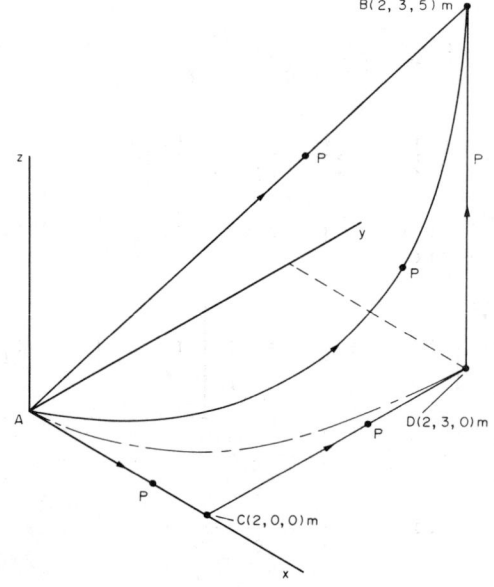

Fig. 4.5.

The various paths are shown in Fig. 4.5.

(a) Let P be a point on the path AB. Then

Solution of Dynamics Problems 173

$$\{R_{PA}\} = \begin{bmatrix} t \\ 3t^2/4 \\ 5t^3/8 \end{bmatrix} \text{ m} , \quad d\{R_{PA}\} = \begin{bmatrix} 1 \\ 3t/2 \\ 15t^2/8 \end{bmatrix} dt \text{ m}$$

and

$$F = \begin{bmatrix} 3t - 3t^2/2 \\ 3t^2/4 + 5t^3/4 \\ -t^2 \end{bmatrix} \text{ N}$$

giving

$$W_{A \to B} = \int_A^B \{F\}^T d\{R_{PA}\}$$

$$= \int_0^2 [\, 3t - \tfrac{3}{2}t^2 \quad \tfrac{3}{4}t^2 + \tfrac{5}{4}t^3 \quad -t^2\,] \begin{bmatrix} 1 \\ 3t/2 \\ 15t^2/8 \end{bmatrix} dt$$

$$= \int_0^2 (3t - \tfrac{3}{2}t^2 + \tfrac{9}{8}t^3 + \tfrac{15}{8}t^4 - \tfrac{15}{8}t^4)\, dt$$

$$= \tfrac{3}{2}t^2 - \tfrac{t^2}{2} + \tfrac{9}{32}t^4 \bigg|_0^2 = \tfrac{17}{2} \text{ J} .$$

(b) Let P be a point on each part of the path. From A to C, for which y = z = 0, and therefore

$$\{F\} = \begin{bmatrix} 3x \\ 0 \\ -x^2 \end{bmatrix} \text{ N} , \quad \{R_{PA}\} = \begin{bmatrix} x \\ 0 \\ 0 \end{bmatrix} \text{ m} \quad \text{and} \quad d\{R_{PA}\} = \begin{bmatrix} dx \\ 0 \\ 0 \end{bmatrix} \text{ m} .$$

The work done is thus

$$W_{A \to B} = \int_0^2 3x\, dx = \tfrac{3}{2}x^2 \bigg|_0^2 = 6 \text{ J}.$$

From C to D, for which x = 2 and z = 0, and therefore

$$\{F\} = \begin{bmatrix} 6 - 2y \\ y \\ -4 \end{bmatrix} \text{ N} , \quad \{R_{PC}\} = \begin{bmatrix} 2 \\ y \\ 0 \end{bmatrix} \text{ m} \quad \text{and} \quad d\{R_{PC}\} = \begin{bmatrix} 0 \\ dy \\ 0 \end{bmatrix} \text{ m}$$

The work done is thus

$$W_{C \to D} = \int_0^3 y \, dy = \left.\frac{y^2}{2}\right|_0^3 = \frac{9}{2} \text{ J.}$$

From D to B, for which $x = 2$ and $y = 3$, and therefore

$$\{F\} = \begin{bmatrix} 0 \\ 3 + 2z \\ -4 \end{bmatrix} \text{ N, } \quad \{R_{PD}\} = \begin{bmatrix} 2 \\ 3 \\ z \end{bmatrix} \text{ m and } d\{R_{PD}\} = \begin{bmatrix} 0 \\ 0 \\ dz \end{bmatrix} \text{ m .}$$

The work done is thus

$$W_{D \to B} = \int_0^5 -4 \, dz = \left.-4z\right|_0^5 = -20 \text{ J.}$$

(c) Let P be a point on the straight line AB. Then

$$\{R_{PA}\} = \begin{bmatrix} 2t \\ 3t \\ 5t \end{bmatrix} \text{ m , } \quad d\{R_{PA}\} = \begin{bmatrix} 2 \\ 3 \\ 5 \end{bmatrix} dt \text{ m and } \{F\} = \begin{bmatrix} 0 \\ 13t \\ -4t^2 \end{bmatrix} \text{ N.}$$

The work done is thus

$$W_{A \to B} = \int_0^1 (39t - 20t^2) \, dt = \left.\frac{39}{2}t^2 - \frac{20}{3}t^3\right|_0^1 = \frac{77}{6} \text{ J .}$$

When a closed path is traced out, such as for example ACDB and the straight line to A, the work done is not zero as it would be in the case of a conservative force, but there is a net expenditure of energy. In the present example

$$W_{A \to C \to D \to B} = 6 + 4.5 - 20 = -21.5 \text{ J,}$$

while for the straight line path between A and B

$$W_{B \to A} = -W_{A \to B} = -\frac{77}{6} \text{ J ,}$$

giving a net expenditure of $-103/3$ J.

Problem 4.6. A particle of mass m rests on a rough horizontal surface. Obtain an expression for the work done by the force which the surface exerts on the particle when there is relative motion and hence show that there is no potential associated with the force. Assume that the friction force is independent of the relative velocity and proportional to the normal force between the particle and the surface (Coulomb friction).

Determine the work done

$$W_{A \to B}$$

when the path of the particle is part of a circle which has the equation

$$x^2 + y^2 = a^2$$

and the points A and B are (0, -a) and (0, a) respectively. Also determine the work done

$$W_{A \to C \to B}$$

when the particle moves along the path A to C and then to B along the x axis such that A, C and B are the following points:

A(a, 0), C(c, 0) and B(b, 0)

where c > b and c > a.

Solution. When the particle moves relative to the surface there are four forces acting on it. These are the weight force mg vertically down, a normal force mg on the particle due to the surface vertically up, a friction force, which is μ times the normal force, along the surface and tangential to the path of the particle on the surface directed so that it opposes the motion and an externally applied force causing the motion. These forces are shown in Fig. 4.6a.

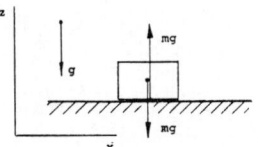

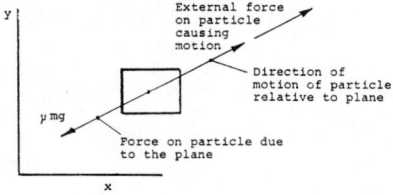

Fig. 4.6a.

Now the work done is given by

$$W_{A \to B} = \int_A^B \{F\}^T d\{R\} = \int_A^B F_x dx + F_y dy = \int_A^B |F| ds .$$

Also, by reference to Fig. 4.6b,

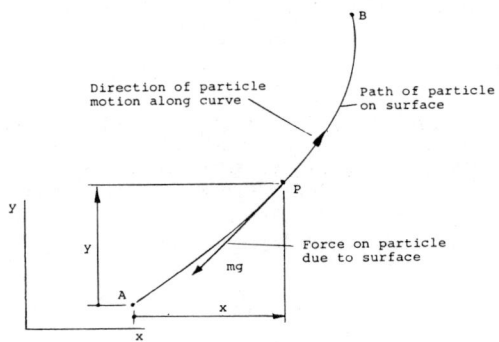

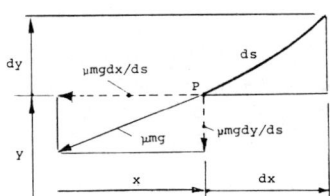

Fig. 4.6b.

$$F_x = - \mu mg dx/ds \quad \text{and} \quad F_y = - \mu mg dy/ds,$$

where

$$ds = \sqrt{(dx)^2 + (dy)^2} = dx\sqrt{1 + (Dy)^2}$$

and

$$Dy = dy/dx,$$

giving

$$W_{A \to B} = - \mu mg \int_A^B \{dx + (Dy) dy\}/\sqrt{1 + (Dy)^2}$$

$$= -\mu mg \int_A^B \{dx + (Dy)^2 dx\}/\sqrt{\{1 + (Dy)^2\}}$$

$$= -\mu mg \int_A^B \sqrt{\{1 + (Dy)^2\}}\, dx. \qquad (1)$$

Since
$$\sqrt{\{1 + (Dy)^2\}}$$

is not an exact differential, the work done will depend on the path between A and B, so that the friction force μmg has no potential associated with it and it is therefore non-conservative.

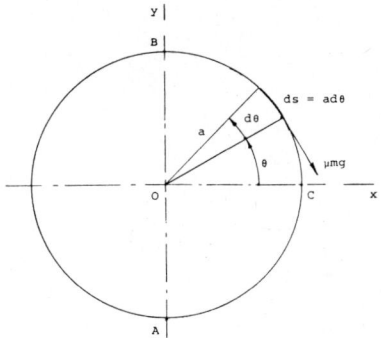

Fig. 4.6c.

For the case in which the particle moves along a circular path as shown in Fig. 4.6c

$$W_{A \to C \to B} = -\mu mg \int_0^0 \sqrt{(1 + x^2/y^2)}\, dx = -\mu mga \int_0^0 \frac{dx}{\sqrt{(a^2 - x^2)}}$$

$$= -\mu mga \sin^{-1}(x/a)\Big|_0^0 = -\mu mga\pi$$

on interpreting the upper and lower limits as the physics of the problem requires (Fig. 4.6d). Of course it is much easier to find the work done from

$$W_{A \to C \to B} = \int |F| ds = -\mu mga \int_{-\pi/2}^{\pi/2} d\theta = -\mu mga\pi .$$

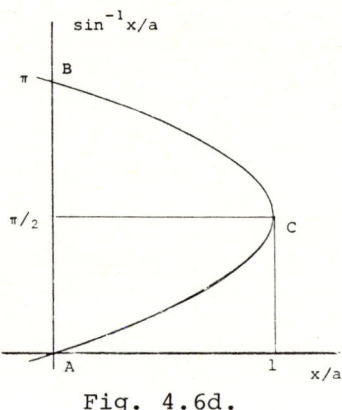

Fig. 4.6d.

Note that, for example,

$$W_{A \to O \to B} = -\mu mg \int_{-a}^{a} dy = -2\mu mga \neq W_{A \to C \to B} .$$

When the particle moves along a path which does not have a finite derivative at all points, the path must be divided into segments which do have finite derivatives at all points. This is a requirement because Dy of Eq. 1 must be finite in the range of integration. Thus in the case of motion along the the path ACB as shown in Fig. 4.6e the work done must be evaluated in two parts

$$W_{A \to C} \quad \text{and} \quad W_{C \to B}$$

Thus

$$W_{A \to C} = \int_{a}^{c} (-\mu mg) dx \quad \text{and} \quad W_{C \to B} = \int_{c}^{b} (+\mu mg) dx$$

giving

$$W_{A \to C \to B} = -\mu mg(c - a) + \mu mg(b - c)$$
$$= -2\mu mgc + \mu mg(a + b)$$
$$= \mu mg(a + b - 2c) .$$

Clearly the work done is not conserved as available mechanical energy.

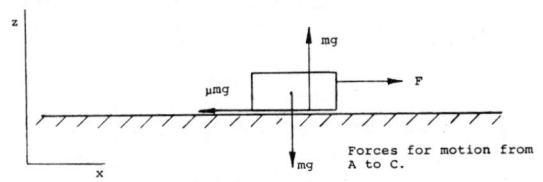

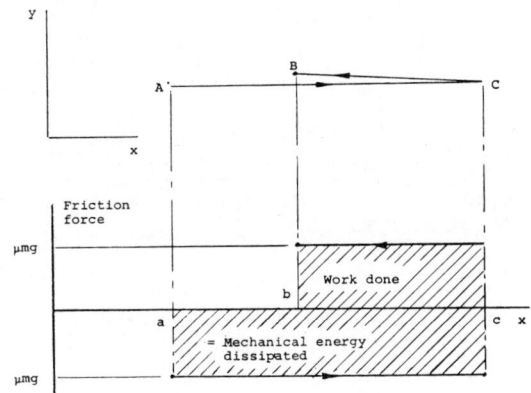

Fig. 4.6e.

Problem 4.7. A particle of mass m is attached to one end of an elastic string which is fixed at the other end. The free length of the string is a and its stiffness is k. Obtain an expression for the potential of the system when it is constrained to move in the vertical plane and the string remains straight.

Solution. The system is shown in Fig. 4.7a. The gravitational potential, using the x axis as datum, is

$$- mgr\cos\theta$$

and the potential due to strain energy is

$$k(r - a)^2/2.$$

The total potential can thus be written

$$V = - mgr\cos\theta + k(r - a)^2/2 + c$$

where c is a constant.

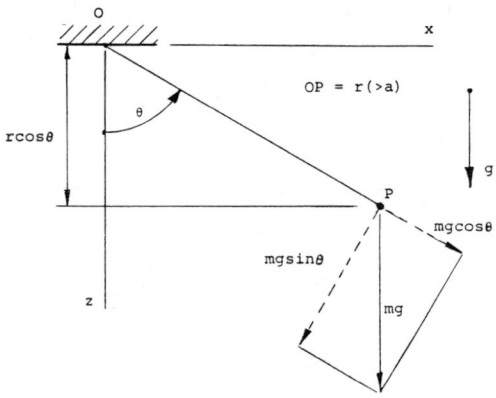

Fig. 4.7a.

In this case the potential of the system is expressed in terms of r and θ rather than z and x. It is interesting therefore to find what significance could be attached to

$$\frac{\partial V}{\partial r} \quad \text{and} \quad \frac{\partial V}{\partial \theta},$$

since

$$\frac{\partial V}{\partial z} = -F_z \quad \text{and} \quad \frac{\partial V}{\partial x} = -F_x.$$

Hence

$$-\frac{\partial V}{\partial r} = mg\cos\theta - k(r - a) = F_r \text{ say},$$

which is clearly the force along r giving rise to the potential. Also

$$-\frac{\partial V}{\partial \theta} = -mgr\sin\theta = F_\theta \text{ say},$$

but this does not have the dimensions of force as might at first be expected, since the differentiation was with respect to an angle and not a length. Nevertheless, it is convenient to regard F_θ as a 'force' which influences θ in the way that F_r influences r. F_θ is in fact called a generalised force.

The reader is invited to show that for the frames specified in Fig. 4.7b

$$\{F\}_1 = \begin{bmatrix} -k\sin\theta(r - a) \\ 0 \\ mg - k\cos\theta(r - a) \end{bmatrix}$$

from the expression for V by substituting

$$r\cos\theta = z \quad \text{and} \quad r = \sqrt{(z^2 + x^2)}$$

and hence finding

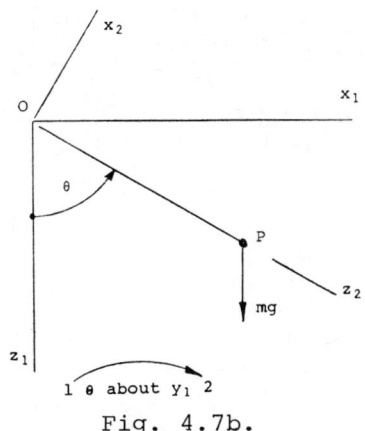

Fig. 4.7b.

$$\{F\}_2 = [\ell_1]_2 \{F\}_1 .$$

How do the terms in $\{F\}_2$ compare with F_r and F_θ?

Problem 4.8. The position of a system can be specified by the position of a point on a rotating z_2 axis as shown in Fig. 4.8 (see Problem 4.7). Show that $\{\nabla\}_2 V$, del V referred to rotating frame 2, is given by

$$\{\nabla\}_2 V = \begin{bmatrix} \dfrac{1}{r}\dfrac{\partial V}{\partial \theta} \\ 0 \\ \dfrac{\partial V}{\partial r} \end{bmatrix} .$$

Solution. Now

$$\{F\}_1^T d\{R\}_{1/1} = -dV$$

or

$$\{\nabla\}_1^T V d\{R\}_{1/1} = dV$$

or

$$\{\nabla\}_2^T V [\ell_1]_2 d\{R\}_{1/1} = dV \tag{1}$$

and

$$[\ell_1]_2 d\{R\}_{1/1} = [\ell_1]_2 \left\{ \{R_r\}_{1/1} dr + \{R_\theta\}_{1/1} d\theta \right\}$$

$$= \{R_r\}_{1/2} dr + \{R_\theta\}_{1/2} d\theta$$

where

$$\{R_r\}_{1/1} = \frac{\partial}{\partial r}\{R\}_{1/1} \quad \text{and} \quad \{R_\theta\}_{1/1} = \frac{\partial}{\partial \theta}\{R\}_{1/1}.$$

Also, since

$$dV = \frac{\partial V}{\partial r} dr + \frac{\partial V}{\partial \theta} d\theta,$$

Eq. 1 can be written

$$[A \quad B \quad C] \left\{ \{R_r\}_{1/2} dr + \{R_\theta\}_{1/2} d\theta \right\}$$

$$= \frac{\partial V}{\partial r} dr + \frac{\partial V}{\partial \theta} d\theta \tag{2}$$

where

$$\{\nabla\}_2 V = \begin{bmatrix} A \\ B \\ C \end{bmatrix}.$$

Now

$$\{R\}_{1/1} = [\ell_2]_1 \{R\}_{2/2} = \begin{bmatrix} r\sin\theta \\ 0 \\ r\cos\theta \end{bmatrix},$$

$$\frac{\partial}{\partial r}\{R\}_{1/1} = \{R_r\}_{1/1} = \begin{bmatrix} \sin\theta \\ 0 \\ \cos\theta \end{bmatrix},$$

$$\{R_r\}_{1/2} = [\ell_1]_2 \{R_r\}_{1/1} = \begin{bmatrix} 0 \\ 0 \\ 1 \end{bmatrix},$$

$$\frac{\partial}{\partial \theta}\{R\}_{1/1} = \{R_\theta\}_{1/1} = \begin{bmatrix} r\cos\theta \\ 0 \\ -r\sin\theta \end{bmatrix}$$

and

$$\{R_\theta\}_{1/2} = [\ell_1]_2 \{R_\theta\}_{1/1} = \begin{bmatrix} r \\ 0 \\ 0 \end{bmatrix}.$$

Substitution of these last results in Eq. 2 gives

$$[A \quad B \quad C] \left[\begin{bmatrix} 0 \\ 0 \\ 1 \end{bmatrix} dr + r \begin{bmatrix} 1 \\ 0 \\ 0 \end{bmatrix} d\theta \right] = \frac{\partial V}{\partial r} dr + \frac{\partial V}{\partial \theta} d\theta$$

Solution of Dynamics Problems 183

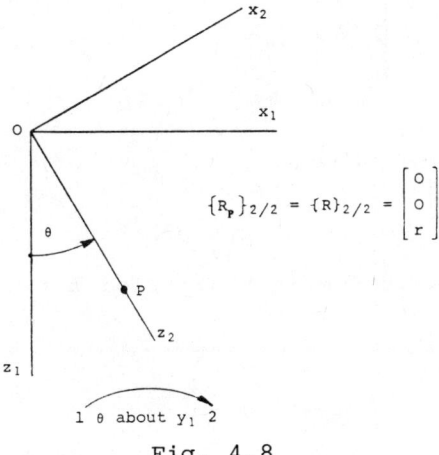

Fig. 4.8

and therefore

$$rA = \frac{\partial V}{\partial \theta} \quad \text{or} \quad A = \frac{1}{r}\frac{\partial V}{\partial \theta}, \quad C = \frac{\partial V}{\partial r} \quad \text{and} \quad B = 0.$$

Hence

$$\{\nabla\}_2 V = \begin{bmatrix} \frac{1}{r}\frac{\partial V}{\partial \theta} \\ 0 \\ \frac{\partial V}{\partial r} \end{bmatrix}$$

as required. Clearly, $\{\nabla\}_1 V$ is readily found from

$$\{\nabla\}_1 V = [\ell_2]_1 \{\nabla\}_2 V .$$

Problem 4.9. A particle, which has a mass of 2 kg, is subjected to a force

$$\{F\}_1 = \begin{bmatrix} 12t^2 \\ 18t - 8 \\ -6t \end{bmatrix} \text{ N}$$

where t is in seconds. When $t = 0$ the particle is at the position

$$\{R\}_{1/1}\big|_{t=0} = \begin{bmatrix} 6 \\ -2 \\ 8 \end{bmatrix} \text{ m}$$

and its velocity is

$$\{V\}_{1/1}\big|_{t=0} = \begin{bmatrix} 3 \\ 8 \\ -4 \end{bmatrix} \text{ m/s.}$$

Determine for the particle
(a) the velocity at any time t,
(b) the position at any time t,
(c) the kinetic energy at t = 2 s,
(d) the work done in the period t = 0 2 to t = 2 s,
(e) the momentum at t = 2 s and
(f) the impulse of the force in the period t = 0 s to t = 2 s.

Solution. Now

$$\{F\}_1 = m\{A\}_{1/1}$$

and therefore

$$\{A\}_{1/1} = \{\dot{V}\}_{1/1} = \tfrac{1}{m}\{F\}_1 = \begin{bmatrix} 6t^2 \\ 9t - 4 \\ -3t \end{bmatrix} \text{ m/s}^2. \tag{1}$$

Hence, by integration of Eq. 1,

$$\{V\}_{1/1} = \{\dot{R}\}_{1/1} = \begin{bmatrix} 2t^2 \\ 4.5t^2 - 4t \\ -1.5t^2 \end{bmatrix} + \begin{bmatrix} v_x \\ v_y \\ v_z \end{bmatrix}$$

and since

$$\{V\}_{1/1}\big|_{t=0} = \begin{bmatrix} 3 \\ 8 \\ -4 \end{bmatrix} \text{ m/s,}$$

$$\{V\}_{1/1} = \begin{bmatrix} 2t^3 + 3 \\ 4.5t^2 - 4t + 8 \\ -1.5t^2 - 4 \end{bmatrix} \text{ m/s .} \tag{2}$$

Integration of Eq. 2 gives

$$\{R\}_{1/1} = \begin{bmatrix} 0.5t^4 + 3t \\ 1.5t^3 - 2t^2 + 8t \\ -0.5t^3 - 4t \end{bmatrix} + \begin{bmatrix} x \\ y \\ z \end{bmatrix}$$

and since

$$\{R\}_{1/1}\big|_{t=0} = \begin{bmatrix} 6 \\ -2 \\ 8 \end{bmatrix} \text{ m ,}$$

Solution of Dynamics Problems

$$\{R\}_{1/1} = \begin{bmatrix} 0.5t^4 + 3t + 6 \\ 1.5t^3 - 2t^2 + 8t - 2 \\ -0.5t^2 - 4t + 8 \end{bmatrix} \text{ m.}$$

From Eq. 2

$$|V|_1^2 \Big|_{t=2} = 19^2 + 18^2 + (-10)^2 = 785 \text{ m}^2/\text{s}^2$$

and therefore

$$T\Big|_{t=2} = \frac{m}{2}|V|_1^2\Big|_{t=2} = 785 \text{ J}$$

Also, since

$$T\Big|_{t=0} = \frac{2}{2}\{3^2 + 8^2 + (-4)^2\} = 89 \text{ J},$$

$$\Delta T = 696 \text{ J}.$$

The work done is given by

$$W = \int \{F\}_1^T d\{R\}_{1/1} = \int \{F\}_1^T \frac{d}{dt}\{R\}_{1/1} dt$$

$$= \int \{F\}_1^T \{V\}_{1/1} dt$$

so that for this case

$$W = \int_0^2 [12t^2 \quad 18t - 8 \quad -6t] \begin{bmatrix} 2t^3 + 3 \\ 4.5t^2 - 4t + 8 \\ -1.5t^2 - 4 \end{bmatrix} dt$$

$$= \int_0^2 (24t^5 + 90t^3 - 72t^2 + 200t - 64) dt$$

$$= (4t^6 + 22.5t^4 - 24t^3 + 100t^2 - 64t) \Big|_0^2$$

$$= 696 \text{ J}$$

and this is the same as the change if kinetic energy.

Now

186 Matrix Methods in Engineering Mechanics

$$\{G\}_{1/1} = m\{V\}_{1/1} = 2\begin{bmatrix} 2t^3 + 3 \\ 4.5t^2 - 4t + 8 \\ -1.5t^2 - 4 \end{bmatrix} \text{ kg m/s}$$

and therefore

$$\{G\}_{1/1}\big|_{t=2} = \begin{bmatrix} 38 \\ 36 \\ -20 \end{bmatrix} \text{ kg m/s.}$$

The impulse of the force is

$$\int_0^2 \{F\}_1 dt = \int_0^2 \begin{bmatrix} 12t^2 \\ 18t - 8 \\ -6t \end{bmatrix} dt = \begin{bmatrix} 4t^3 \\ 9t^2 - 8t \\ -3t^2 \end{bmatrix}\bigg|_0^2$$

$$= \begin{bmatrix} 32 \\ 10 \\ -12 \end{bmatrix} \text{ kg m/s}$$

and since

$$\{G\}_{1/1}\big|_{t=0} = \begin{bmatrix} 6 \\ 16 \\ -8 \end{bmatrix} \text{ kg m/s },$$

$$\Delta\{G\}_{1/1} = \{G\}_{1/1}\big|_{t=2} - \{G\}_{1/1}\big|_{t=0} = \begin{bmatrix} 32 \\ 10 \\ -12 \end{bmatrix} \text{ kg m/s}$$

which is the impulse of the force.

Problem 4.10. A body of mass m = 4 kg is free to move along a smooth vertical rod. One end of a spring is attached to the body at A by a smooth pin joint and the other end of the spring is attached to a fixed point O by a similar joint as shown in Fig. 4.10. The free length of the spring, measured between the centres of the pin joints, is a = 10 cm and the stiffness of the spring is k = 0.5 kN/m. Find the velocity of the mass if it moves from rest when OA is horizontal to the position in which A is h = 15 cm below its original position.

Solution. If the gravitational potential is measured relative to the x_1 axis as datum then

$$V_A = k(r_A - a)^2/2 \quad \text{and} \quad V_B = k(r_B - a)^2/2 - mgh.$$

Solution of Dynamics Problems

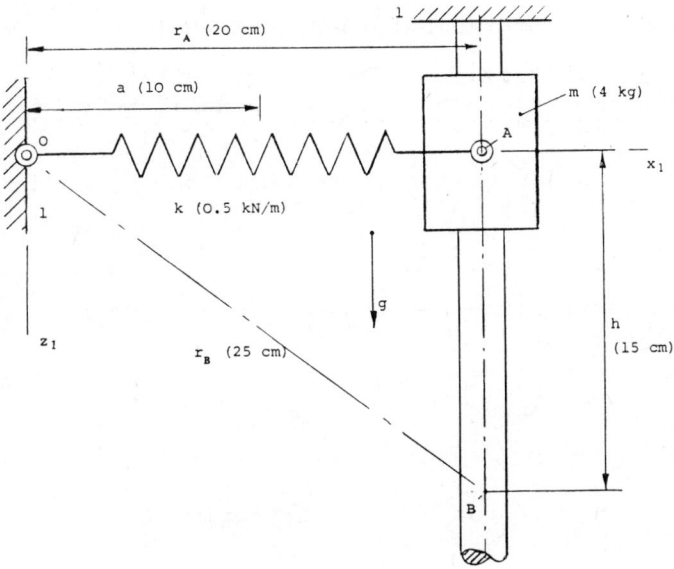

Fig. 4.10.

Hence

$$\Delta V = V_B - V_A = 250\{(25-10)^2 - (20-10)^2\} - 4 \times 981 \times 15$$
$$= -27610 \text{ N cm}$$
$$= -276.1 \text{ J}.$$

Since energy is conserved

$$T|_A + V_A = T|_B + V_B .$$

In this case $T|_A = 0$ and therefore

$$T|_B = V_A - V_B = -\Delta V$$

$$\frac{4}{2}|v|_1^2 = 276.1 .$$

Hence

$$|v|_1 = 11.75 \text{ m/s}$$

and

$$\{V\}_{1/1} = \begin{bmatrix} 0 \\ 0 \\ 11.75 \end{bmatrix} \text{ m/s} .$$

Problem 4.11. A particle of mass m is constrained to move along a smooth helical path, of radius a and lead angle α, which has its axis vertical.

Determine the velocity of the particle when it moves from rest through the vertical distance h.

Also find the magnitude of the force which the path exerts on the particle in terms of the magnitude of the velocity of the particle.

Solution. Let the magnitude of the particle velocity $\{V\}_1$ be

$$|V|_1 = v.$$

The kinetic energy is thus

$$T = mv^2/2.$$

The potential energy of the particle relative to the position in which the kinetic energy was zero is

$$V = -mgh$$

and since the system is conservative

$$0 = T + V,$$

giving

$$v = \sqrt{(2gh)}.$$

Let the particle be treated as a moving point in the rotating frame 2 as shown in Fig. 4.11. Then

$$\{R\}_{1/1} = [\ell_2]_1 \{R\}_{2/2}$$

where

$$\{R\}_{2/2} = \begin{bmatrix} a \\ 0 \\ z \end{bmatrix}.$$

Hence

$$\{V\}_{1/1} = [\omega_2]_{1/1}[\ell_2]_1\{R\}_{2/2} + [\ell_2]_1\{\dot{R}\}_{2/2}$$

or

$$\{V\}_{1/2} = [\omega_2]_{1/2}\{R\}_{2/2} + \{\dot{R}\}_{2/2}$$

$$= \begin{bmatrix} 0 & -\omega & 0 \\ \omega & 0 & 0 \\ 0 & 0 & 0 \end{bmatrix} \begin{bmatrix} a \\ 0 \\ z \end{bmatrix} + \begin{bmatrix} 0 \\ 0 \\ \dot{z} \end{bmatrix} = \begin{bmatrix} 0 \\ a\omega \\ \dot{z} \end{bmatrix}$$

where $\omega = \dot{\gamma}$.

Solution of Dynamics Problems

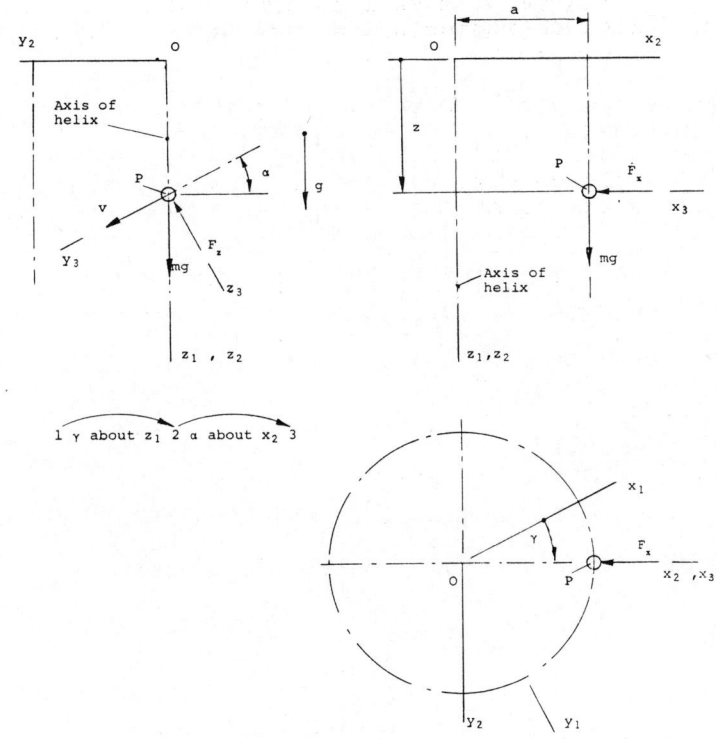

Fig. 4.11.

Since the lead angle of the helix is α,

$$\tan\alpha = \dot{z}/(a\omega) \, ,$$

$$\{V\}_{1/2} = \begin{bmatrix} 0 \\ \dot{z}/\tan\alpha \\ \dot{z} \end{bmatrix}$$

and

$$v = \dot{z}\sqrt{\{(1/\tan^2\alpha) + 1\}} = \dot{z}/\sin\alpha.$$

Now

$$\{V\}_{1/1} = [\ell_2]_1 \{V\}_{1/2}$$

and therefore

$$\{A\}_{1/1} = [\omega_2]_{1/1}[\ell_2]_1\{V\}_{1/2} + [\ell_2]_1 \frac{d}{dt}\{V\}_{1/2}$$

or

$$\{A\}_{1/2} = [\omega_2]_{1/2}\{V\}_{1/2} + \frac{d}{dt}\{V\}_{1/2}$$

$$= \begin{bmatrix} 0 & -\omega & 0 \\ \omega & 0 & 0 \\ 0 & 0 & 0 \end{bmatrix} \begin{bmatrix} 0 \\ a\omega \\ \dot{z} \end{bmatrix} + \begin{bmatrix} 0 \\ a\dot{\omega} \\ \ddot{z} \end{bmatrix}$$

$$= \begin{bmatrix} -\omega^2 a \\ \dot{\omega}a \\ \ddot{z} \end{bmatrix} = \begin{bmatrix} -\dot{z}^2/a\tan^2\alpha \\ \ddot{z}/\tan\alpha \\ \ddot{z} \end{bmatrix} .$$

Let the force on the particle due to the path which is referred to frame 3 be

$$\{F\}_3 = \begin{bmatrix} -F_x \\ 0 \\ -F_z \end{bmatrix} .$$

The equation of motion for the particle is thus

$$[\ell_3]_2\{F\}_3 + \{W\}_2 = m\{A\}_{1/2}$$

$$\begin{bmatrix} 1 & 0 & 0 \\ 0 & c\alpha & -s\alpha \\ 0 & s\alpha & c\alpha \end{bmatrix} \begin{bmatrix} -F_x \\ 0 \\ -F_z \end{bmatrix} + \begin{bmatrix} 0 \\ 0 \\ mg \end{bmatrix} = m \begin{bmatrix} -\dot{z}^2/a\tan\alpha \\ \ddot{z}/\tan\alpha \\ \ddot{z} \end{bmatrix}$$

$$\begin{bmatrix} F_x \\ F_z \sin\alpha \\ -F_z \cos\alpha + mg \end{bmatrix} = \begin{bmatrix} -m\dot{z}^2/a\tan^2\alpha \\ m\ddot{z}/\tan\alpha \\ m\ddot{z} \end{bmatrix} .$$

Eliminating F_z between the y and z component equations gives

$$\ddot{z} = g\sin^2\alpha$$

and since the particle is released from rest at time $t = 0$,

$$\dot{z} = g\sin^2\alpha \, t$$

and

$$z = g\sin^2\alpha \, t^2/2.$$

The time to move vertically through the distance h is thus given by

$$t = \sqrt{(2h/g\sin^2\alpha)} .$$

Also

$$F_x = mv^2\cos^2\alpha/a$$

and

$$F_z = mg\cos\alpha ,$$

Solution of Dynamics Problems

giving

$$|F| = mg\cos\alpha \sqrt{\{(v^2\cos^2\alpha/a^2g^2) + 1\}} .$$

Problem 4.12. A particle is released from rest relative to the earth at a height h above the earth, which is much less than the radius of the earth, and at a place where the latitude is α. Obtain an expression for its position measured in and referred to a set of axes fixed in the earth. Neglect the effects of air resistance.

Solution. Select a set of axes as shown in Fig. 3.10, where O is vertically below the point at which the particle is released.

Neglecting terms in ω_e^2 and Ω^2

$$\{A\}_{1/4} = \begin{bmatrix} \ddot{x} - 2\omega_e(\dot{y}\cos\alpha - \dot{z}\sin\alpha) \\ \ddot{y} + 2\omega_e \dot{x}\cos\alpha \\ \ddot{z} - 2\omega_e \dot{x}\sin\alpha \end{bmatrix}$$

as obtained in Problem 3.10.

The equation of motion for the particle is

$$\{F\}_4 = m\{A\}_{1/4}$$

$$\begin{bmatrix} 0 \\ -mg \\ 0 \end{bmatrix} = m \begin{bmatrix} \ddot{x} - 2\omega_e(\dot{y}\cos\alpha - \dot{z}\sin\alpha) \\ \ddot{y} + 2\omega_e \dot{x}\cos\alpha \\ \ddot{z} - 2\omega_e \dot{x}\sin\alpha \end{bmatrix} \qquad (1)$$

The y component equation can be integrated to give

$$\dot{y} + 2\omega_e x\cos\alpha = -gt + A$$

and since $\dot{y} = 0$ and $\dot{x} = 0$ when $t = 0$, $A = 0$, giving

$$\dot{y} + 2\omega_e x\cos\alpha = -gt . \qquad (2)$$

The z component equation can be similarly integrated to give

$$\dot{z} - 2\omega_e x\sin\alpha = B$$

and since $\dot{z} = 0$ and $x = 0$ when $t = 0$, $B = 0$, giving

$$\dot{z} - 2\omega_e x\sin\alpha = 0 . \qquad (3)$$

Substituting Eqs. 2 and 3 in the x component equation of Eqs. 1 gives

$$\ddot{x} - 2\omega_e \{\cos\alpha(-gt - 2\omega_e x\cos\alpha) + 2\omega_e x\sin^2\alpha\} = 0$$

which reduces to

$$\ddot{x} + 2\omega_e gt\cos\alpha = 0 \tag{4}$$

when terms in ω_e^2 are neglected. Equation 4 integrates to give

$$\dot{x} + \omega_e gt^2\cos\alpha = C$$

and since $\dot{x} = 0$ when $t = 0$, $C = 0$, giving

$$\dot{x} = -\omega_e gt^2\cos\alpha . \tag{5}$$

Equation 5 integrates to give

$$x = -(\omega_e gt^3\cos\alpha)/3 \tag{6}$$

since the constant of integration is zero. Substituting Eq. 6 in Eqs. 2 and 3 gives

$$\dot{y} = -gt + (2\omega_e^2 gt^3\cos^2\alpha)/3$$

and

$$\dot{z} = -(2\omega_e^2 gt^3\sin\alpha\cos\alpha)/3 ,$$

which reduce to

$$\dot{y} = -gt$$

and

$$\dot{z} = 0$$

on neglecting terms in ω_e^2. Hence

$$y = -(gt^2)/2 + h \tag{7}$$

and

$$z = 0 \tag{8}$$

and the path of the particle drawn on frame 4 is given by

$$\{R\}_{4/4} = \begin{bmatrix} -(\omega_e gt^3\cos\alpha)/3 \\ -(gt^2)/2 + h \\ 0 \end{bmatrix} .$$

Problem 4.13. A straight rigid rod is constrained to rotate at a constant rate ω with its longitudinal axis in the horizontal plane as shown in Fig. 4.13a. A particle, of mass m, moves along the rod without appreciable frictional constraint under the action of a spring of stiffness k.

For the case in which the spring force is zero when $x = \delta$, obtain an expression for the motion of the mass along the rod when it is released from rest relative to the rod and x is then equal to δ. Assume that the motion is always controlled by the elastic characteristics of the spring.

Solution of Dynamics Problems

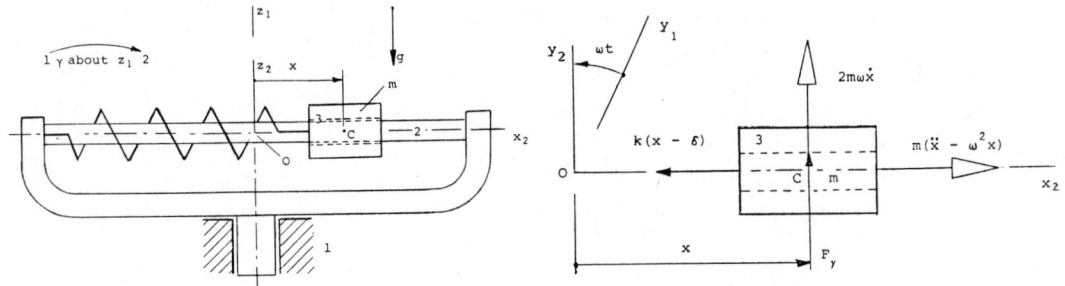

Fig. 4.13a. Fig. 4.13b.

Solution. Let the particle be treated as a point moving along the x_2 axis of the rotating frame 2. Then

$$\{R_c\}_{1/1} = [\ell_2]_1 \{R_c\}_{2/2}$$

and therefore

$$\{V_c\}_{1/1} = [\omega_2]_{1/1}[\ell_2]_1 \{R_c\}_{2/2} + [\ell_2]_1 \{\dot R_c\}_{2/2}$$

or

$$\{V_c\}_{1/2} = [\omega_2]_{1/2}\{R_c\}_{2/2} + \{\dot R_c\}_{2/2}$$

$$= \begin{bmatrix} 0 & -\omega & 0 \\ \omega & 0 & 0 \\ 0 & 0 & 0 \end{bmatrix} \begin{bmatrix} x \\ 0 \\ 0 \end{bmatrix} + \begin{bmatrix} \dot x \\ 0 \\ 0 \end{bmatrix} = \begin{bmatrix} \dot x \\ \omega x \\ 0 \end{bmatrix}.$$

Now

$$\{V_c\}_{1/1} = [\ell_2]_1 \{V_c\}_{1/2}$$

and therefore

$$\{A_c\}_{1/1} = [\omega_2]_{1/1}[\ell_2]_1 \{V_c\}_{1/2} + [\ell_2]_1 \frac{d}{dt}\{V_c\}_{1/2}$$

or

$$\{A_c\}_{1/2} = [\omega_2]_{1/2}\{V_c\}_{1/2} + \frac{d}{dt}\{V_c\}_{1/2}$$

$$= \begin{bmatrix} 0 & -\omega & 0 \\ \omega & 0 & 0 \\ 0 & 0 & 0 \end{bmatrix} \begin{bmatrix} \dot x \\ \omega x \\ 0 \end{bmatrix} + \begin{bmatrix} \ddot x \\ \omega \dot x \\ 0 \end{bmatrix} = \begin{bmatrix} -\omega^2 x + \ddot x \\ 2\omega \dot x \\ 0 \end{bmatrix}$$

where ω is constant.

The equation of motion is thus, by reference to Fig. 4.13b,

$$\{W_3\}_2 + \{F_{32}\}_2 = m\{A_c\}_{1/2}$$

$$\begin{bmatrix} 0 \\ 0 \\ -mg \end{bmatrix} + \begin{bmatrix} -k(x-\delta) \\ F_y \\ F_z \end{bmatrix} = m \begin{bmatrix} -\omega^2 x + \ddot{x} \\ 2\omega\dot{x} \\ 0 \end{bmatrix}.$$

From the x component equation

$$m(\ddot{x} - \omega^2 x) = -k(x-\delta)$$

$$\ddot{x} + \omega_n^2 x = \omega_o^2 \delta$$

where

$$\omega_n^2 = \omega_o^2 - \omega^2 \quad \text{and} \quad \omega_o^2 = k/m.$$

Taking Laplace transforms, writing

$$L\{x(t)\} = X(s),$$

$$s^2 X - s\delta + \omega_n^2 X = \frac{\omega_o^2 \delta}{s}$$

or

$$X = \omega_o^2 \delta \frac{1}{s(s^2 + \omega_n^2)} + \delta \frac{s}{s^2 + \omega_n^2}.$$

By reference to tables of transform pairs

$$x = \frac{\omega_o^2 \delta}{\omega_n^2}(1 - \cos\omega_n t) + \delta \cos\omega_n t$$

$$= \frac{\delta}{1-r^2}(1 - r^2 \cos\omega_n t)$$

where

$$r = \omega/\omega_o = \omega\sqrt{(m/k)}.$$

Problem 4.14. A simple pendulum, of length b, hangs at rest from a support C. C is made to move in a horizontal circular path drawn on an inertial reference body at the uniform rate

$$\omega = \sqrt{(g/2b)}.$$

Find the position of the pendulum bob as a function of time. Assume that the radius of the path of C, a, is much less than b and that therefore the vertical movement of the bob can be neglected.

Solution of Dynamics Problems   195

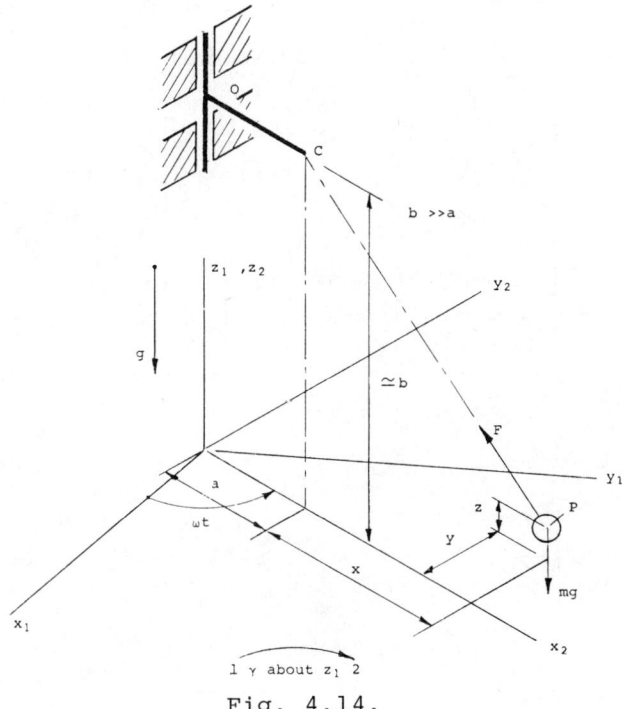

Fig. 4.14.

Solution. The system is as shown in Fig. 4.14. Let the pendulum bob be treated as a moving point in the rotating frame 2. Then, writing

$$\{R_p\}_1 = \{R\}_1,$$

$$\{R\}_{1/1} = [\ell_2]_1 \{R\}_{2/2}$$

where

$$\{R\}_{2/2} = \begin{bmatrix} a + x \\ y \\ 0 \end{bmatrix}$$

since $z \simeq 0$. Hence

$$\{V\}_{1/2} = [\omega_2]_{1/2} \{R\}_{2/2} + \{\dot{R}\}_{2/2}$$

$$= \begin{bmatrix} 0 & -\omega & 0 \\ \omega & 0 & 0 \\ 0 & 0 & 0 \end{bmatrix} \begin{bmatrix} a + x \\ y \\ 0 \end{bmatrix} + \begin{bmatrix} \dot{x} \\ \dot{y} \\ 0 \end{bmatrix}$$

$$= \begin{bmatrix} -\omega y + \dot{x} \\ \omega(a + x) + \dot{y} \\ 0 \end{bmatrix} \qquad (1)$$

and

$$\{\dot{A}\}_{1/2} = [\omega_2]_{1/2}\{V\}_{1/2} + \frac{d}{dt}\{V\}_{1/2}$$

$$= \begin{bmatrix} 0 & -\omega & 0 \\ \omega & 0 & 0 \\ 0 & 0 & 0 \end{bmatrix} \begin{bmatrix} -\omega y + \dot{x} \\ \omega(a+x) + \dot{y} \\ 0 \end{bmatrix} + \begin{bmatrix} -\omega\dot{y} + \ddot{x} \\ \omega\dot{x} + \ddot{y} \\ 0 \end{bmatrix}$$

$$= \begin{bmatrix} -\omega^2(a+x) - 2\omega\dot{y} + \ddot{x} \\ -\omega^2 y + 2\omega\dot{x} + \ddot{y} \\ 0 \end{bmatrix} . \qquad (2)$$

The force on the bob due to the string is

$$\{F\}_2 = F\{R_{CP}\}_{2/2}/|R_{CP}|$$

$$= F \begin{bmatrix} -x/b \\ -y/b \\ 1 \end{bmatrix}$$

and the equation of motion can be written as

$$\{F\}_2 + \{W\}_2 = m\{\dot{A}\}_{1/2}$$

$$F \begin{bmatrix} -x/b \\ -y/b \\ 1 \end{bmatrix} + mg \begin{bmatrix} 0 \\ 0 \\ -1 \end{bmatrix} = m \begin{bmatrix} -\omega^2(a+x) - 2\omega\dot{y} + \ddot{x} \\ -\omega^2 y + 2\omega\dot{x} + \ddot{y} \\ 0 \end{bmatrix} \qquad (3)$$

From the z component equation

$$F = mg .$$

With this result, and writing

$$g/b = 4\omega^2,$$

the x component equation becomes

$$-3\omega^2 x = -\omega^2 a - 2\omega\dot{y} + \ddot{x} . \qquad (4)$$

Similarly, the y component equation becomes

$$-3\omega^2 y = 2\omega\dot{x} + \ddot{y} . \qquad (5)$$

Taking Laplace transforms of Eqs. 4 and 5, the initial conditions being zero,

$$(s^2 + 3\omega^2)X = \frac{\omega^2 a}{s} + 2\omega s Y \qquad (6)$$

and

$$(s^2 + 3\omega^2)Y = -2\omega sX \,. \tag{7}$$

Eliminating X between these last two equations gives

$$Y = -\frac{2\omega^3 a}{(s^2 + 9\omega^2)(s^2 + \omega^2)}$$

and by reference to tables of transform pairs

$$y = \frac{a}{12}(\sin 3\omega t - 3\sin\omega t)\,.$$

Also

$$X = \frac{\omega^2 a(s^2 + 3\omega^2)}{s(s^2 + 9\omega^2)(s^2 + \omega^2)}$$

$$= \omega^2 a \left\{ \frac{s}{(s^2 + 9\omega^2)(s^2 + \omega^2)} + \frac{3\omega^2}{s(s^2 + 9\omega^2)(s^2 + \omega^2)} \right\}$$

and

$$x = \omega^2 a \left\{ \frac{1}{8\omega^2}(\cos\omega t - \cos 3\omega t) + \frac{1}{3\omega^2} - \frac{3}{8\omega^2}\cos\omega t + \frac{1}{24\omega^2}\cos 3\omega t \right\}$$

$$= \frac{a}{12}(4 - 3\cos\omega t - \cos 3\omega t)\,.$$

Hence

$$\{R\}_{2/2} = \frac{a}{12}\begin{bmatrix} 16 - 3\cos\omega t - \cos 3\omega t \\ \sin 3\omega t - 3\sin\omega t \\ 0 \end{bmatrix}$$

and

$$\{R\}_{1/1} = \frac{a}{12}\begin{bmatrix} \cos\omega t & -\sin\omega t & 0 \\ \sin\omega t & \cos\omega t & 0 \\ 0 & 0 & 1 \end{bmatrix}\begin{bmatrix} 16 - 3\cos\omega t - \cos 3\omega t \\ \sin 3\omega t - 3\sin\omega t \\ 0 \end{bmatrix}$$

$$= \frac{a}{12}\begin{bmatrix} 16\cos\omega t - \cos 2\omega t - 3 \\ 16\sin\omega t - \sin 2\omega t \\ 0 \end{bmatrix}\,.$$

198   Matrix Methods in Engineering Mechanics

Problem 4.15. Derive an expression for the motion of the bob of a long simple pendulum taking into account the rotation of the earth about its axis.

Solution.

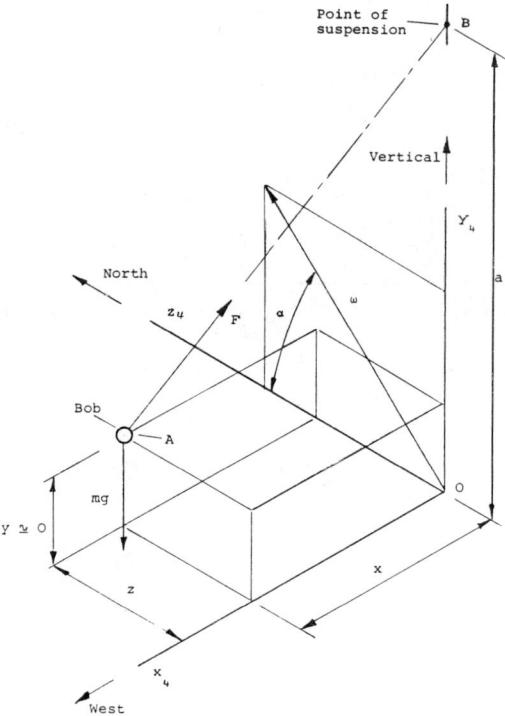

Fig. 4.15a.

From Problem 3.10, the acceleration of a point relative to an inertial set of axes referred to frame 4 fixed in the earth, as shown in Fig. 3.10, is given by

$$\{A\}_{1/4} = \begin{bmatrix} \ddot{x} + 2\omega\dot{z}\sin\alpha \\ \ddot{y} + 2\omega\dot{x}\cos\alpha \\ \ddot{z} - 2\omega\dot{x}\sin\alpha \end{bmatrix} \tag{1}$$

for the case in which $\dot{y}$ can be neglected and $\omega \equiv \omega_e$.

By reference to Fig. 4.15a,

$$\{R_B\}_{4/4} = \begin{bmatrix} 0 \\ a \\ 0 \end{bmatrix}, \quad \{R_A\}_{4/4} = \begin{bmatrix} x \\ y \\ z \end{bmatrix} \quad \text{and} \quad \{R_{BA}\}_{4/4} \simeq \begin{bmatrix} -x \\ a \\ -z \end{bmatrix}$$

when y is much less than a.

Solution of Dynamics Problems

The force on the bob due to the string is

$$F\{R_{BA}\}_4/4/|R_{BA}| = F\begin{bmatrix} -x/a \\ 1 \\ -z/a \end{bmatrix}.$$

The equations of motion for the bob are thus

$$\begin{bmatrix} 0 \\ -mg \\ 0 \end{bmatrix} + F\begin{bmatrix} -x/a \\ 1 \\ -z/a \end{bmatrix} = m\begin{bmatrix} \ddot{x} + 2\omega\dot{z}\sin\alpha \\ \ddot{y} + 2\omega\dot{x}\cos\alpha \\ \ddot{z} - 2\omega\dot{x}\sin\alpha \end{bmatrix} \quad (2)$$

and these can be rewritten

$$\frac{F}{m} = 2\omega\dot{x}\cos\alpha + g \simeq g \quad (3)$$

$$\ddot{x} + 2\omega\dot{z}\sin\alpha + \omega_n^2 x = 0 \quad (4)$$

$$\ddot{z} - 2\omega\dot{x}\sin\alpha + \omega_n^2 z = 0 \quad (5)$$

where $\omega_n^2 = g/a$.

If the initial conditions are

$$x(0) = A, \; \dot{x}(0) = 0, \; z(0) = 0 \; \text{and} \; \dot{z}(0) = 0,$$

then the Laplace transforms of Eqs. 4 and 5 become

$$s^2 X - sA + 2\omega\sin\alpha \, sX + \omega_n^2 X = 0 \quad (6)$$

$$s^2 Z - 2\omega\sin\alpha(sX - A) + \omega_n^2 Z = 0. \quad (7)$$

On writing

$$2\omega\sin\alpha = B,$$

$$X = \frac{As(s^2 + \omega_n^2 + B^2)}{s^2 + (2\omega_n^2 + B^2)s + \omega_n^4} \quad (8)$$

and

$$Z = -\frac{AB}{s^2 + (2\omega_n^2 + B^2)s + \omega_n^4}. \quad (9)$$

The roots of the denominators of Eqs 8 and 9 are

$$s = \pm j\omega_n \sqrt{(1 + 2\omega\sin\alpha/\omega_n)}$$

$$s = \pm j\omega_n \sqrt{(1 - 2\omega\sin\alpha/\omega_n)}$$

and since ω is much less than ω_n these roots can be written approximately as

and
$$s = \pm j(\omega_n + \omega\sin\alpha) = +j\omega_1$$
$$s = \pm j(\omega_n - \omega\sin\alpha) = +j\omega_2.$$

The inverse transforms of Eqs 8 and 9 can thus be determined giving

$$x = \frac{A}{2}(\cos\omega_1 t + \cos\omega_2 t) \tag{10}$$

and

$$z = \frac{A}{2}(\sin\omega_1 t + \sin\omega_2 t) \tag{11}$$

using the fact that ω is much less than ω_n. The reader is invited to complete the intermediate steps required to determine the inverse transforms.

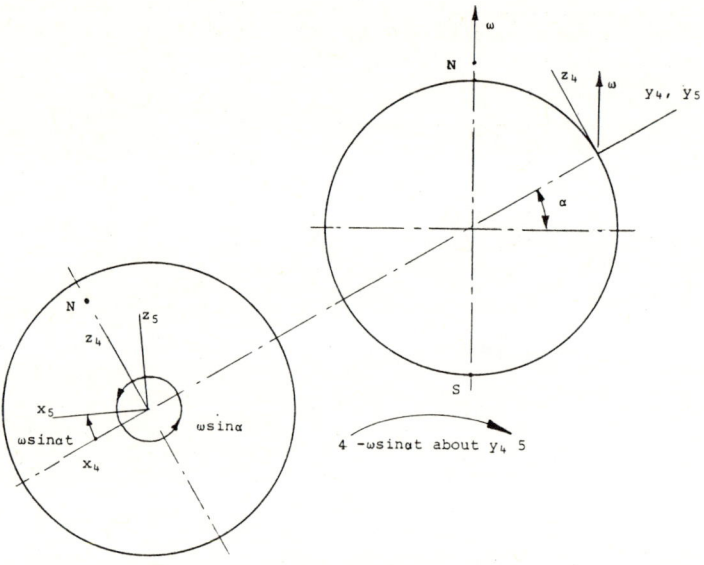

Fig. 4.15b.

Equations 10 and 11 can be rewritten

$$x = A\cos\omega_n t \cos(\omega\sin\alpha t) \tag{12}$$

and

$$z = A\cos\omega_n t \sin(\omega\sin\alpha t) \tag{13}$$

by use of appropriate trigonometric identities. Thus

$$\{R_A\}_{4/4} = A\cos\omega_n t \begin{bmatrix} \cos(\omega\sin\alpha t) \\ 0 \\ \sin(\omega\sin\alpha t) \end{bmatrix}$$

and if this vector is referred to frame 5 positioned relative to frame 4 as shown in Fig. 4.15b (that is frame 5 rotates 'against' the direction of rotation of the earth), then

$$\{R_A\}_{5/5} = [\ell_4]_5\{R_A\}_{4/4} = A\cos\omega_n t \begin{bmatrix} 1 \\ 0 \\ 0 \end{bmatrix}.$$

The pendulum bob is thus seen to move with simple harmonic motion along a line which rotates relative to the earth with the angular velocity $\omega\sin\alpha$ as shown in Fig. 4.14b. Of course the bob is moving along a line fixed in inertial space while the earth rotates underneath it.

Problem 4.16. A homogeneous solid is in the form of a truncated sector of a sphere as shown in Fig. 4.16a. Find the position of the centre of this solid when $a = 10$ cm, $a\cos\alpha - h = 2$ cm and $\alpha = 15°$.

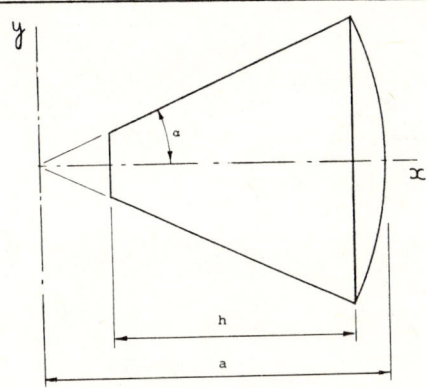

Fig. 4.16a.

Solution. The reference frame chosen is as shown in Fig. 4.16b. The centre of mass is clearly on the x axis. Consider the solid as being made up of two parts, one of which is body 2, a truncated cone, and the other body 3, a spherical cap. The composite body is designated 4. By reference to Fig. 4.16b

$$m_2 = \int_{a\cos\alpha - h}^{a\cos\alpha} \rho\pi y^2 dx ,$$

where ρ is the density of the material of the solid, and since

$$y = x\tan\alpha ,$$

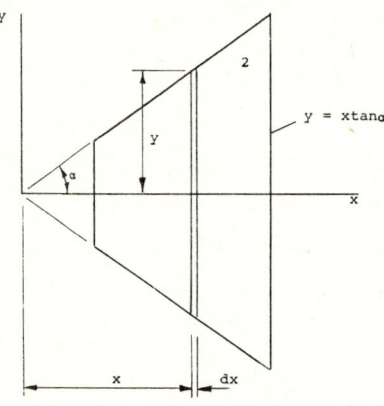
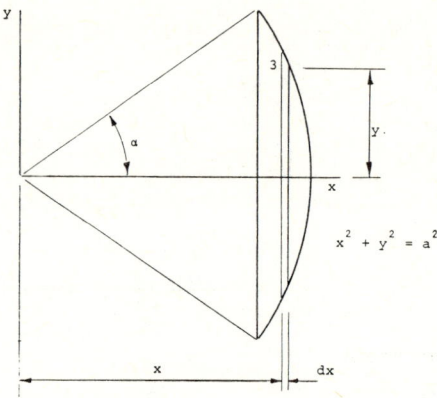

Fig. 4.16b. Fig. 4.16c.

$$m_2 = \rho\pi\tan^2\alpha \int_{a\cos\alpha - h}^{a\cos\alpha} x^2 dx$$

$$= \frac{\rho\pi\tan^2\alpha}{3}\{a^3\cos^3\alpha + (a\cos\alpha - h)^3\} . \tag{1}$$

Also

$$m_2\{R_2^x\} = \int_{a\cos\alpha - h}^{a\cos\alpha} (\rho\pi y^2 dx)x = \rho\pi\tan^2\alpha \int_{a\cos\alpha - h}^{a\cos\alpha} x^3 dx$$

$$= \frac{\rho\pi\tan^2\alpha}{4}\{a^4\cos^4\alpha - (a\cos\alpha - h)^4\}. \tag{2}$$

Therefore

$$\{R_2^x\} = \frac{3\{a^4\cos^4\alpha - (a\cos\alpha - h)^4\}}{4\{a^3\cos^3\alpha - (a\cos\alpha - h)^3\}} . \tag{3}$$

By reference to Fig. 4.16c,

$$m_3 = \int_{a\cos\alpha}^{a} \rho\pi y^2 dx = \rho\pi \int_{a\cos\alpha}^{a} (a^2 - x^2) dx$$

$$= \frac{\rho\pi a^3}{3}\{2 - \cos^2\alpha(3 - \cos^2\alpha)\} \tag{4}$$

since
$$y^2 = a^2 - x^2.$$

Also
$$m_3\{R_3^x\} = \int_{a\cos\alpha}^{a} (\rho\pi y^2 dx) x$$

$$= \frac{\rho\pi a^4}{4}\{1 - \cos^2\alpha(2 - \cos^2\alpha)\}. \tag{5}$$

Therefore
$$\{R_3^x\} = \frac{3a\{1 - \cos^2\alpha(2 - \cos^2\alpha)\}}{4\{2 - \cos\alpha(3 - \cos^2\alpha)\}}. \tag{6}$$

By Eq. 1
$$\frac{m_3}{\rho\pi} = 0.0718\{10^3 \times 0.8705 - 8\} = 20.785 \text{ cm}^3$$

and by Eq. 4
$$\frac{m_3}{\rho\pi} = \frac{10^3}{3}\{2 - 0.9659(3 - 0.93301)\} = 1.15 \text{ cm}^3.$$

By Eq. 2
$$\frac{m_2\{R_2^x\}}{\rho\pi} = \frac{0.0718\{10^4 \times 0.7587 - 16\}}{4} = 132.33 \text{ cm}^4$$

and by Eq. 5
$$\frac{m_3\{R_3^x\}}{\rho\pi} = \frac{10^4}{4}\{1 - 0.93301(2 - 0.93301)\} = 11.217 \text{ cm}^4.$$

Hence
$$\{R_4^x\} = \frac{m_2\{R_2^x\} + m_3\{R_3^x\}}{m_2 + m_3}$$

$$= \frac{132.33 + 11.217}{20.785 + 1.15}$$

$$= 6.54 \text{ cm}.$$

Problem 4.17. Figure 4.17 shows a machine part made from a homogeneous material. Locate its centre of mass.

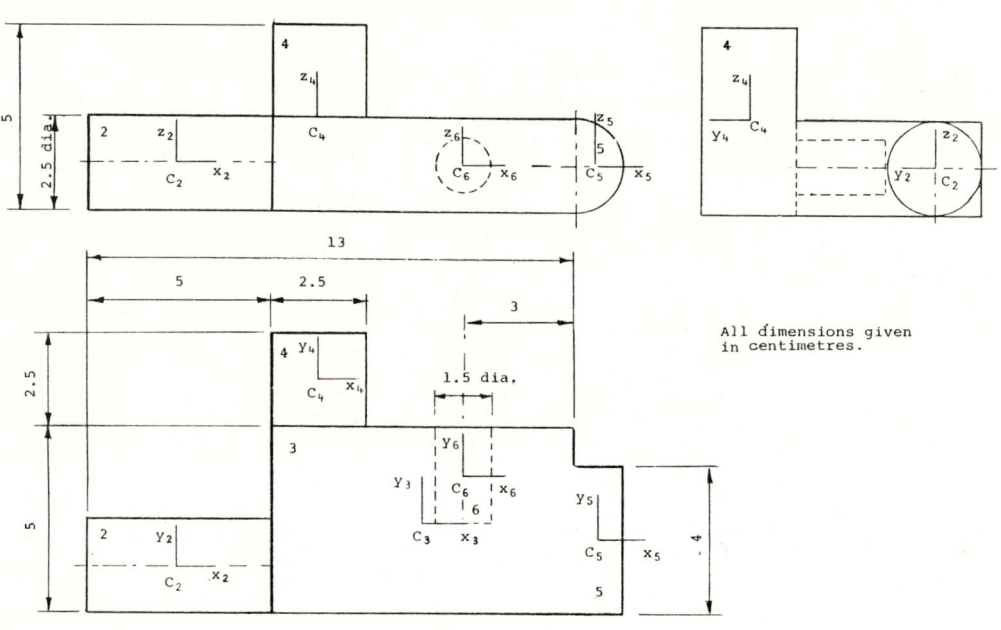

Fig. 4.17.

Solution. Treat the machine part as the composite body 7, being made up of the bodies 2, 3, 4 and 5 and the hole 6. If the density of the material is ρ, then the masses of the various parts are

$$m_2 = (\rho \pi \times 2.5^2 \times 5)/4 = 24.543\rho,$$

$$m_3 = \rho \times 2.5 \times 8 \times 5 = 100\rho,$$

$$m_4 = \rho \times 2.5 \times 2.5 \times 5 = 31.25\rho,$$

$$m_5 = (\rho \times 2.5^2 \times 4)/(4 \times 2) = 9.82\rho$$

and

$$m_6 = -(\rho \pi \times 1.5^2 \times 2.5)/4 = -4.42\rho$$

so that

$$m_7 = 161.2\rho .$$

The positions of the centres of mass of bodies 3, 4, 5 and 6 relative to the centre of mass of body 2 are

$$\{R_{C_3C_2}\} = \begin{bmatrix} 6.5 \\ 1.25 \\ 0 \end{bmatrix} \text{cm}, \quad \{R_{C_4C_2}\} = \begin{bmatrix} 3.75 \\ 5 \\ 1.25 \end{bmatrix} \text{cm},$$

All dimensions given in centimetres.

$$\{R_{C_5C_2}\} = \begin{bmatrix} 11.03 \\ 0.75 \\ 0 \end{bmatrix} \text{cm} \quad \text{and} \quad \{R_{C_6C_2}\} = \begin{bmatrix} 7.5 \\ 2.5 \\ 0 \end{bmatrix} \text{cm}.$$

(It is easy to show that the centre of mass of a semi-circular cylinder of radius a is $4a/3\pi$ from the plane of the diameteral surface). Hence, the position of the centre of mass of the composite body 7 relative to the position of the centre of mass of body 2 is given by

$$\{R_{C_7C_2}\} = \frac{m_3\{R_{C_3C_2}\} + m_4\{R_{C_4C_2}\} + m_5\{R_{C_5C_2}\} - m_6\{R_{C_6C_2}\}}{m_7}$$

$$= \frac{100\begin{bmatrix} 6.5 \\ 1.25 \\ 0 \end{bmatrix} + 31.25\begin{bmatrix} 3.75 \\ 5 \\ 1.25 \end{bmatrix} + 9.82\begin{bmatrix} 11.03 \\ 0.75 \\ 0 \end{bmatrix} - 4.42\begin{bmatrix} 7.5 \\ 2.5 \\ 0 \end{bmatrix}}{161.2}$$

$$= \begin{bmatrix} 5.22 \\ 1.72 \\ 0.24 \end{bmatrix} \text{cm}.$$

Problem 4.18. Show that the moment of a force about a given point is independent of the position vector chosen, provided that it starts at the given point and terminates on the line of action of the force.

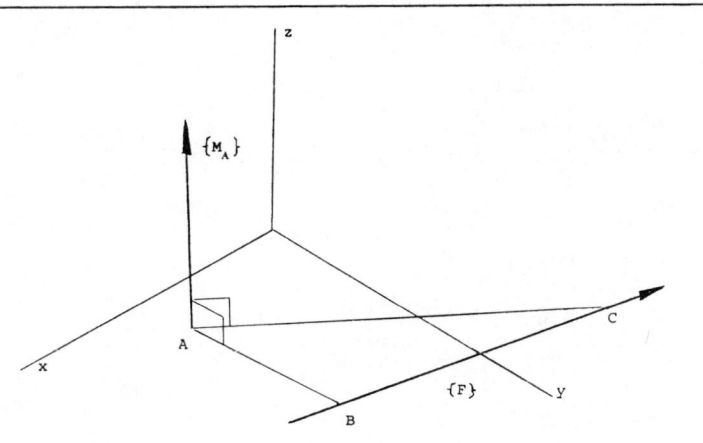

Fig. 4.18.

Solution. Refer to Fig. 4.18. Let A be the point about which the moment of {F} is to be determined. If B and C are any points on the line of action of {F}, then

$$\{M_A\} = [R_{CA}]\{F\}$$

and

$$\{R_{CA}\} = \{R_{BA}\} + \{R_{CB}\} .$$

Hence

$$\{M_A\} = \left[[R_{BA}] + [R_{CB}]\right]\{F\}$$

and since

$$\{R_{CB}\} = \lambda\{F\}$$

where λ is a scalar,

$$\{M_A\} = [R_{BA}]\{F\} + \lambda[F]\{F\} = [R_{BA}]\{F\}$$

because the vector product of the force vector with itself is a null vector. Thus, since B is any point on the line of action of the force vector, the moment vector is independent of the position vector chosen.

Problem 4.19. A vertical mast, which is supported at the ground in a smooth spherical cup, is held erect by guy ropes as shown in Fig. 4.19. If the resultant force on the mast due to the guy ropes is to be vertically down, find the guy rope tensions in AD and AB and the force on the mast at A in terms of the tension in the guy rope AC.

Solution. Let body 6 be the device to which the ropes are attached and which transmits the resultant force to the mast, body 2. Then the forces on body 6 are

$$\{F_{63}\} = F_D \frac{\{R_{DA}\}}{|R_{DA}|} = \frac{F_D}{25}\begin{bmatrix} 0 \\ -15 \\ -20 \end{bmatrix} = F_D \begin{bmatrix} 0 \\ -0.6 \\ -0.8 \end{bmatrix} = F_D \begin{bmatrix} a_1 \\ b_1 \\ c_1 \end{bmatrix},$$

$$\{F_{64}\} = F_B \frac{\{R_{BA}\}}{|R_{BA}|} = \frac{F_B}{21.47}\begin{bmatrix} 5 \\ 6 \\ -20 \end{bmatrix} = F_B \begin{bmatrix} 0.233 \\ 0.279 \\ -0.931 \end{bmatrix} = F_B \begin{bmatrix} a_2 \\ b_2 \\ c_2 \end{bmatrix}$$

and

Solution of Dynamics Problems

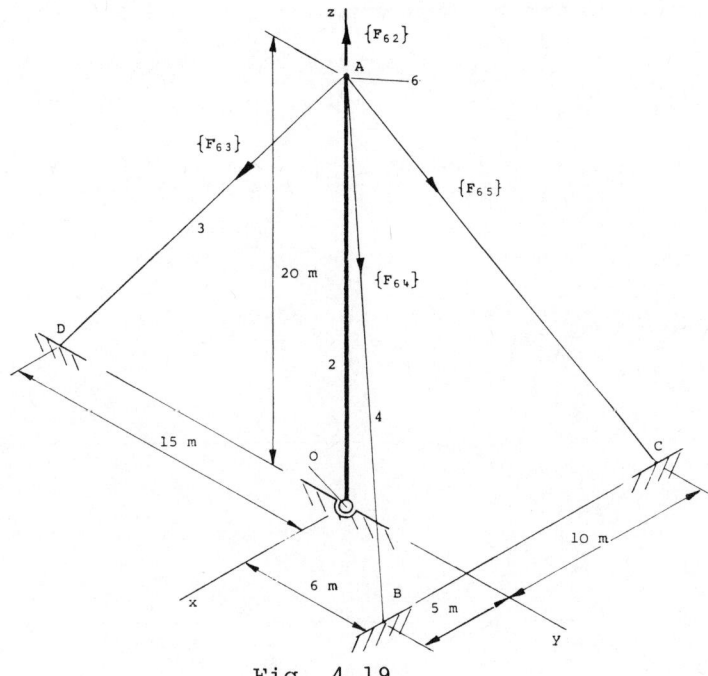

Fig. 4.19.

$$\{F_{65}\} = F_C \frac{\{R_{CA}\}}{|R_{CA}|} = \frac{F_C}{23.15}\begin{bmatrix} -10 \\ 6 \\ -20 \end{bmatrix} = F_C \begin{bmatrix} -0.432 \\ 0.259 \\ -0.864 \end{bmatrix} = F_C \begin{bmatrix} a_3 \\ b_3 \\ c_3 \end{bmatrix}$$

due to the guy ropes, and

$$\{F_{62}\} = \begin{bmatrix} 0 \\ 0 \\ F \end{bmatrix}$$

due to the mast.

For body 6, since its mass acceleration is zero,

$$\{F_{63}\} + \{F_{64}\} + \{F_{65}\} + \{F_{62}\} = \{0\}$$

$$F_D \begin{bmatrix} a_1 \\ b_1 \\ c_1 \end{bmatrix} + F_B \begin{bmatrix} a_2 \\ b_2 \\ c_2 \end{bmatrix} + F_C \begin{bmatrix} a_3 \\ b_3 \\ c_3 \end{bmatrix} + F \begin{bmatrix} 0 \\ 0 \\ 1 \end{bmatrix} = \begin{bmatrix} 0 \\ 0 \\ 0 \end{bmatrix}. \tag{1}$$

From the x component equation

$$F_B = -\frac{a_3}{a_2} F_C = -\frac{(-0.432)}{0.233} F_C = 1.85 F_C$$

and from the y component equation

$$F_D = -\frac{b_3}{b_1} F_C - \frac{b_3}{b_1} F_B = \frac{b_3}{b_1}\left(\frac{b_2 a_3}{b_3 a_2} - 1\right) F_C$$

$$= \frac{0.259}{(-0.6)}\left(\frac{0.279 \times (-0.432)}{0.259 \times 0.233} - 1\right) F_C = 1.294 F_C \;.$$

Also, from the z component equation,

$$F = \left\{-\frac{c_1 b_3}{b_1}\left(\frac{b_2 a_3}{b_3 a_2} - 1\right) + \frac{c_2 a_3}{a_2} - c_3\right\} F_C$$

$$= \left\{-(-0.8) \times 1.294 + \frac{(-0.931)(-0.432)}{0.233} - (-0.864)\right\} F_C$$

$$= 3.625 F_C \;.$$

Problem 4.20. A uniform rectangular trapdoor, of mass m, is connected to fixed points at O and D in a horizontal plane by smooth hinges as shown in Fig.4.20a. A string is attached to the mid point of the edge parallel to the hinged edge and is used to support the door at an angle θ to the horizontal. The string is also attached to a fixed point A.

Find, for a range of values of θ from $0°$ to $90°$ at $10°$ intervals, the string tension and hinge forces if the hinge at D is not capable of exerting a force in the direction of the hinge axis.

Solution. By reference to Fig. 4.20b,

$$\{F_{23}\} = F\frac{\{R_{BA}\}}{|R_{BA}|} \;, \quad \{F_{24}\} = \begin{bmatrix} F_4 \\ 0 \\ F_5 \end{bmatrix} \;,$$

$$\{F_{25}\} = \begin{bmatrix} F_1 \\ F_2 \\ F_3 \end{bmatrix} \quad \text{and} \quad \{W_2\} = mg\begin{bmatrix} 0 \\ 0 \\ -1 \end{bmatrix} \;.$$

Now

$$\{R_A\} = \begin{bmatrix} -1 \\ 0 \\ 2 \end{bmatrix} m \;, \quad \{R_A\} = \begin{bmatrix} \cos\theta \\ 0.7 \\ \sin\theta \end{bmatrix} = \begin{bmatrix} P \\ 0.7 \\ Q \end{bmatrix} m \;,$$

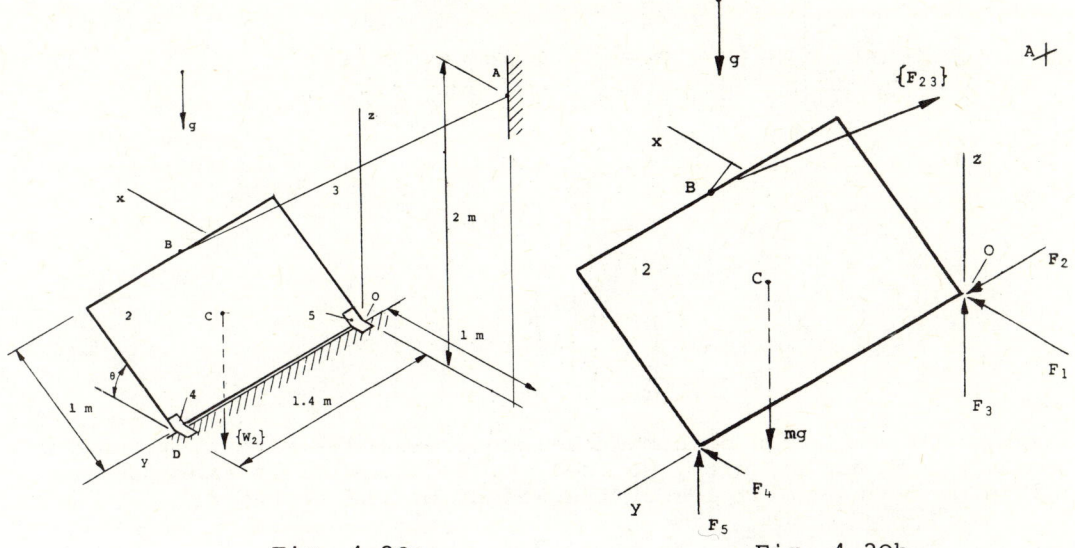

Fig. 4.20a.   Fig. 4.20b.

$$\{R_D\} = \begin{bmatrix} 0 \\ 1.4 \\ 0 \end{bmatrix} \text{ m}, \quad \{R_C\} = \begin{bmatrix} 0.5\cos\theta \\ 0.7 \\ 0.5\sin\theta \end{bmatrix} = \begin{bmatrix} 0.5P \\ 0.7 \\ 0.5Q \end{bmatrix} \text{ m},$$

$$\{R_{BA}\} = \begin{bmatrix} -(1+\cos\theta) \\ -0.7 \\ 2-\sin\theta \end{bmatrix} = \begin{bmatrix} -(1+P) \\ -0.7 \\ 2-Q \end{bmatrix} \text{ m}$$

and

$$|R_{BA}| = R = \sqrt{\{(1+P)^2 + 0.7^2 + (2-Q)^2\}}.$$

Taking moments about O, which for equilibrium gives a null vector as a resultant,

$$[R_D]\{F_{24}\} + [R_A]\{F_{23}\} + [R_C]\{W_2\} = \{0\}$$

$$\begin{bmatrix} 0 & 0 & 1.4 \\ 0 & 0 & 0 \\ -1.4 & 0 & 0 \end{bmatrix} \begin{bmatrix} F_4 \\ 0 \\ F_5 \end{bmatrix} + \frac{F}{R} \begin{bmatrix} 0 & -2 & 0 \\ 2 & 0 & 1 \\ 0 & -1 & 0 \end{bmatrix} \begin{bmatrix} -(1-P) \\ -0.7 \\ 2-Q \end{bmatrix}$$

$$+ mg \begin{bmatrix} 0 & -0.5Q & 0.7 \\ 0.5Q & 0 & -0.5P \\ 0.7 & 0.5P & 0 \end{bmatrix} \begin{bmatrix} 0 \\ 0 \\ -1 \end{bmatrix} = \begin{bmatrix} 0 \\ 0 \\ 0 \end{bmatrix}$$

$$\begin{bmatrix} 1.4F_5 \\ 0 \\ -1.4F_4 \end{bmatrix} + \frac{F}{R}\begin{bmatrix} 1.4 \\ -(2P+Q) \\ 1+P \end{bmatrix} = mg\begin{bmatrix} 0 \\ -0.5P \\ 0 \end{bmatrix} \qquad (1)$$

Also, for equilibrium, the sum of the forces is a null vector

$$\{F_{24}\} + \{F_{23}\} + \{F_{25}\} + \{W_2\} = \{0\}$$

$$\begin{bmatrix} F_4 \\ 0 \\ F_5 \end{bmatrix} + \frac{F}{R}\begin{bmatrix} -(1+P) \\ -0.7 \\ 2-P \end{bmatrix} + \begin{bmatrix} F_1 \\ F_2 \\ F_3 \end{bmatrix} + mg\begin{bmatrix} 0 \\ 0 \\ -1 \end{bmatrix} = \begin{bmatrix} 0 \\ 0 \\ 0 \end{bmatrix} \qquad (2)$$

Equations 1 and 2 can be combined into the single martrix given below when mg is taken as unity, so that the forces will be expressed as multiples or sub-multiples of the weight of the trapdoor.

$$\begin{bmatrix} 1 & 0 & 0 & 1 & 0 & -(1+P)/R \\ 0 & 1 & 0 & 0 & 0 & -0.7/R \\ 0 & 0 & 1 & 0 & 1 & (2-Q)/R \\ 0 & 0 & 0 & 0 & 1.4 & 1.4/R \\ 0 & 0 & 0 & 0 & 0 & -(2P+Q)/R \\ 0 & 0 & 0 & -1.4 & 0 & (1+P)/R \end{bmatrix}\begin{bmatrix} F_1 \\ F_2 \\ F_3 \\ F_4 \\ F_5 \\ F \end{bmatrix} = \begin{bmatrix} 0 \\ 0 \\ 1 \\ 0.7 \\ -0.5P \\ 0 \end{bmatrix}$$

which is of the form

$$[A]\{F\} = \{B\}$$

and therefore

$$\{F\} = [A]^{-1}\{B\} = [C]\{B\} .$$

A programme to effect the necessary computation, written in a Basic language, together with the print out for the 0°, 50° and 90° positions are given below.

It is always advisable to check the result of a particular computation, which does not involve any special conditions such as for example the 90° position, for which the string tension is zero and $F_3 = F_5 = 0.5mg$, by longhand methods to ensure that the programming is correct. The reader is invited to do this. It is however reassuring to find that the computed results give the correct values for the 90° position.

Solution of Dynamics Problems

Programme

```
100  DIM A(6,6), C(6,6), F(6,1), B(6,1)
110  FOR T1 = 0 TO 90 STEP 10
120  PRINT "TRAPDOOR ANGLE = ";T1
130  PRINT
140  T = T1/57.296
150  P = COS(T)
160  Q = SIN(T)
170  R = SQR((1+P)↑2 + .7↑2 + (2-Q)↑2)
180  MATA = ZER
190  A(1,6) = -(1+P)/R
200  A(2,6) = -.7/R
210  A(3,6) = (2-Q)/R
220  A(4,6) = 1.4/R
230  A(5,6) = -(2*P+Q)/R
240  A(6,6) = (1+P)/R
250  A(1,1) = 1
260  A(2,2) = 1
270  A(3,3) = 1
280  A(1,4) = 1
290  A(3,5) = 1
300  A(4,5) = 1.4
310  A(6,4) = -1.4
320  MATB = ZER
330  B(3,1) = 1
340  B(4,1) = .7
350  B(5,1) = -.5*P
360  MATC = INV(A)
370  MATF = C*B
380  PRINT "F1=";F(1,1);"F2=";F(2,1);"F3=";F(3,1)
390  PRINT "F4=";F(4,1);"F5=";F(5,1);"F6=";F(6,1)
400  PRINT
410  PRINT
420  NEXT T1
430  END
```

Print out

```
TRAPDOOR ANGLE = 0

F1 = .142857  F2 = .175  F3 = .25
F4 = .357143  F5 = .25   F = .72844

TRAPDOOR ANGLE = 50

F1 = 7.35285E-02  F2 = .109658  F3 = .46335
F4 = .183821  F5 = .343346  F = .34003

TRAPDOOR ANGLE = 90

F1 = 8.63530E-07  F2 = 2.11564E-06  F3 = .5
F4 = 2.15883E-06  F4 = .499997  F = 4.76918E-06
```

Problem 4.21. A straight uniform rod, of length 2a and mass m, rests on the ground and the top horizontal edge of a wall as shown in Fig. 4.21. If no slip occurs at the point of contact between the rod and the ground, obtain an expression for θ when slip is about to occur at the point of contact between the rod and the wall.

Also find the force at A when slip is about to occur.

Solution. Refer to the ground as body 1, the wall as body 2 and the rod as body 3. The rod is in equilibrium under the action of

$\{F_{31}\}$ at A, $\{W_3\}$ vertically down through C and

$\{F_{32}\}$ at B.

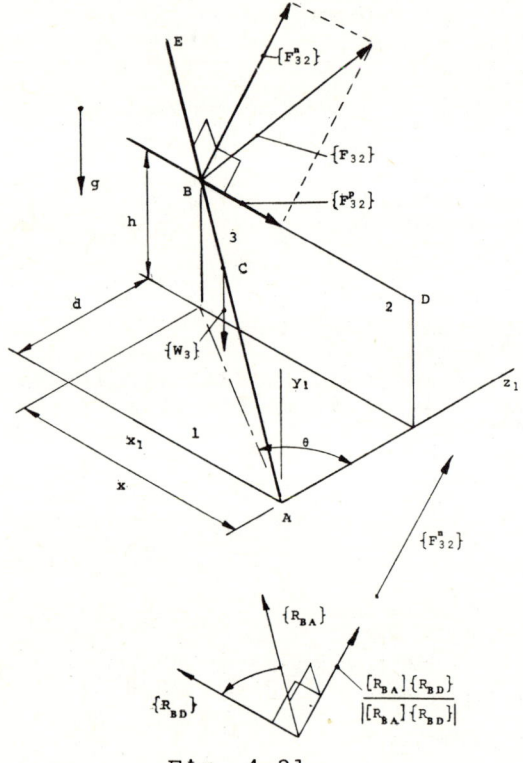

Fig. 4.21.

When slip is about to occur at B, $\{F_{32}\}$ is the vector sum of a force $\{F_{32}^n\}$ which is normal to the plane containing AB and BD, and a force $\{F_{32}^p\}$ along BD which is directed such as to oppose the relative motion between bodies 2 and 3. Unit vector perpendicular to $\{R_{BD}\}$ and $\{R_{BA}\}$ and in the direction of $\{F_{32}^n\}$ is given by

Solution of Dynamics Problems

$$[R_{BA}]\{R_{BD}\}/|[R_{BA}]\{R_{BD}\}| \ .$$

Hence
$$\{F^n_{32}\} = F[R_{BA}]\{R_{BD}\}/|[R_{BA}]\{R_{BD}\}|$$

and since
$$[R_{BA}]\{R_{BD}\} = \begin{bmatrix} 0 & -d & h \\ d & 0 & -x \\ -h & x & 0 \end{bmatrix} \begin{bmatrix} x \\ 0 \\ 0 \end{bmatrix} = x \begin{bmatrix} 0 \\ d \\ -h \end{bmatrix}$$

and
$$|[R_{BA}]\{R_{BD}\}| = x\sqrt{(d^2 + h^2)} \ ,$$

$$\{F^n_{32}\} = \frac{F}{\sqrt{(d^2 + h^2)}} \begin{bmatrix} 0 \\ d \\ -h \end{bmatrix} \ .$$

Also
$$\{F^p_{32}\} = \mu F \begin{bmatrix} -1 \\ 0 \\ 0 \end{bmatrix} \ .$$

For equilibrium, the sum of the moments of the forces on body 3 about A is zero

$$[R_{CA}]\{W_3\} + [R_{BA}]\{F^n_{32}\} + [R_{BA}]\{F^p_{32}\} = \{0\}$$

and since
$$\{R_{CA}\} = |R_{CA}| \frac{\{R_{BA}\}}{|R_{BA}|} = \frac{a}{\sqrt{(x^2 + d^2 + h^2)}} \begin{bmatrix} x \\ d \\ h \end{bmatrix} \ ,$$

$$\begin{bmatrix} 0 & -d & h \\ d & 0 & -x \\ -h & x & 0 \end{bmatrix} \left[\frac{mga}{\sqrt{(x^2 + d^2 + h^2)}} \begin{bmatrix} 0 \\ -1 \\ 0 \end{bmatrix} + \frac{F}{\sqrt{(d^2 + h^2)}} \begin{bmatrix} 0 \\ d \\ -h \end{bmatrix} \right.$$
$$\left. + \mu F \begin{bmatrix} -1 \\ 0 \\ 0 \end{bmatrix} \right] = \begin{bmatrix} 0 \\ 0 \\ 0 \end{bmatrix} \ .$$

From the y component equation, writing $d/h = \lambda$,

$$\frac{xh}{\sqrt{(d^2 + h^2)}} = \mu d \quad \text{or} \quad \tan\theta = \frac{x}{d} = \mu\sqrt{(1 + \lambda^2)}$$

and from the x component equation

$$F\sqrt{(d^2 + h^2)} = \frac{mgad}{\sqrt{(x^2 + d^2 + h^2)}},$$

which reduces to

$$F = \frac{mga\lambda}{h(1 + \lambda^2)\sqrt{(1 + \mu^2\lambda^2)}}$$

by use of the result from the y component equation.

Also, for equilibrium, the sum of the forces on the rod is zero

$$\{W_3\} + \{F_{32}\} + \{F_{31}\} = \{0\}$$

$$mg\begin{bmatrix} 0 \\ -1 \\ 0 \end{bmatrix} + \frac{F}{\sqrt{(d^2 + h^2)}}\begin{bmatrix} 0 \\ d \\ -h \end{bmatrix} + \mu F\begin{bmatrix} -1 \\ 0 \\ 0 \end{bmatrix} + \begin{bmatrix} F_x \\ F_y \\ F_z \end{bmatrix} = \begin{bmatrix} 0 \\ 0 \\ 0 \end{bmatrix}$$

and therefore

$$F_x = \frac{\mu mg\lambda}{(1 + \lambda^2)\sqrt{(1 + \mu^2\lambda^2)}}$$

$$F_y = mg\left\{1 - \frac{\lambda}{(1 + \lambda^2)^{3/2}\sqrt{(1 + \mu^2\lambda^2)}}\right\}$$

and

$$F_z = \frac{mg\lambda}{(1 + \lambda^2)^{3/2}\sqrt{(1 + \mu^2\lambda^2)}}.$$

If slip is to occur before the rod falls down the face of the wall, then

$$x^2 + d^2 + h^2 \leqq 4a^2$$

or

$$(1 + \lambda^2)(1 + \mu^2\lambda^2) \leqq 4a^2/h^2$$

requiring

$$\mu \leqq \frac{1}{\lambda}\sqrt{\{(4a^2/h^2)/(1 + \lambda^2) - 1\}}.$$

Problem 4.22. Any given system of forces and couples can be reduced to a single force {F} through a given point O and a couple {L} as shown in Fig. 4.22a. The force {F} is independent of the choice of the point O but the couple {L} is not. In general the force and couple vectors will not be parallel. Consider the problem of reducing the {F} and {L} system to a force {F} through some point A, to be determined, and a couple {L^p} which is that component of {L} parallel to {F}.

Solution of Dynamics Problems

Find $\{L^p\}$ and the point at which the line of action of $\{F\}$ cuts the xz plane for the case in which

$$\{F\} = \begin{bmatrix} 10 \\ 6 \\ 4 \end{bmatrix} \text{ kN} \quad \text{and} \quad \{L\} = \begin{bmatrix} 6 \\ 3 \\ -6 \end{bmatrix} \text{ kN m}.$$

Solution.

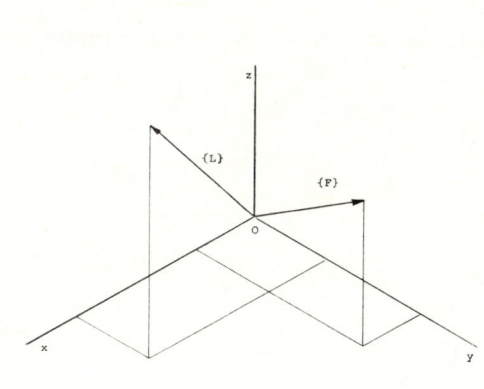

Fig. 4.22a.

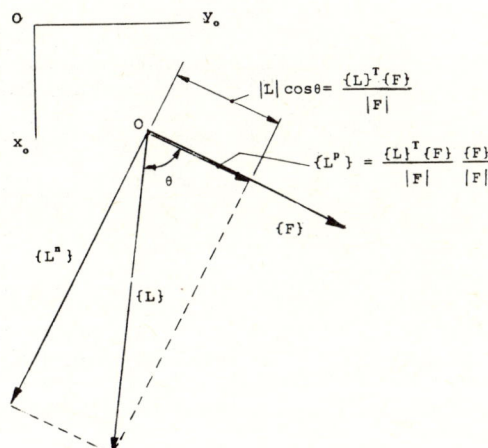

Fig. 4.22b.

The couple vector which is parallel to the force vector is given by

$$\{L^p\} = \frac{\{L\}^T\{F\}}{|F|} \frac{\{F\}}{|F|} = \frac{\{L\}^T\{F\}\{F\}}{|F|^2} = \lambda\{F\}$$

as illustrated in Fig. 4.22b. Thus

$$\{L\} = \{L^p\} + \{L^n\}$$

or

$$\{L^n\} = \{L\} - \{L^p\}$$

where $\{L^n\}$ is normal to the force vector $\{F\}$. If $\{L^n\}$ is to be replaced by moving the line of action of $\{F\}$ so that it passes through some point A, then the position of A relative to O is given by

$$\{L^n\} = [R_{AO}]\{F\}$$

and therefore

$$\{L\} - \{L^p\} = [R_{AO}]\{F\}$$
$$\{L\} - \lambda\{F\} = -[F]\{R_{AO}\}$$

Premultiplying this equation by $[F]$ gives

$$[F]\{L\} - \lambda[F]\{F\} = -[F][F]\{R_{AO}\}$$

and therefore

$$-[F]\{L\} = (\{F\}^T\{R_{AO}\})\{F\} - (\{F\}^T\{F\})\{R_{AO}\}$$

from which

$$\{R_{AO}\} = \frac{(\{F\}^T\{R_{AO}\})}{|F|^2}\{F\} + \frac{[F]\{L\}}{|F|^2} \quad .$$

One of the points satisfying this equation is on a line through O perpendicular to the plane containing {F} and {L} which is

$$\left|\frac{[F]\{L\}}{|F|^2}\right|$$

long and in the direction of

$$[F]\{L\} \quad .$$

The force {F} through this point A, together with $\{L^P\}$ is the *wrench* equivalent to {F} through O and {L}.

For the particular system given in the problem

$$\frac{[F]\{L\}}{|F|^2} = \frac{1}{152}\begin{bmatrix} 0 & -4 & 6 \\ 4 & 0 & -10 \\ -6 & 10 & 0 \end{bmatrix}\begin{bmatrix} 6 \\ 3 \\ -6 \end{bmatrix} = \begin{bmatrix} -0.316 \\ 0.553 \\ -0.0395 \end{bmatrix} \text{ m} \quad .$$

The equation to the line along which {F} lies is thus

$$\frac{x - x_A}{\ell} = \frac{y - y_A}{m} = \frac{z - z_A}{n}$$

where

$$\{R_P\} = \begin{bmatrix} x \\ y \\ z \end{bmatrix}, \quad \{R_A\} = \begin{bmatrix} x_A \\ y_A \\ z_A \end{bmatrix}$$

and ℓ, m and n are the direction cosines of {F}. Thus, for $y = 0$

$$x = x_A - \ell y_A/m = -0.316 - 10 \times 0.553/6 = -1.237 \text{ m}$$

and

$$z = z_A - n y_A/m = -0.0395 - 4 \times 0.553/6 = -0.407 \text{ m}.$$

giving

$$\{R_P\}\Big|_{y=0} = \begin{bmatrix} -1.237 \\ 0 \\ -0.407 \end{bmatrix} \text{ m}.$$

Problem 4.23. Prove the perpendicular axis and parallel axis theorems for a plane lamina.

Solution.

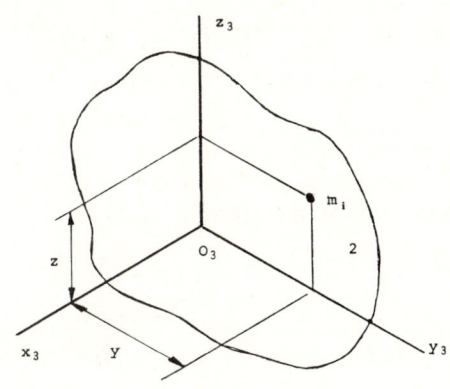 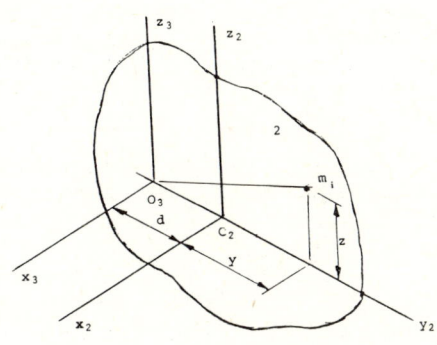

Fig. 4.23a.   Fig. 4.23b.

Perpendicular axis theorem.
Refer to Fig.4.23a showing a lamina, body 2, in the x_3z_3 plane with O_3 at any point in the plane. The moment of inertia of the lamina about the x_3 axis is given by

$$I_{x_3 x_3} = \sum m_i (z^2 + y^2)$$
$$= \sum m_i z^2 + \sum m_i y^2$$
$$= I_{y_3 y_3} + I_{z_3 z_3} \, .$$

Thus, the moment of inertia of a lamina about an axis perpendicular to its plane through any point O is the sum of the moments of inertia about any two mutually perpendicular axes in the plane of the lamina which pass through O. If in particular

$$I_{y_3 y_3} = I_{z_3 z_3} \, ,$$

then

$$I_{x_3 x_3} = 2I_{y_3 y_3} = 2I_{z_3 z_3} \, .$$

Parallel axis theorem.
Refer to Fig. 4.23b showing a lamina, body 2, in the x_2z_2 plane with the origin of frame 2 at the centre of mass of the lamina. The moment of inertia of the lamina about the x_3 axis is

$$I_{x_3 x_3} = \sum m_i \{(d + y)^2 + z^2\}$$
$$= \sum m_i d^2 + 2d \sum m_i y + \sum m_i (y^2 + z^2)$$

$$= m_2 d^2 + I_{x_2 x_2}$$

since $\sum m_i y = 0$.

Thus, the moment of inertia of a lamina about any axis exceeds the inertia about a parallel axis through the centre of mass of the lamina by the product of the mass of the lamina and the square of the distance between the parallel axes.

Problem 4.24. Find, for the thin uniform circular disc, body2, of Fig. 4.24

$$[I_2]_{2/2} \quad \text{and} \quad [I_2]_{2/3} \ .$$

Solution. The moment of inertia of the element shown about the y_2 axis is

$$\rho \ r \ d\theta \ dr \ r^2 \ ,$$

where ρ is the mass per unit area of the disc. The total moment of inertia is thus

$$I_{y_2 y_2} = \int_0^a r^3 \, dr \int_0^{2\pi} d\theta = 2\pi\rho \int_0^a r^3 \, dr = \rho \pi a^4 / 2$$

and since $m_2 = \rho \pi a^2$,

$$I_{y_2 y_2} = m_2 a^2 / 2 \ .$$

By the perpendicular axis theorem of Problem 4.23

$$I_{x_2 x_2} = I_{z_2 z_2} = m_2 a^2 / 4$$

and therefore, since the product of inertia terms are zero by symmetry

$$[I_2]_{2/2} = \frac{m_2 a^2}{4} \begin{bmatrix} 1 & 0 & 0 \\ 0 & 2 & 0 \\ 0 & 0 & 1 \end{bmatrix} \ .$$

Now

$$[I_2]_{2/3} = [\ell_2]_3 [I_2]_{2/2} [\ell_2]_3^T$$

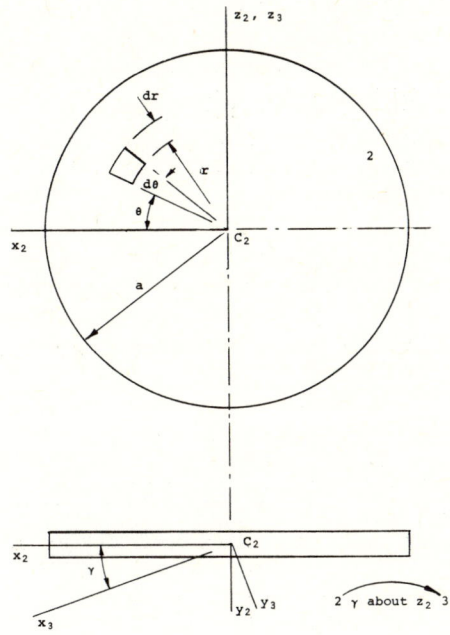

Fig. 4.24.

and therefore

$$[I_2]_{2/3} = \frac{m_2 a^2}{4} \begin{bmatrix} c\gamma & s\gamma & 0 \\ -s\gamma & c\gamma & 0 \\ 0 & 0 & 1 \end{bmatrix} \begin{bmatrix} 1 & 0 & 0 \\ 0 & 2 & 0 \\ 0 & 0 & 1 \end{bmatrix} \begin{bmatrix} c\gamma & -s\gamma & 0 \\ s\gamma & c\gamma & 0 \\ 0 & 0 & 1 \end{bmatrix}$$

$$= \frac{m_2 a^2}{4} \begin{bmatrix} 2 - \cos^2\gamma & \sin\gamma\cos\gamma & 0 \\ \sin\gamma\cos\gamma & 2 - \sin^2\gamma & 0 \\ 0 & 0 & 1 \end{bmatrix}.$$

Problem 4.25. Obtain the inertia matrix for a three-bladed airscrew referred to a set of axes which are fixed in the engine and which have their origin on the srew axis and in the plane of rotation of the screw.

Solution. Refer to the engine fixed frame as 0 and number the blades 2, 3 and 4 as shown in Fig.4.25. Frames 2, 3 and 4 have their origins coincident with that of frame 0, thus departing from the usual convention of positioning the origin of a frame numbered to correspond with that of the body at the centre of mass of that body.

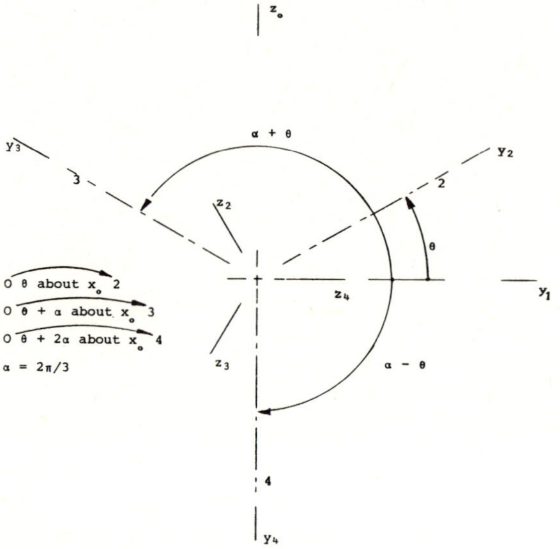

Fig. 4.25.

If the blades are treated as simple straight rods, then

$$[I_2]_{2/2} = [I_3]_{3/3} = [I_4]_{4/4} \simeq \begin{bmatrix} I & 0 & 0 \\ 0 & J & 0 \\ 0 & 0 & I \end{bmatrix}$$

and if the whole screw is body 5, then

$$[I_5]_{5/0} = [I_2]_{2/0} + [I_3]_{3/0} + [I_4]_{4/0}$$

$$= [\ell_2]_0 [I_2]_{2/2} [\ell_2]_0^T + [\ell_3]_0 [I_3]_{3/3} [\ell_3]_0^T$$

$$+ [\ell_4]_0 [I_4]_{4/4} [\ell_4]_0^T .$$

Now

$$[\ell_2]_0 = \begin{bmatrix} 1 & 0 & 0 \\ 0 & c\theta & -s\theta \\ 0 & s\theta & c\theta \end{bmatrix} , \quad [\ell_3]_0 = \begin{bmatrix} 1 & 0 & 0 \\ 0 & c(\theta+\alpha) & -s(\theta+\alpha) \\ 0 & s(\theta+\alpha) & c(\theta+\alpha) \end{bmatrix}$$

$$[\ell_4]_0 = \begin{bmatrix} 1 & 0 & 0 \\ 0 & c(\alpha-\theta) & s(\alpha-\theta) \\ 0 & -s(\alpha-\theta) & c(\alpha-\theta) \end{bmatrix} = \begin{bmatrix} 1 & 0 & 0 \\ 0 & c(\theta-\alpha) & -s(\theta-\alpha) \\ 0 & s(\theta-\alpha) & c(\theta-\alpha) \end{bmatrix}$$

and therefore

$$[I_2]_{2/0} = \begin{bmatrix} I & 0 & 0 \\ 0 & Jc^2\theta + Is^2\theta & -(I-J)s\theta c\theta \\ 0 & -(I-J)s\theta c\theta & Js^2\theta + Ic^2\theta \end{bmatrix},$$

$$[I_3]_{3/0} = \begin{bmatrix} I & 0 & 0 \\ 0 & Jc^2(\theta+\alpha) + Is^2(\theta+\alpha) & -(I-J)s(\theta+\alpha)c(\theta+\alpha) \\ 0 & -(I-J)s(\theta+\alpha)c(\theta+\alpha) & Js^2(\theta+\alpha) + Ic^2(\theta+\alpha) \end{bmatrix}$$

and

$$[I_4]_{4/0} = \begin{bmatrix} I & 0 & 0 \\ 0 & Jc^2(\theta-\alpha) + Is^2(\theta-\alpha) & -(I-J)s(\theta-\alpha)c(\theta-\alpha) \\ 0 & -(I-J)s(\theta-\alpha)c(\theta-\alpha) & Js^2(\theta-\alpha) + Ic^2(\theta-\alpha) \end{bmatrix}.$$

For $\alpha = 2\pi/3$

$$s\theta c\theta + s(\theta+\alpha)c(\theta+\alpha) + s(\theta-\alpha)c(\theta-\alpha)$$
$$= 0.5\{s2\theta + s2(\theta+\alpha) + s2(\theta-\alpha)\}$$
$$= 0,$$

$$c^2\theta + c^2(\theta+\alpha) + c^2(\theta-\alpha)$$
$$= 0.5\{1 + c2\theta + 1 + c2(\theta+\alpha) + 1 + c2(\theta-\alpha)\}$$
$$= 1.5$$

and, similarly

$$s^2\theta + s^2(\theta+\alpha) + s^2(\theta-\alpha)$$
$$= 1.5$$

for all values of θ. Hence

$$[I_5]_{5/0} = \begin{bmatrix} 3I & 0 & 0 \\ 0 & 1.5(I+J) & 0 \\ 0 & 0 & 1.5(I+J) \end{bmatrix}.$$

Problem 4.26. The uniform rectangular parallelepiped shown in Fig.4.26 a is machined from a steel forging and fixed to a light shaft with its axis along BD. The shaft runs in bearings, which can be considered inertial, at a constant rate of 1 000 rev/min. Determine the angular momentum and the rate of change of angular momentum of the parallelepiped referred to a frame fixed in it.

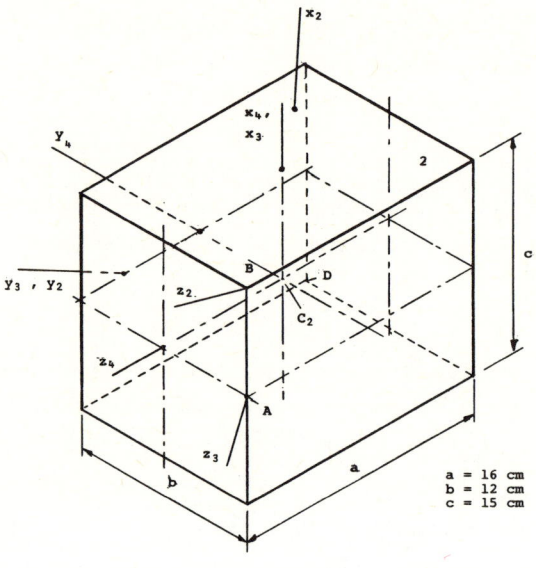

Fig. 4.26a.

Solution.

Since the density of steel is about 7.8 g/cm³ , the mass of the body is

$$m_2 = 16 \times 12 \times 15 \times 7.8 \times 10^{-3} = 22.46 \text{ kg} .$$

Now

$$[I_2]_{4/4} = [I_2]_{2/2} = \frac{m_2}{12} \begin{bmatrix} a^2 + b^2 & 0 & 0 \\ 0 & a^2 + c^2 & 0 \\ 0 & 0 & b^2 + c^2 \end{bmatrix}$$

$$= \begin{bmatrix} 748.7 & 0 & 0 \\ 0 & 900.3 & 0 \\ 0 & 0 & 690.6 \end{bmatrix} \text{ kg m}^2 .$$

To refer $[I_2]_{2/4}$ to frame 2 it is necessary to find $[\ell_4]_2$. Frame 4 can be aligned with frame 2 by the sequence of rotations shown in Fig.4.26b where

$$\cos\alpha = 0.8, \quad \sin\alpha = 0.6, \quad \cos\beta = 10/12.5 = 0.8$$

and

$$\sin\beta = 7.5/12.5 = 0.6 .$$

Hence

$$[\ell_4]_2 = [\ell_3]_2 [\ell_4]_3 = \begin{bmatrix} c\beta & 0 & -s\beta \\ 0 & 1 & 0 \\ s\beta & 0 & c\beta \end{bmatrix} \begin{bmatrix} 1 & 0 & 0 \\ 0 & c\alpha & s\alpha \\ 0 & -s\alpha & c\alpha \end{bmatrix}$$

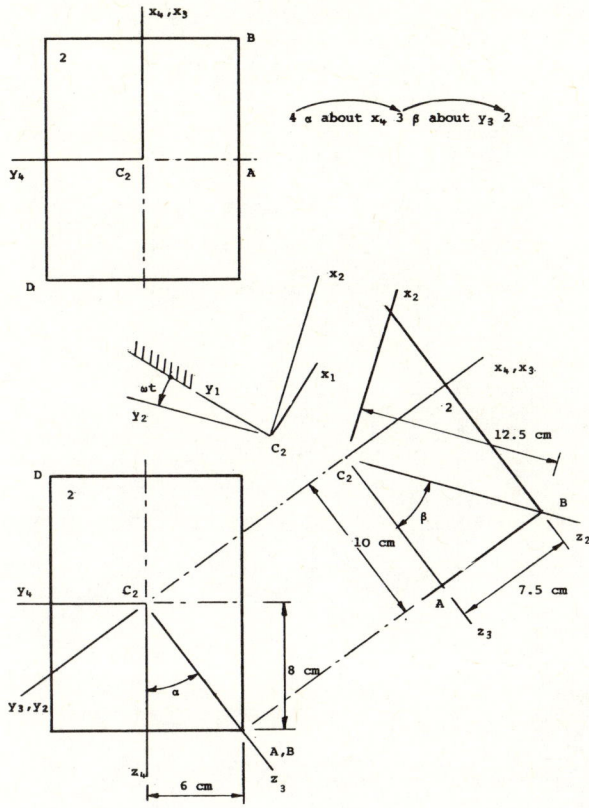

Fig. 4.26b.

$$= \begin{bmatrix} c\beta & s\alpha s\beta & -c\alpha s\beta \\ 0 & c\alpha & s\alpha \\ s\beta & -s\alpha c\beta & c\alpha c\beta \end{bmatrix} = \begin{bmatrix} 0.8 & 0.36 & -0.48 \\ 0 & 0.8 & 0.6 \\ 0.6 & -0.48 & 0.64 \end{bmatrix}$$

Therefore

$$[I_2]_{2/2} = [\ell_4]_2 [I_2]_{2/4} [\ell_4]_2^T$$

$$= \begin{bmatrix} 755 & 60.4 & -8.35 \\ 60.4 & 825 & -80.5 \\ -8.35 & -80.5 & 760 \end{bmatrix} \text{ kg m}^2.$$

Also

$$|\omega_2|_1 = 2\pi \times 1\,000/60 = 104.7 \text{ rad/s}$$

and

$$\{\omega_2\}_{1/1} = \{\omega_2\}_{1/2} = 104.7 \begin{bmatrix} 0 \\ 0 \\ 1 \end{bmatrix} \text{ rad/s,}$$

giving

$$\{H_2\}_{1/2} = [I_2]_{2/2}\{\omega_2\}_{1/2}$$

$$= 104.7 \begin{bmatrix} 755 & 60.4 & -8.35 \\ 60.4 & 825 & -80.5 \\ -8.35 & -80.5 & 760 \end{bmatrix} \begin{bmatrix} 0 \\ 0 \\ 1 \end{bmatrix}$$

$$= \begin{bmatrix} -874 \\ -8\,428 \\ 79\,572 \end{bmatrix} \text{ kg m}^2/\text{s}$$

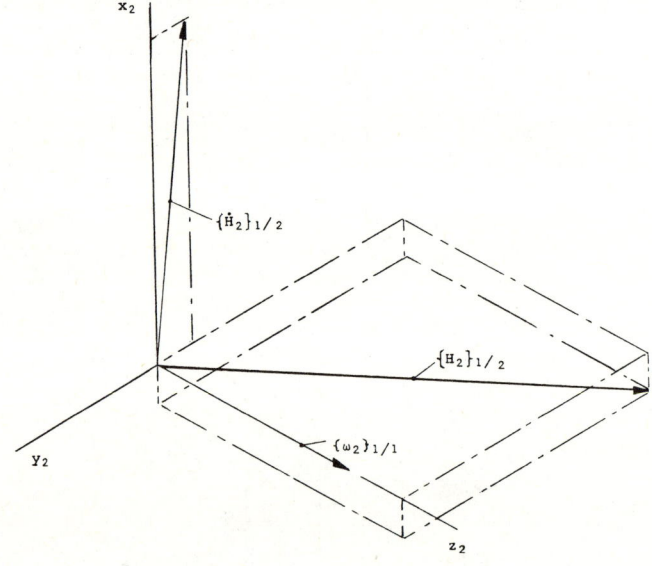

Fig. 4.26c.

Since $\{\dot{\omega}_2\}_{1/2}$ is a null matrix,

$$\{\dot{H}_2\}_{1/2} = \lfloor \omega_2 \rfloor_{1/2} [I_2]_{2/2}\{\omega_2\}_{1/2} = \lfloor \omega_2 \rfloor_{1/2}\{H_2\}_{1/2}$$

$$= 104.7 \begin{bmatrix} 0 & -1 & 0 \\ 1 & 0 & 0 \\ 0 & 0 & 0 \end{bmatrix} \begin{bmatrix} -874 \\ -8\,428 \\ 79\,572 \end{bmatrix} = \begin{bmatrix} 882 \\ -91.5 \\ 0 \end{bmatrix} \text{ kN m.}$$

Problem 4.27. The uniform rotor, body 3, of Fig. 4.27a runs on body 2 with a constant angular velocity of magnitude ω, while body 2 rotates relative to an inertial frame 1 with a constant angular velocity of magnitude Ω. Find, for the axes shown

$$[I_3]_{3/1}, \quad \{H_3\}_{1/1}, \quad \{\dot{H}_3\}_{1/1}, \quad \{H_3\}_{1/2} \text{ and } \{\dot{H}_3\}_{1/2}.$$

Solution.

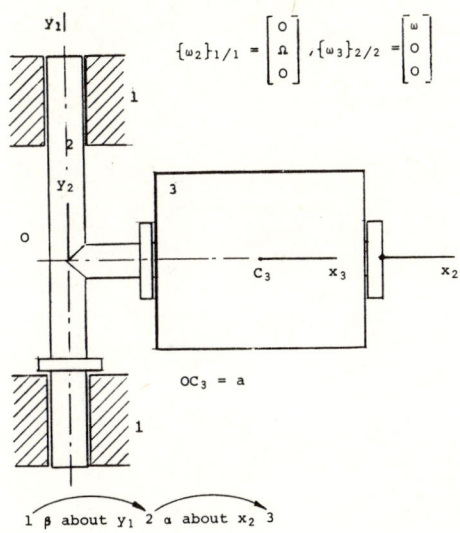

$$\{\omega_2\}_{1/1} = \begin{bmatrix} 0 \\ \Omega \\ 0 \end{bmatrix}, \{\omega_3\}_{2/2} = \begin{bmatrix} \omega \\ 0 \\ 0 \end{bmatrix}$$

Fig. 4.27a.

By reference to Fig. 4.27b

$$\{\omega_3\}_{1/1} = \{\omega_2\}_{1/1} + [\ell_2]_1 \{\omega_3\}_{2/2}$$

$$= \Omega \begin{bmatrix} 0 \\ 1 \\ 0 \end{bmatrix} + \begin{bmatrix} c\beta & 0 & s\beta \\ 0 & 1 & 0 \\ -s\beta & 0 & c\beta \end{bmatrix} \begin{bmatrix} \omega \\ 0 \\ 0 \end{bmatrix} = \begin{bmatrix} \omega\cos\beta \\ \Omega \\ -\omega\sin\beta \end{bmatrix} \quad (1)$$

and

$$\{\omega_3\}_{1/2} = [\ell_1]_2 \{\omega_3\}_{1/1}$$

$$= \begin{bmatrix} c\beta & 0 & -s\beta \\ 0 & 1 & 0 \\ s\beta & 0 & c\beta \end{bmatrix} \begin{bmatrix} \omega c\beta \\ \Omega \\ -\omega s\beta \end{bmatrix} = \begin{bmatrix} \omega \\ \Omega \\ 0 \end{bmatrix} \quad (2)$$

Equations 1 and 2 are illustrated in Fig. 4.27c. Differentiation of Eq. 1 gives

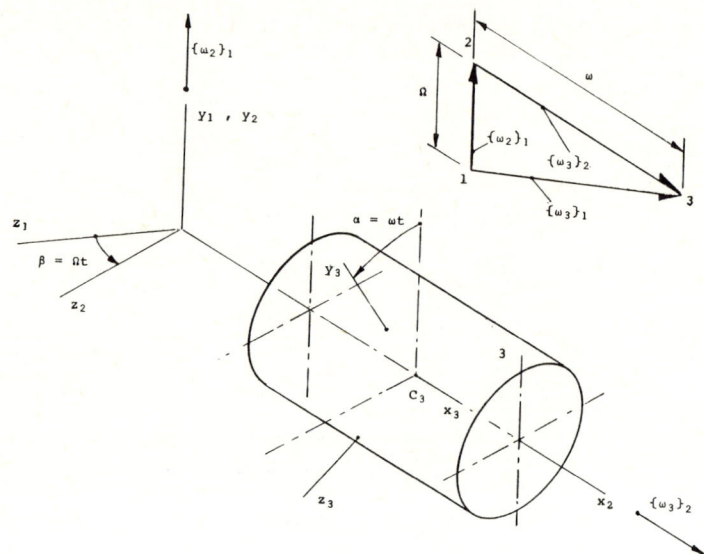

Fig. 4.27b.

$$\{\dot{\omega}_3\}_{1/1} = \begin{bmatrix} -\omega\dot{\beta}s\beta \\ 0 \\ -\omega\dot{\beta}c\beta \end{bmatrix} = -\omega\Omega \begin{bmatrix} \sin\beta \\ 0 \\ \cos\beta \end{bmatrix} \qquad (3)$$

and

$$\{\dot{\omega}_3\}_{1/2} = [\ell_1]_2 \{\dot{\omega}_3\}_{1/1} = -\omega\Omega \begin{bmatrix} c\beta & 0 & -s\beta \\ 0 & 1 & 0 \\ s\beta & 0 & c\beta \end{bmatrix} \begin{bmatrix} s\beta \\ 0 \\ c\beta \end{bmatrix} = \begin{bmatrix} 0 \\ 0 \\ -\omega\Omega \end{bmatrix} . (4)$$

Alternatively,

$$\{\dot{\omega}_3\}_{1/2} = [\omega_2]_{1/2}\{\omega_3\}_{1/2} + \frac{d}{dt}\{\omega_3\}_{1/2}$$

$$= \Omega \begin{bmatrix} 0 & 0 & 1 \\ 0 & 0 & 0 \\ -1 & 0 & 0 \end{bmatrix} \begin{bmatrix} \omega \\ \Omega \\ 0 \end{bmatrix} + \begin{bmatrix} 0 \\ 0 \\ 0 \end{bmatrix} = \begin{bmatrix} 0 \\ 0 \\ -\omega\Omega \end{bmatrix} .$$

Since body 3 is a solid of revolution, the axis of generation being the x_3 axis

$$[I_3]_{3/3} = \begin{bmatrix} J & 0 & 0 \\ 0 & I & 0 \\ 0 & 0 & I \end{bmatrix} .$$

Solution of Dynamics Problems

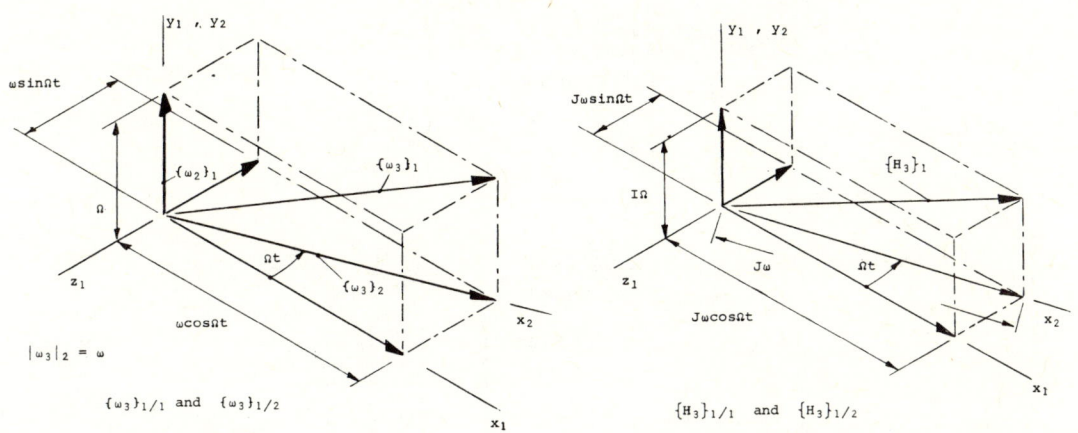

Fig. 4.27c.   Fig. 4.27d

Therefore

$$[I_3]_{3/2} = [\ell_3]_2 [I_3]_{3/3} [\ell_3]_2^T$$

$$= \begin{bmatrix} 1 & 0 & 0 \\ 0 & c\alpha & -s\alpha \\ 0 & s\alpha & c\alpha \end{bmatrix} \begin{bmatrix} J & 0 & 0 \\ 0 & I & 0 \\ 0 & 0 & I \end{bmatrix} \begin{bmatrix} 1 & 0 & 0 \\ 0 & c\alpha & s\alpha \\ 0 & -s\alpha & c\alpha \end{bmatrix}$$

$$= \begin{bmatrix} J & 0 & 0 \\ 0 & I & 0 \\ 0 & 0 & I \end{bmatrix} \quad . \tag{5}$$

Also

$$[I_3]_{3/1} = [\ell_2]_1 [I_3]_{3/2} [\ell_2]_1^T$$

$$= \begin{bmatrix} c\beta & 0 & s\beta \\ 0 & 1 & 0 \\ -s\beta & 0 & c\beta \end{bmatrix} \begin{bmatrix} J & 0 & 0 \\ 0 & I & 0 \\ 0 & 0 & I \end{bmatrix} \begin{bmatrix} c\beta & 0 & -s\beta \\ 0 & 1 & 0 \\ s\beta & 0 & c\beta \end{bmatrix}$$

$$= \begin{bmatrix} J\cos^2\beta + I\sin^2\beta & 0 & (I-J)\sin\beta\cos\beta \\ 0 & I & 0 \\ (I-J)\sin\beta\cos\beta & 0 & J\sin^2\beta + I\cos^2\beta \end{bmatrix} \quad . \tag{6}$$

Now

$$\{H_3\}_{1/1} = [I_3]_{3/1} \{\omega_3\}_{1/1} = [\ell_2]_1 [I_3]_{3/2} [\ell_2]_1^T \{\omega_3\}_{1/1}$$

$$= \begin{bmatrix} c\beta & 0 & s\beta \\ 0 & 1 & 0 \\ -s\beta & 0 & c\beta \end{bmatrix} \begin{bmatrix} J & 0 & 0 \\ 0 & I & 0 \\ 0 & 0 & I \end{bmatrix} \begin{bmatrix} c\beta & 0 & -s\beta \\ 0 & 1 & 0 \\ s\beta & 0 & c\beta \end{bmatrix} \begin{bmatrix} \omega c\beta \\ \Omega \\ -\omega s\beta \end{bmatrix}$$

$$= \begin{bmatrix} J\omega\cos\beta \\ I\Omega \\ -J\omega\sin\beta \end{bmatrix} \qquad (7)$$

and

$$\{H_3\}_{1/2} = [\ell_1]_2 \{H_3\}_{1/1}$$

$$= \begin{bmatrix} c\beta & 0 & -s\beta \\ 0 & 1 & 0 \\ s\beta & 0 & c\beta \end{bmatrix} \begin{bmatrix} J\omega c\beta \\ I\Omega \\ -J\omega s\beta \end{bmatrix} = \begin{bmatrix} J\omega \\ I\Omega \\ 0 \end{bmatrix}, \qquad (8)$$

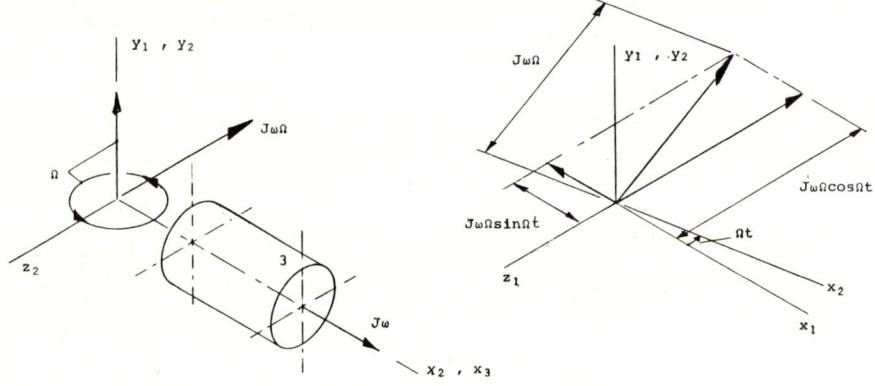

$\{\dot{H}_3\}_{1/2}$ and $\{\dot{H}_3\}_{1/1}$

Fig. 4.27e.

or alternatively

$$\{H_3\}_{1/2} = [I_3]_{3/2} \{\omega_3\}_{1/2} \quad .$$

Equations 7 and 8 are illustrated in Fig. 4.27d.

Differentiation of Eq. 7 gives

$$\{\dot{H}_3\}_{1/1} = \begin{bmatrix} -J\omega\Omega s\beta \\ 0 \\ -J\omega\Omega c\beta \end{bmatrix} = -J\omega\Omega \begin{bmatrix} \sin\beta \\ 0 \\ \cos\beta \end{bmatrix} \qquad (9)$$

and

Solution of Dynamics Problems

$$\{\dot H_3\}_{1/2} = [\ell_1]_2 \{\dot H_3\}_{1/1}$$

$$-J\omega\Omega \begin{bmatrix} c\beta & 0 & -s\beta \\ 0 & 1 & 0 \\ s\beta & 0 & c\beta \end{bmatrix} \begin{bmatrix} s\beta \\ 0 \\ c\beta \end{bmatrix} = -J\omega\Omega \begin{bmatrix} 0 \\ 0 \\ 1 \end{bmatrix} \quad . \tag{10}$$

Alternatively

$$\{\dot H_3\}_{1/2} = [\omega_2]_{1/2}\{H_3\}_{1/2} + \frac{d}{dt}\{H_3\}_{1/2}$$

or

$$\{\dot H_3\}_{1/2} = [\omega_3]_{1/2}[I_3]_{3/2}\{\omega_3\}_{1/2} + [I_3]_{3/2}\{\dot\omega_3\}_{1/2}$$

The reader is asked to evaluate these two alternative expressions for the rate of change of angular momentum of body 3, measured in frame 1 and referred to the rotating frame 2. Equations 9 and 10 are illustrated in Fig. 4.27e.

Problem 4.28. Body 2 has the inertia matrix

$$[I_2]_{2/2} = \begin{bmatrix} 7 & 0 & 0 \\ 0 & 25 & 0 \\ 0 & 0 & 32 \end{bmatrix} \text{ kg m}^2$$

and is constrained so that it is free to rotate about its centre of mass. At time $t = 0$ its angular velocity is

$$\{\omega_2\}_{1/2} = \Omega \begin{bmatrix} 4/5 \\ 0 \\ 3/5 \end{bmatrix} \text{ rad/s .}$$

Obtain an expression for $\{\omega_2\}_{1/2}$ at any subsequent time t, for the case in which there are no external couples acting on the body and the only external force has a line of action through the centre of mass of the body.

Solution. Let the angular velocity of the body at any time t be

$$\{\omega_2\}_{1/2} = \begin{bmatrix} \omega_x \\ \omega_y \\ \omega_z \end{bmatrix}$$

and therefore

$$\{\dot{\omega}_2\}_{1/2} = \frac{d}{dt}\{\omega_2\}_{1/2} = \begin{bmatrix} \dot{\omega}_x \\ \dot{\omega}_y \\ \dot{\omega}_z \end{bmatrix}.$$

For the motion subsequent to t = 0, since there are no external couples

$$\{0\} = \{\dot{H}_2\}_{1/2} = \lfloor \omega_2 \rfloor_{1/2} [I_2]_{2/2} \{\omega_2\}_{1/2} + [I_2]_{2/2} \{\dot{\omega}_2\}_{1/2}$$

$$\begin{bmatrix} 0 \\ 0 \\ 0 \end{bmatrix} = \begin{bmatrix} 0 & -\omega_z & \omega_y \\ \omega_z & 0 & -\omega_x \\ -\omega_y & \omega_x & 0 \end{bmatrix} \begin{bmatrix} 7 & 0 & 0 \\ 0 & 25 & 0 \\ 0 & 0 & 32 \end{bmatrix} \begin{bmatrix} \omega_x \\ \omega_y \\ \omega_z \end{bmatrix}$$

$$+ \begin{bmatrix} 7 & 0 & 0 \\ 0 & 25 & 0 \\ 0 & 0 & 32 \end{bmatrix} \begin{bmatrix} \dot{\omega}_x \\ \dot{\omega}_y \\ \dot{\omega}_z \end{bmatrix}.$$

These equations reduce to

$$\dot{\omega}_x + \omega_y \omega_z = 0, \tag{1}$$

$$\dot{\omega}_y + \omega_x \omega_z = 0 \tag{2}$$

and

$$16\dot{\omega}_z + 9\omega_x \omega_y = 0. \tag{3}$$

If Eq. 1 is multiplied by ω_x and Eq. 2 by ω_y then

$$\omega_x \dot{\omega}_x + \omega_x \omega_y \omega_z = 0$$

$$\omega_y \dot{\omega}_y - \omega_x \omega_y \omega_z = 0$$

and these equations sum to give

$$\omega_x \dot{\omega}_x + \omega_y \dot{\omega}_y = 0. \tag{4}$$

Hence

$$\int \omega_x \frac{d}{dt}\omega_x \, dt + \int \omega_y \frac{d}{dt}\omega_y \, dt = 0$$

or

$$\omega_x^2 + \omega_y^2 = A.$$

When t = 0

$\omega_x = 4\Omega/5$ and $\omega_y = 0$ giving $A = 16\Omega^2/25$

and therefore

Solution of Dynamics Problems

$$\omega_x^2 + \omega_y^2 = 16\Omega^2/25 \qquad (5)$$

If Eq. 2 is multiplied by $9\omega_y$ and Eq. 3 is multiplied by ω_z then

$$9\omega_y \dot{\omega}_y - 9\omega_x \omega_y \omega_z = 0$$
$$16\omega_z \dot{\omega}_z + 9\omega_x \omega_y \omega_z = 0$$

and these equations sum to give

$$9\omega_y \dot{\omega}_y + 16\omega_z \dot{\omega}_z = 0$$

which integrates to

$$9\omega_y^2 + 16\omega_z^2 = 144\Omega^2/25 . \qquad (6)$$

From Eq. 2

$$\frac{d}{dt}\omega_y = \omega_x \omega_z \quad \text{or} \quad dt = \frac{d\omega_y}{\omega_x \omega_z}$$

and by substitution from Eqs. 5 and 6

$$dt = \frac{d\omega_y}{\sqrt{\{(144\Omega^2/25) - 9\omega_y^2\}\{(16\Omega^2/25) - \omega_y^2\}/16}}$$

$$= \frac{4d\omega_y}{3\{(4\Omega/5)^2 - \omega_y^2\}} .$$

This equation integrates to

$$t = \frac{5}{3\Omega} \tanh^{-1}(5\omega_y/4\Omega)$$

since $\omega_y = 0$ when $t = 0$. Therefore

$$\omega_y = \frac{4\Omega}{5} \tanh(3\Omega t/5)$$

and by Eqs. 5 and 6

$$\omega_x = \frac{4\Omega}{5}\sqrt{\{1 - \tanh^2(3\Omega t/5)\}}$$

$$\omega_z = \frac{3\Omega}{5}\sqrt{\{1 - \tanh^2(3\Omega t/5)\}} .$$

It is thus seen that while the component angular velocity along the y_2 axis (the axis corresponding to the intermediate principal moment of inertia) is initially zero, finally the angular velocity is wholly along this axis. A graph showing the variation of the component angular velocities with time is shown in Fig. 4.28.

The reader will have noted that a special set of initial conditions

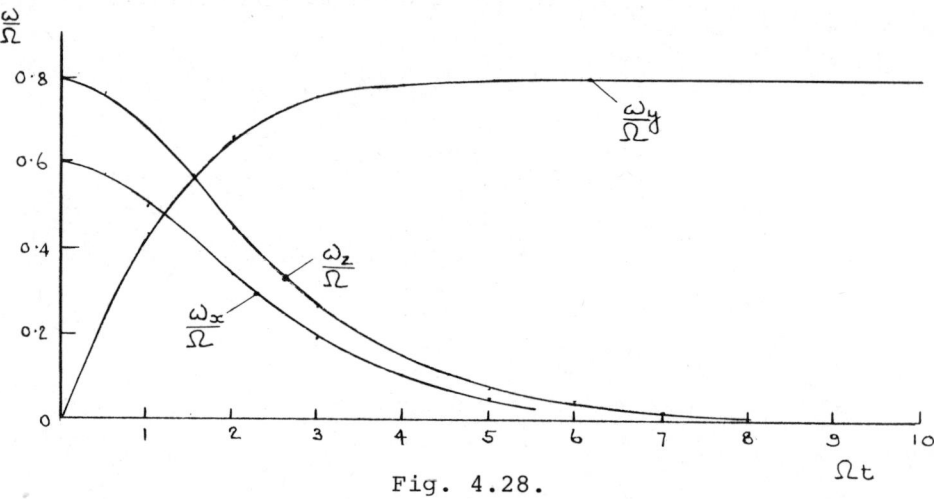

Fig. 4.28.

have been chosen to give a solution in a closed form. Had an arbitrary set of initial conditions been chosen, an integral of the form

$$\int \frac{d\omega}{\sqrt{(1 - a\omega^2)(1 - b\omega^2)}}$$

would have to be effected. This is one form of elliptic integral which does not have a closed form solution.

Problem 4.29. Body 2 is moving in free space relative to an inertial frame 1 such that

$$\{\omega_2\}_{1/2} = \begin{bmatrix} \Omega \\ 0 \\ 0 \end{bmatrix} \quad \text{and} \quad [I_2]_{2/2} = \begin{bmatrix} A & 0 & 0 \\ 0 & B & 0 \\ 0 & 0 & C \end{bmatrix},$$

where A, B and C are not equal th each other, when it is subjected to a small external impulsive couple. Determine whether or not the subsequent motion is stable.

Solution. Let the angular velocity of the body after the impulse be

$$\{\omega_2\}_{1/2} = \begin{bmatrix} \Omega + \omega_x \\ \omega_y \\ \omega_z \end{bmatrix}$$

Solution of Dynamics Problems

where ω_x, ω_y and ω_z are small compared with Ω. The equation of motion after the impulse is

$$\{0\} = \{\dot{H}_2\}_{1/2} = [\omega_2]_{1/2}[I_2]_{2/2}\{\omega_2\}_{1/2} + [I_2]_{2/2}\{\dot{\omega}_2\}_{1/2}$$

$$\begin{bmatrix} 0 \\ 0 \\ 0 \end{bmatrix} = \begin{bmatrix} 0 & \omega_z & \omega_y \\ -\omega_z & 0 & -(\Omega+\omega_x) \\ \omega_y & \Omega+\omega_x & 0 \end{bmatrix} \begin{bmatrix} A & 0 & 0 \\ 0 & B & 0 \\ 0 & 0 & C \end{bmatrix} \begin{bmatrix} \Omega+\omega_x \\ \omega_y \\ \omega_z \end{bmatrix}$$

$$+ \begin{bmatrix} A & 0 & 0 \\ 0 & B & 0 \\ 0 & 0 & C \end{bmatrix} \begin{bmatrix} \dot{\omega}_x \\ \dot{\omega}_y \\ \dot{\omega}_z \end{bmatrix}$$

and these equations reduce to

$$(C - B)\omega_y \omega_z + A\dot{\omega}_x = 0$$

$$(A - C)(\Omega + \omega_x)\omega_z + B\dot{\omega}_y = 0$$

and

$$(B - A)(\Omega + \omega_x)\omega_y + C\dot{\omega}_z = 0.$$

Now

$$\omega_x \omega_z \ll \Omega \omega_z \quad \text{and} \quad \omega_x \omega_y \ll \Omega \omega_y$$

so that the equations can be further reduced to

$$(C - B)\omega_y \omega_z + A\dot{\omega}_x = 0 \tag{1}$$

$$(A - C)\Omega \omega_z + B\dot{\omega}_y = 0 \tag{2}$$

$$(B - A)\Omega \omega_y + C\dot{\omega}_z = 0. \tag{3}$$

Differentiating Eq. 2 with respect to time gives

$$\dot{\omega}_z = - \frac{B}{\Omega(A - C)} \ddot{\omega}_y$$

and this result in Eq. 3 leads to

$$\ddot{\omega}_y + \frac{(A - C)(A - B)\Omega^2}{BC} \omega_y = 0.$$

If

$$(A - C)(A - B) < 0$$

then the motion will be unstable. Consider the three possible alternatives.
Case (i) If $A - C > 0$ and $A - B > 0$, or A is greater than either B or C then

$$(A - C)(B - C) > 0$$

and the motion is stable with a natural frequency of

233

$$\Omega \sqrt{\{(A - C)(A - B)/BC\}} \ .$$

Case(ii) If $A - C < 0$ and $A - B < 0$, or A is less than either B or C then

$$(A - C)(A - B) > 0$$

and the motion is stable.

Case(iii) If $A - C < 0$ and $A - B > 0$, or A lies between B and C then

$$(A - C)(A - B) < 0$$

and the motion is unstable.

Thus, if the body is rotating steadily about a principal axis for which the inertia is intermediate between the other principal moments of inertia, then any small external disturbance will give rise to an unstable motion. If the motion is about either of the other axes then the motion will be stable.

Problem 4.30. A body which has axial symmetry as shown in Fig. 4.30a is moving without constraint in deep space. Examine its angular motion.

Solution. One set of axes which are convenient for this study are as shown in Fig.4.30b. Frame 4 is fixed in the body with its origin at the centre of mass and the z_4 axis is along the axis of symmetry. Frame 4 is positioned relative to an inertial frame 1 through the intermediate frames 2 and 3 as follows:

(i) frame 2 rotates through the angle ψ about an axis parallel to the z_1 axis,

(ii) frame 3 rotates through the angle θ about the x_2 axis and

(iii) frame 4 rotates through the angle ϕ about the z_3 axis.

The angular velocity of body 4 relative to the inertial reference 1 is thus given by

$$\{\omega_4\}_1 = \{\omega_2\}_1 + \{\omega_3\}_2 + \{\omega_4\}_3$$

where

$$\{\omega_2\}_{1/1} = \{\omega_2\}_{1/2} = \begin{bmatrix} 0 \\ 0 \\ \dot{\psi} \end{bmatrix}, \quad \{\omega_3\}_{2/2} = \{\omega_3\}_{2/3} = \begin{bmatrix} \dot{\theta} \\ 0 \\ 0 \end{bmatrix}$$

and

$$\{\omega_4\}_{3/3} = \begin{bmatrix} \dot{\phi} \\ 0 \\ 0 \end{bmatrix} \ .$$

Solution of Dynamics Problems

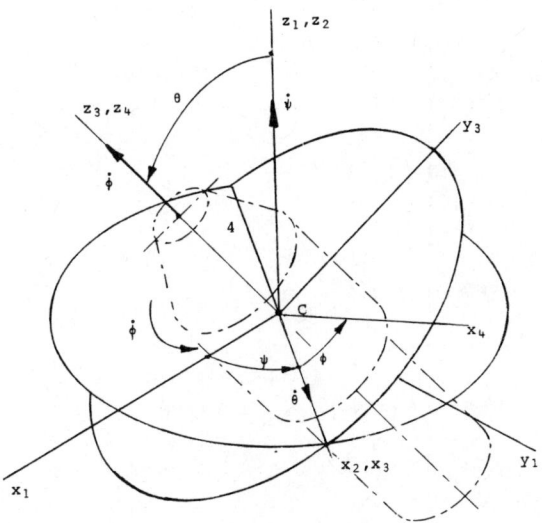

Fig. 4.30a.

The component angular velocity $\dot{\psi}$ is the angular velocity of precession. The component angular velocity $\dot{\theta}$ is the angular velocity of nutation and the x_2 axis is the line of nodes. The z_3 axis is the spin axis, the angular velocity of spin being $\dot{\psi}\cos\theta + \dot{\phi}$.

The angular velocity vector for body 4 measured in frame 1 and referred to frame 3 is given by

$$\{\omega_4\}_{1/3} = \{\omega_2\}_{1/3} + \{\omega_3\}_{2/3} + \{\omega_4\}_{3/3}$$

$$= [\ell_2]_3\{\omega_2\}_{1/2} + \{\omega_3\}_{2/2} + \{\omega_4\}_{3/3}$$

$$= \begin{bmatrix} 1 & 0 & 0 \\ 0 & c\theta & s\theta \\ 0 & -s\theta & c\theta \end{bmatrix}\begin{bmatrix} 0 \\ 0 \\ \dot{\psi} \end{bmatrix} + \begin{bmatrix} \dot{\theta} \\ 0 \\ 0 \end{bmatrix} + \begin{bmatrix} 0 \\ 0 \\ \dot{\phi} \end{bmatrix}$$

$$= \begin{bmatrix} \dot{\theta} \\ \dot{\psi}\sin\theta \\ \dot{\psi}\cos\theta + \dot{\phi} \end{bmatrix} \qquad (1)$$

Also, if the angular velocity vector for body 4 measured in frame 1 and referred to frame 4 is given by

$$\{\omega_4\}_{1/4} = \begin{bmatrix} \omega_x \\ \omega_y \\ \omega_z \end{bmatrix} \qquad (2)$$

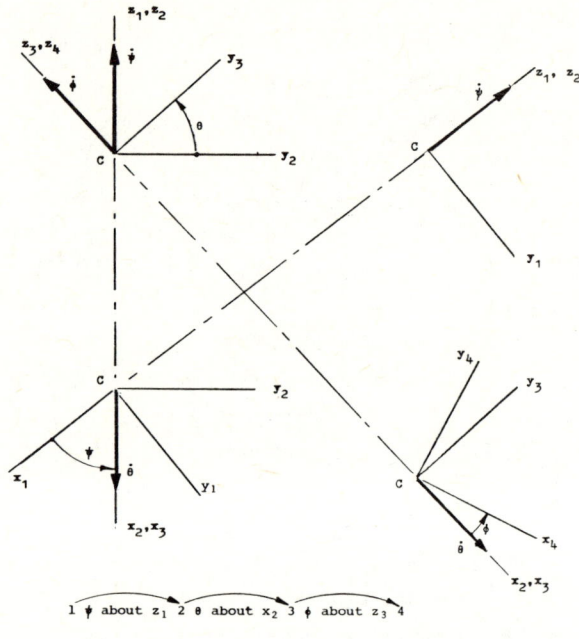

1 ψ about z_1 2 θ about x_2 3 ϕ about z_3 4

Fig. 4.30b.

then

$$\{\omega_4\}_{1/3} = [\ell_4]_3\{\omega_4\}_{1/4} = \begin{bmatrix} c\phi & -s\phi & 0 \\ s\phi & c\phi & 0 \\ 0 & 0 & 1 \end{bmatrix}\begin{bmatrix} \omega_x \\ \omega_y \\ \omega_z \end{bmatrix}$$

$$= \begin{bmatrix} \omega_x \cos\phi - \omega_y \sin\phi \\ \omega_x \sin\phi + \omega_y \cos\phi \\ \omega_z \end{bmatrix} \quad (3)$$

and

$$\frac{d}{dt}\{\omega_4\}_{1/4} = \{\dot{\omega}_4\}_{1/4} = \begin{bmatrix} \dot{\omega}_x \\ \dot{\omega}_y \\ \dot{\omega}_z \end{bmatrix} \quad (4)$$

The angular momentum and rate of change of angular momentum of body 4 measured in frame 1 and referred to frame 4 are, respectively

$$\{H_4\}_{1/4} = [I_4]_{4/4}\{\omega_4\}_{1/4} = \begin{bmatrix} I & 0 & 0 \\ 0 & I & 0 \\ 0 & 0 & J \end{bmatrix}\begin{bmatrix} \omega_x \\ \omega_y \\ \omega_z \end{bmatrix} = \begin{bmatrix} I\omega_x \\ I\omega_y \\ J\omega_z \end{bmatrix} \quad (5)$$

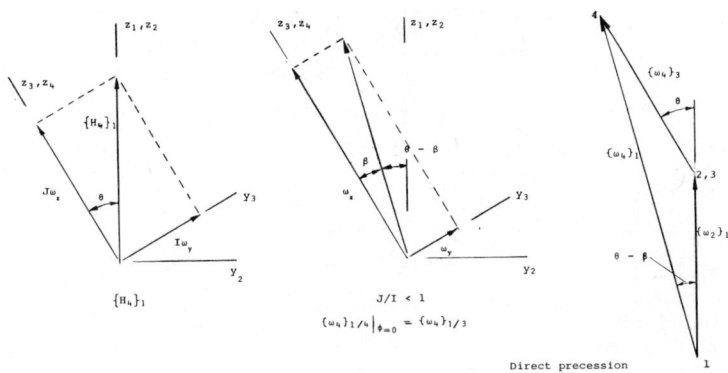

Fig. 4.30c.

and

$$\{\dot{H}_4\}_{1/4} = [\omega_4]_{1/4}[I_4]_{4/4}\{\omega_4\}_{1/4} + [I_4]_{4/4}\{\dot{\omega}_4\}_{1/4}$$

$$= \begin{bmatrix} I\dot{\omega}_x - (I - J)\omega_y\omega_z \\ I\dot{\omega}_y + (I - J)\omega_x\omega_z \\ J\dot{\omega}_z \end{bmatrix} . \quad (6)$$

The reader is invited to show that the result obtained from

$$\{\dot{H}_4\}_{1/4} = [\omega_4]_{1/4}\{H_4\}_{1/4} + \frac{d}{dt}\{H_4\}_{1/4}$$

is the same as that above.

Since there are no external couples on the body $\{\dot{H}_4\}_1$ is a null vector, one immediate result of which is that $\dot{\omega}_z = 0$. Also, the $\{H_4\}_1$ vector is of constant magnitude and fixed directionally in inertial space. As a matter of convenience, let the $\{H_4\}_1$ vector be directed along the z_1 axis as shown in Fig. 4.30c. If $\{H_4\}_1$ is fixed in inertial space then $\dot{\theta}$ must be zero, θ being given by

$$\tan\theta = I\omega_y/J\omega_z \quad (7)$$

The angular velocity vector $\{\omega_4\}_1$ is inclined at the angle β to the z_3 axis given by

$$\tan\beta = \omega_y/\omega_z \quad (8)$$

From the angular velocity vector diagram of Fig. 4.30d, since

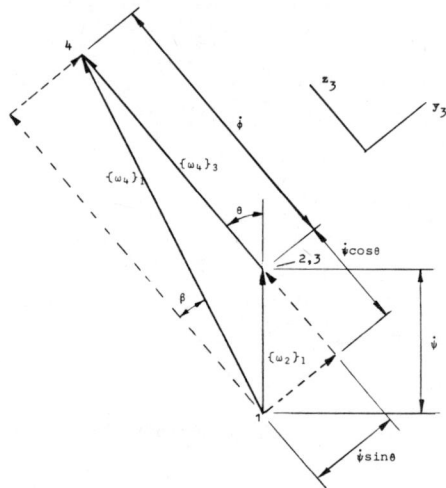

Fig. 4.30d.

$$\tan\beta = J/I\tan\theta , \qquad (9)$$

$$\frac{\dot{\psi}}{\dot{\psi}\cos\theta + \dot{\phi}} = \frac{J}{I\cos\theta}$$

and therefore

$$\dot{\psi} = \frac{J\dot{\phi}}{(I - J)\cos\theta} , \qquad (10)$$

or alternatively

$$\frac{\omega}{\sin\theta} = \frac{\dot{\phi}}{\sin(\theta - \beta)} = \frac{\dot{\psi}}{\sin\beta} \qquad (11)$$

where $|\omega_4|_1 = \omega$.

The angular velocity vector diagram of Fig. 4.30d can be related to the taper roller thrust bearing of Problem 3.14. The angular motion of an axially symmetric free body can thus be seen to be equivalent to rolling of a cone fixed in the body, the body cone, on a cone fixed in inertial space, the space cone, as shown in Fig. 4.30e.

To this point the body has been assumed to be rod-shaped and therefore such that $J/I < 1$. In this case $\beta < \theta$ and the precessional motion is said to be *direct*. The body and space cone configuration corresponding to
$$1 < J/I < 2$$
for a disc-shaped body is as shown in Fig. 3.40f. Here $\beta > \theta$ and the

Solution of Dynamics Problems

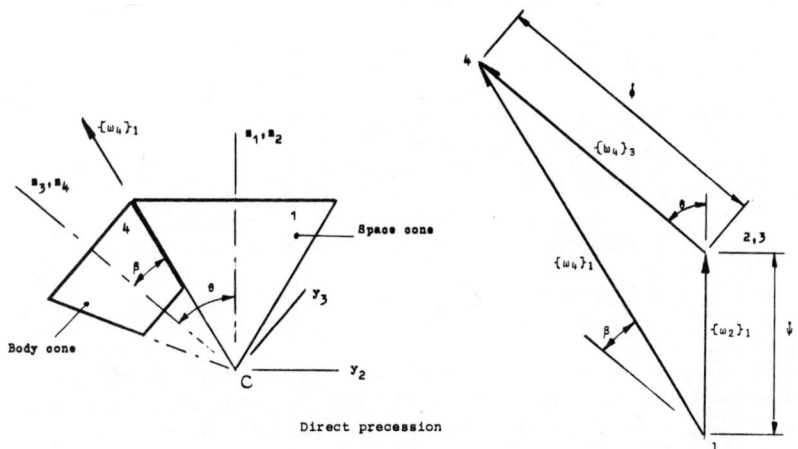

Fig. 4.30e.

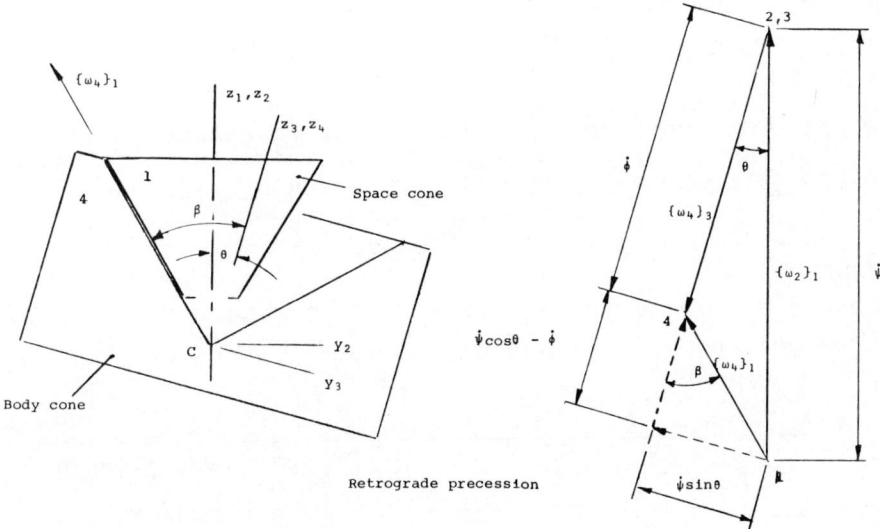

Fig. 4.30f.

precessional motion is said to be *retrograde*.

The motion can also be examined by effecting a direct solution of the equations of motion, Eqs. 6, which can be written

$$\dot{\omega}_x - A\omega_y = 0 \tag{12}$$

$$\dot{\omega}_y + A\omega_x = 0 \tag{13}$$

where $A = (I - J)\omega_z/J$, a constant. Multiplication of these equations by ω_x and ω_y respectively and adding gives

$$\omega_x \dot{\omega}_x + \omega_y \dot{\omega}_y = 0.$$

This equation integrates to

$$\omega_x^2 + \omega_y^2 = \text{constant} \tag{14}$$

and since ω_z is constant

$$\omega_x^2 + \omega_y^2 + \omega_z^2 = \text{constant} \tag{15}$$

which, as might be expected, indicates that the rotational kinetic energy is constant. The result obtained by differentiating Eq. 13 with respect to time, when substituted in Eq. 12 is

$$\ddot{\omega}_y + A^2 \omega_y = 0 \tag{16}$$

and therefore

$$\omega_y = P\cos At + Q\sin At \tag{17}$$

and

$$\omega_x = P\sin At + Q\cos At \tag{18}$$

where P and Q are constant of integration. To relate these equations for ω_y and ω_x to the previous result, the initial conditions must be selected to correspond to the axis system chosen. By Eqs. 1, 2 and 10

$$A = \frac{(I-J)}{J}\omega_z = \frac{(I-J)}{J}(\dot{\psi}\cos\theta + \dot{\phi}) = \dot{\phi}.$$

Also, when $\phi = 0$, $\omega_y = \dot{\psi}\sin\theta$ and $\omega_x = 0$, giving $P = \dot{\psi}\sin\theta$, $Q = 0$, $\omega_y = \dot{\psi}\sin\theta\cos\phi$ and $\omega_x = \dot{\psi}\sin\theta\sin\phi$.

Problem 4.31. A uniform rod, body 3, having a circular cross section of radius a and length b, is mounted in a frame, body 2, as shown in Fig. 4.31. The motion of body 3 relative to body 2 about the central pivot can be considered free from fictional constraint. The frame is driven at a constant rate Ω relative to an inertial body 1.

Find, for the case in which $b/a > \sqrt{3}$, the frequency of small oscillations of body 3 when it is positioned such that $\alpha = \pi/2$.

Also find for the case in which body 3 is released from rest relative to body 2 when $\alpha = 0$, $\dot{\alpha}$ when $\alpha = \pi/2$.

Solution of Dynamics Problems

Solution. Now

$$\{\omega_3\}_1 = \{\omega_2\}_1 + \{\omega_3\}_2$$

and therefore

$$\{\omega_3\}_{1/3} = [\ell_2]_3\{\omega_2\}_{1/2} + [\ell_2]_3\{\omega_3\}_{2/2}$$

$$= \begin{bmatrix} 1 & 0 & 0 \\ 0 & c\alpha & s\alpha \\ 0 & -s\alpha & c\alpha \end{bmatrix} \begin{bmatrix} \dot\alpha \\ \Omega \\ 0 \end{bmatrix} = \begin{bmatrix} \dot\alpha \\ \Omega\cos\alpha \\ -\Omega\sin\alpha \end{bmatrix} = \begin{bmatrix} \omega_x \\ \omega_y \\ \omega_z \end{bmatrix}$$

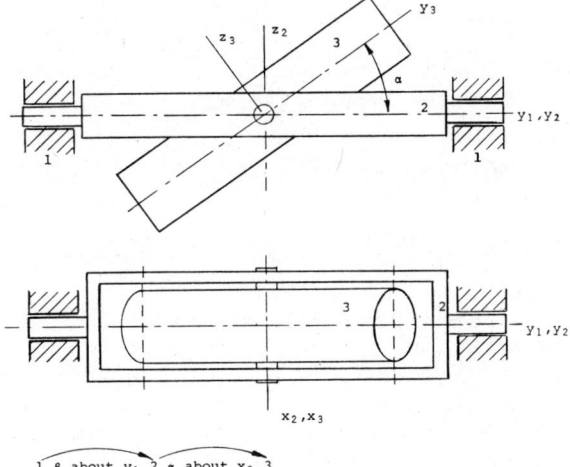

Fig. 4.31.

giving

$$\{\dot\omega_3\}_{1/3} = \frac{d}{dt}\{\omega_3\}_{1/3} = \begin{bmatrix} \ddot\alpha \\ -\dot\alpha\Omega\sin\alpha \\ -\dot\alpha\Omega\cos\alpha \end{bmatrix} = \begin{bmatrix} \dot\omega_x \\ \dot\omega_y \\ \dot\omega_z \end{bmatrix}.$$

The equation of motion for body 3 is

$$\{L_{32}\}_3 = \{\dot H_3\}_{1/3} = \lfloor\omega_3\rfloor_{1/3}[I_3]_{3/3}\{\omega_3\}_{1/3} + [I_3]_{3/3}\{\dot\omega_3\}_{1/3}$$

$$\begin{bmatrix} 0 \\ L_y \\ L_z \end{bmatrix} = \begin{bmatrix} (I - J)\omega_y\omega_z + I\dot\omega_x \\ J\dot\omega_z \\ -(I - J)\omega_y\omega_x + I\dot\omega_z \end{bmatrix}$$

where

$\{L_{32}\}_3$ is the couple which body 2 exerts on body 3 referred to frame,

$$[I_3]_{3\,3} = \begin{bmatrix} I & 0 & 0 \\ 0 & J & 0 \\ 0 & 0 & I \end{bmatrix}, \quad I = m_3\{(b^2/12) + (a^2/4)\}$$

and

$$J = m_3 a^2/2.$$

The x component equation of motion can be written

$$I\ddot{\alpha} - (I - J)\Omega^2 \sin\alpha\cos\alpha = 0.$$

To consider small motions of body 3 relative to the position in which $\alpha = \alpha_o$, the $\sin\alpha\cos\alpha$ product can be written

$$\sin(\alpha_o + \alpha)\cos(\alpha_o + \alpha) = \tfrac{1}{2}\sin 2(\alpha_o + \alpha)$$

where α is now redefined as a small displacement from the α_o position. If $\alpha_o = \pi/2$, then

$$\tfrac{1}{2}\sin 2(\alpha_o + \alpha) = \tfrac{1}{2}\sin(\pi + 2\alpha) = -\tfrac{1}{2}\sin 2\alpha \approx -\alpha$$

and the equation of motion for small movements relative to the $\alpha_o = \pi/2$ position can be written

$$\ddot{\alpha} + \frac{(I - J)\Omega^2}{I}\alpha = 0$$

so that the natural frequency of small vibrations about this position is

$$\omega_n = \Omega\sqrt{(I - J)/I}$$

provided

$$I > J \text{ or } \frac{b^2}{12} + \frac{a^2}{4} > \frac{a^2}{2}, \text{ i.e. } \frac{b}{a} > \sqrt{3}.$$

Now

$$\ddot{\alpha} = \frac{(I - J)\Omega^2}{I}\sin\alpha\cos\alpha \quad \text{or} \quad \dot{\alpha}\frac{d\dot{\alpha}}{d\alpha} = \omega_n^2 \sin\alpha\cos\alpha$$

and

$$\int_0^{\dot{\alpha}} \dot{\alpha}\,d\dot{\alpha} = \omega_n^2 \int_0^{\pi/2} \sin\alpha\,d(\sin\alpha)$$

$$\dot{\alpha}^2 = \omega_n^2 (\sin^2\alpha)\Big|_0^{\pi/2}$$

$$\dot{\alpha} = \omega_n.$$

Solution of Dynamics Problems

It is suggested that the reader finds

$$\{L_{32}\}_3, \{L_{32}\}_2, \{L_{32}\}_1, \{L_{21}\}_2 \text{ and } \{L_{21}\}_1$$

for the case in which $\alpha_o = \pi/2$ and a steady state vibration

$$\alpha = A\sin\omega_n t$$

is taking place.

Problem 4.32. A uniform ring is pivoted to a shaft which rotates in fixed bearings as shown in Fig. 4.32a. The motion of the ring about its pivot relative to the shaft is controlled by a spring which exerts no torque on the ring when the axis of the ring is at 70° to the axis of the shaft.

Determine the stiffness of the spring if the axis of the ring is to be 30° to the axis of the shaft when the shaft speed is 200 rad/s. What will be the attitude of the ring when the shaft speed is 100 rad/s?

The ring, which is of brass (density 8 200 kg/m^3), has the following dimensions: external diameter 7 cm, internal diameter 6 cm and length 1 cm.

Solution.

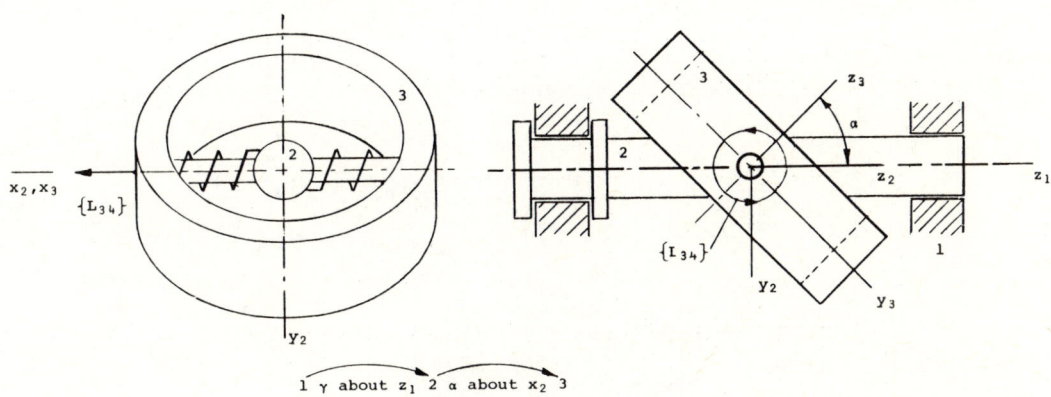

Fig. 4.32a.

The reader is left to show that

$$m_3 = 0.0837 \text{ kg}$$

and

$$[I_3]_{3/3} = \begin{bmatrix} I & 0 & 0 \\ 0 & I & 0 \\ 0 & 0 & J \end{bmatrix} = 10^{-6} \begin{bmatrix} 45 & 0 & 0 \\ 0 & 45 & 0 \\ 0 & 0 & 89 \end{bmatrix} \text{kg m}^2 .$$

If the spring exerts no torque on the ring when the axis of the ring is at 70° to the shaft axis then it will exert a torque about the x_2 axis equal to

$$k\theta/57.3$$

when the ring turns through the angle θ (in degrees) as indicated in Fig. 4.32b, where k is the spring stiffness.

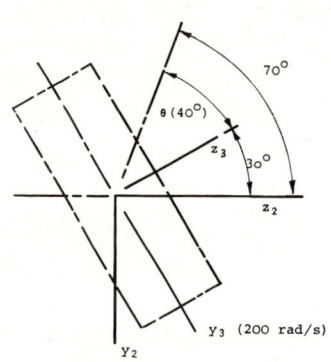

Fig. 4.32b.

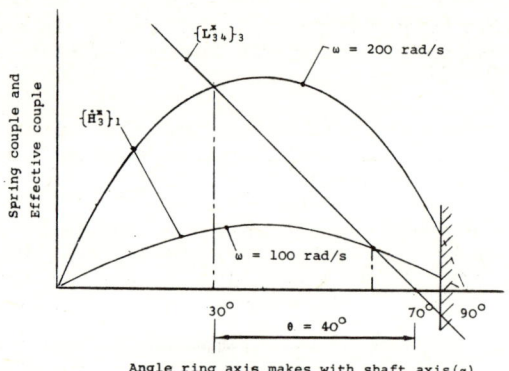

Fig. 4.32c

When the shaft is running at a constant speed and the ring is at rest relative to the shaft

$$\{\omega_2\}_{1/1} = \{\omega_3\}_{1/1} = \{\omega_3\}_{1/2} = \begin{bmatrix} 0 \\ 0 \\ \omega \end{bmatrix},$$

$$\{\omega_3\}_{1/3} = [\ell_2]_3 \{\omega_3\}_{1/2} = \begin{bmatrix} 0 \\ \omega \sin\alpha \\ \omega \cos\alpha \end{bmatrix}$$

and $\{\dot{\omega}_3\}_{1/1}$ is a null matrix. The equation of motion for the ring

$$\{L_{34}\}_3 = \{\dot{H}_3\}_{1/3} = [\omega_3]_{1/3} [I_3]_{3/3} \{\omega_3\}_{1/3}$$

thus becomes

$$\frac{k\theta}{57.3} \begin{bmatrix} 1 \\ 0 \\ 0 \end{bmatrix} = \omega^2 (J - I) \sin\alpha \cos\alpha \begin{bmatrix} 1 \\ 0 \\ 0 \end{bmatrix} .$$

Solution of Dynamics Problems

This equation is illustrated in Fig. 4.32c. On substitution of the values appropriate to a speed of 200 rad/s, the equation becomes

$$\frac{40k}{57.3} = 4 \times 10^4 \times 10^{-6} (89 - 45) \times 0.5 \times 0.866$$

giving
$$k = 1.09 \text{ N m/rad}.$$

When $\omega = 100$ rad/s

$$\frac{1.09\theta}{57.3} = \frac{10^4 \times 10^{-6} (89 - 45) \sin(140 - 2\theta)}{2}.$$

By trial and error $\theta \simeq 10°$.

The reader is invited to determine the natural frequency of the ring at each speed. Reference to Problem 4.31 will help in an approach to this problem.

Problem 4.33. Figure 4.33a shows a part section of a ball thrust race. Determine the angular velocity of the upper track relative to the lower fixed track when there is slip between the ball and the track at each point of contact.

Solution.

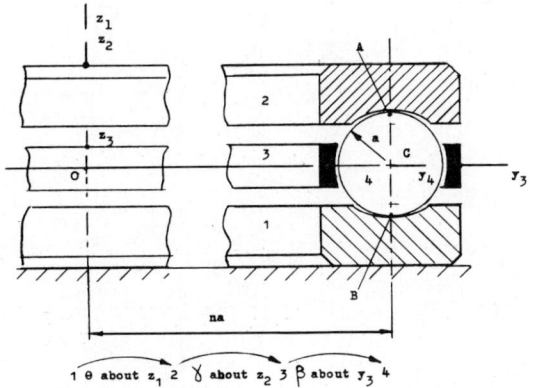

Fig. 4.33a.

From Problem 3.13

$$\{A_4\}_{1/3} = \frac{na\omega^2}{4} \begin{bmatrix} 0 \\ -1 \\ 0 \end{bmatrix} \quad \text{and} \quad \{\dot{\omega}_4\}_{1/3} = \frac{n\omega^2}{4} \begin{bmatrix} 1 \\ 0 \\ 0 \end{bmatrix}.$$

Now

$$\{\dot{H}_4\}_{1/3} = [I_4]_{4/3}\{\dot{\omega}_4\}_{1/3} = \frac{2ma^2}{5} \frac{n\omega^2}{4} \begin{bmatrix} 1 & 0 & 0 \\ 0 & 1 & 0 \\ 0 & 0 & 1 \end{bmatrix} \begin{bmatrix} 1 \\ 0 \\ 0 \end{bmatrix}$$

$$= \frac{mna^2\omega^2}{10} \begin{bmatrix} 1 \\ 0 \\ 0 \end{bmatrix} .$$

Taking moments about B, by reference to Fig. 4.33b,

$$[R_{AB}]_{3/3}\{F_{42}\}_3 = m[R_{CB}]_{3/3}\{A_4\}_{1/3} + \{\dot{H}_4\}_{1/3} .$$

$$2a \begin{bmatrix} 0 & -1 & 0 \\ 1 & 0 & 0 \\ 0 & 0 & 0 \end{bmatrix} \begin{bmatrix} F_1 \\ F_2 \\ F_3 \end{bmatrix} = \frac{mna^2\omega^2}{4} \begin{bmatrix} 0 & -1 & 0 \\ 1 & 0 & 0 \\ 0 & 0 & 0 \end{bmatrix} \begin{bmatrix} 0 \\ -1 \\ 0 \end{bmatrix} + \frac{mna^2\omega^2}{10} \begin{bmatrix} 1 \\ 0 \\ 0 \end{bmatrix}$$

from which

$$F_2 = -\frac{7mna\omega^2}{40} .$$

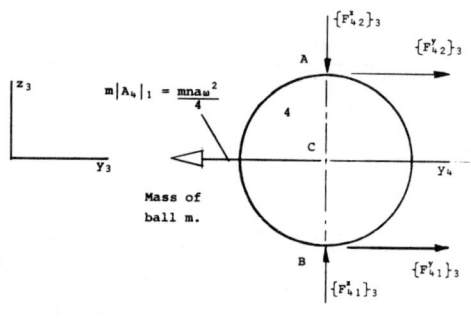

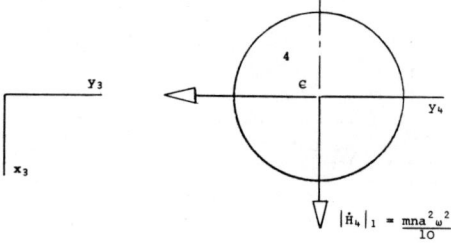

Fig. 4.33b.

Solution of Dynamics Problems

Equating applied and effective forces

$$\{F_{42}\}_3 + \{F_{41}\}_3 = m\{A_4\}_{1/3}, \begin{bmatrix} F_1 \\ F_2 \\ F_3 \end{bmatrix} + \begin{bmatrix} F_4 \\ F_5 \\ F_6 \end{bmatrix} = \frac{mna\omega^2}{4} \begin{bmatrix} 0 \\ -1 \\ 0 \end{bmatrix}$$

from which

$$F_5 = -\frac{3mna\omega^2}{40}.$$

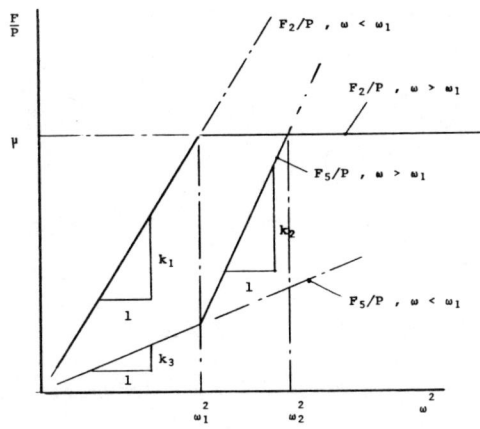

Fig. 4.33c.

If P is the axial load on each ball then

$$\frac{F_2}{P} = \frac{7mna\omega^2}{40P} = k_1\omega^2 \quad \text{and} \quad \frac{F_5}{P} = \frac{3mna\omega^2}{40P} = k_3\omega^2.$$

When slip is about to occur, first at A,

$$\frac{F_2}{P} = \mu = k_1\omega_1^2$$

as shown in Fig. 4.33c. In general

$$F_2 + F_5 = \frac{mna\omega^2}{4}$$

and when $\omega > \omega_1$, $F_2 = \mu P$. In this case

$$\mu P + F_5 = \frac{mna\omega^2}{4}$$

or

$$\frac{F_5}{P} = \frac{mna\omega^2}{4P} - \mu = k_2\omega^2 - \mu.$$

When slip occurs at B

$$\frac{F_5}{P} = \mu \quad \text{and} \quad 2\mu = k_2 \omega_2^2$$

giving

$$\omega_2^2 = \frac{8\mu P}{mna} .$$

Problem 4.34. A rotor, body 4, is mounted in gimbals as shown in Fig. 4.34. The inertia matrices for the elements are given by

$$[I_2]_{2/2} = \begin{bmatrix} A_2 & 0 & 0 \\ 0 & B_2 & 0 \\ 0 & 0 & C_2 \end{bmatrix}, [I_3]_{3/3} = \begin{bmatrix} A_3 & 0 & 0 \\ 0 & B_3 & 0 \\ 0 & 0 & C_3 \end{bmatrix}$$

and

$$[I_4]_{4/4} = \begin{bmatrix} J & 0 & 0 \\ 0 & I & 0 \\ 0 & 0 & I \end{bmatrix} .$$

Obtain expressions for γ and θ for the case in which the system moves from rest when subjected to each of the following disturbances

(i) a couple

$$\{L_2\}_2 = \begin{bmatrix} 0 \\ 0 \\ \ell \end{bmatrix}$$

applied suddenly and

(ii) a small mass m is placed on the inner gimbal at B such that

$$\{R_{BC}\}_{3/3} = \begin{bmatrix} a \\ 0 \\ b \end{bmatrix} .$$

Assume that friction effects can be neglected, the rotor runs at a constant high speed ω, the motion of the inner gimbal about the y_2 axis is small, the angular velocity products due to $\dot{\gamma}$ and $\dot{\theta}$ can be neglected and the mass acclerations due to m can be similarly neglected.

Solution. Angular velocities and accelerations are given by

$$\{\omega_2\}_{1/2} = \begin{bmatrix} 0 \\ 0 \\ \dot{\gamma} \end{bmatrix}, \{\dot{\omega}_2\}_{1/2} = \begin{bmatrix} 0 \\ 0 \\ \ddot{\gamma} \end{bmatrix}$$

Solution of Dynamics Problems

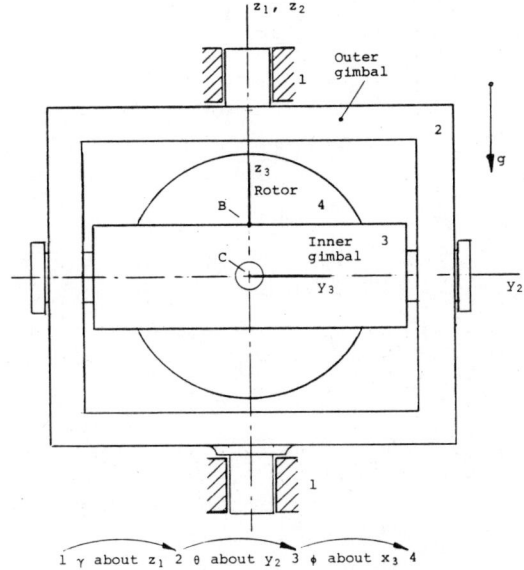

1 γ about z_1 2 θ about y_2 3 ϕ about x_3 4

Fig. 4.34.

$$\{\omega_3\}_{1/3} = \begin{bmatrix} -\dot{\gamma}\sin\theta \\ \dot{\theta} \\ \dot{\gamma}\cos\theta \end{bmatrix} = \begin{bmatrix} \omega_x \\ \omega_y \\ \omega_z \end{bmatrix} \;,$$

$$\{\dot{\omega}_3\}_{1/3} = \begin{bmatrix} \dot{\omega}_x \\ \dot{\omega}_y \\ \dot{\omega}_z \end{bmatrix} = \begin{bmatrix} -\ddot{\gamma}\sin\theta - \dot{\gamma}\dot{\theta}\cos\theta \\ \ddot{\theta} \\ \ddot{\gamma}\cos\theta - \dot{\gamma}\dot{\theta}\sin\theta \end{bmatrix} \;,$$

$$\{\omega_4\}_{1/3} = \begin{bmatrix} \omega_x + \omega \\ \omega_y \\ \omega_z \end{bmatrix} \text{ and } \{\dot{\omega}_4\}_{1/3} = \begin{bmatrix} \dot{\omega}_x \\ \dot{\omega}_y + \omega\omega_z \\ \dot{\omega}_z + \omega\omega_y \end{bmatrix} \;.$$

Rates of change of angular momenta are given by

$$\{\dot{H}_2\}_{1/2} = \begin{bmatrix} 0 \\ 0 \\ C_2\ddot{\gamma} \end{bmatrix}, \{\dot{H}_3\}_{1/3} = \begin{bmatrix} A_3\dot{\omega}_x + (C_3 - B_3)\omega_y\omega_z \\ B_3\dot{\omega}_y + (A_3 - C_3)\omega_x\omega_z \\ C_3\dot{\omega}_z + (B_3 - A_3)\omega_x\omega_y \end{bmatrix}$$

and

$$\{\dot{H}_4\}_{1/3} = \begin{bmatrix} J\dot{\omega}_x \\ I\dot{\omega}_y + h\omega_z + (J - I)\omega_x\omega_z \\ I\dot{\omega}_z - h\omega_y - (J - I)\omega_x\omega_y \end{bmatrix}$$

where $h = J\omega$. The reader is advised to justify these results.

Taking moments about C, for body 2,

$$\{L_2\}_2 + \{L_{23}\}_2 = \{\dot{H}_2\}_{1/2} \,, \qquad (1)$$

for body 3,

$$\{L_{32}\}_3 + \{L_{34}\}_3 = \{\dot{H}_3\}_{1/3} \qquad (2)$$

and for body 4

$$\{L_{43}\}_3 = \{\dot{H}_4\}_{1/3} \,. \qquad (3)$$

Combining Eqs. 2 and 3

$$\{L_{32}\}_3 = \{\dot{H}_4\}_{1/3} + \{\dot{H}_3\}_{1/3} \,. \qquad (4)$$

Combining Eqs. 4 and 1

$$\{L_2\}_2 = [\ell_3]_2 \left(\{\dot{H}_4\}_{1/3} + \{\dot{H}_3\}_{1/3} \right) + \{\dot{H}_2\}_{1/2}. \qquad (5)$$

On neglecting the terms involving products of $\dot{\gamma}$ and $\dot{\theta}$, the approximate equations of motion become, from Eq. 4

$$\begin{bmatrix} L_1 \\ L_2 \\ L_3 \end{bmatrix} = \begin{bmatrix} (J + A_3)\dot{\omega}_x \\ (I + B_3)\dot{\omega}_y + h\omega_z \\ (I + C_3)\dot{\omega}_z - h\omega_y \end{bmatrix} \qquad (6)$$

and from Eq. 5

$$\begin{bmatrix} L_4 \\ L_5 \\ L_6 \end{bmatrix} = \begin{bmatrix} c\theta & 0 & s\theta \\ 0 & 1 & 0 \\ -s\theta & 0 & c\theta \end{bmatrix} \begin{bmatrix} (J + A_3)\dot{\omega}_x \\ (I + B_3)\dot{\omega}_y + h\omega_z \\ (I + C_3)\dot{\omega}_z - h\omega_y \end{bmatrix} + \begin{bmatrix} 0 \\ 0 \\ C_2\ddot{\gamma} \end{bmatrix} \,. \qquad (7)$$

Taking the y equation from Eqs. 6 and the z equation from Eqs. 7, for θ small

$$L_2 = A\ddot{\theta} + h\dot{\gamma} \qquad (8)$$

$$L_6 = B\ddot{\gamma} - h\dot{\theta} \qquad (9)$$

where $A = I + B_3$ and $B = I + C_2 + C_3$.

For the case in which the system moves from rest when a couple of magnitude ℓ is applied to body 2, the Laplace transform of these equations, for zero initial conditions is

$$0 = As\Theta + h\Gamma$$

and

$$\frac{\ell}{s} = Bs^2\Gamma - hs\Theta$$

where the Laplace transform of $\{\theta(t)\} = \Theta(s) = \Theta$ and the Laplace transform of $\{\gamma(t)\} = \Gamma(s) = \Gamma$. Hence

$$\Gamma = \frac{\ell}{B} \frac{1}{s(s^2 + \omega_n^2)}$$

and
$$\theta = -\frac{h\ell}{AB}\frac{1}{s^2(s^2+\omega_n^2)}$$

where $\omega_n^2 = h^2/AB$. Therefore

$$\gamma = A\ell(1 - \cos\omega_n t)/h^2$$

and

$$\theta = -\ell(\omega_n t - \sin\omega_n t)/h\omega_n .$$

Since θ increases without limit, these solutions cease to be valid shortly after the motion starts.

For the case in which the system moves from rest after the mass m is placed on body 3 as described

$$\frac{mga}{s} = As^2\theta + hs\Gamma$$

and

$$0 = Bs\Gamma - h\theta .$$

Hence

$$\Gamma = \frac{mgah}{AB}\frac{1}{s^2(s^2+\omega_n^2)}$$

and

$$\theta = \frac{mgaB}{h}\frac{1}{s(s^2+\omega_n^2)} .$$

Therefore

$$\gamma = mga(\omega_n t - \sin\omega_n t)/h\omega_n$$

and

$$\theta = mgaB(1 - \cos\omega_n t)/h^2 .$$

Problem 4.35. Body 4 is in motion relative to an inertial reference, body 1. Vectors can be referred to either frame 2 or frame 3, so that

$$2T_{4\,rot} = \{\omega_4\}^T_{1/n} [I_4]_{4/n} \{\omega_4\}_{1/n}$$

and

$$2T_{4\,trans} = m_4 \{V_4\}^T_{1/n} \{V_4\}_{1/n}$$

where n is either 2 or 3. Show that

$$\begin{bmatrix} \partial T_4 \text{rot}/\partial \omega_x \\ \partial T_4 \text{rot}/\partial \omega_y \\ \partial T_4 \text{rot}/\partial \omega_z \end{bmatrix} = \{\nabla_\omega\}_n T_4 \text{rot} = \{H_4\}_{1/n}$$

and

$$\begin{bmatrix} \partial T_4 \text{trans}/\partial v_x \\ \partial T_4 \text{trans}/\partial v_y \\ \partial T_4 \text{trans}/\partial v_z \end{bmatrix} = \{\nabla_v\}_n T_4 \text{trans} = \{G_4\}_{1/n}$$

where

$$\{\omega_4\}_{1/n} = \begin{bmatrix} \omega_x \\ \omega_y \\ \omega_z \end{bmatrix} \quad \text{and} \quad \{V_4\}_{1/n} = \begin{bmatrix} v_x \\ v_y \\ v_z \end{bmatrix}.$$

In the system of Fig. 4.35 body 2 turns at a constant rate Ω_2 relative to body 1 and body 5 turns at a constant rate Ω_5 relative to body 1. Body 4 rolls without slip on body 5. Obtain expressions for

$$\{H_4\}_{1/3}, \quad \{G_4\}_{1/3}, \quad \{H_4\}_{1/2} \quad \text{and} \quad \{G_4\}_{1/2}$$

by determining

$$\{\nabla_\omega\}_n T_4 \text{rot} \quad \text{and} \quad \{\nabla_v\}_n T_4 \text{trans}.$$

Also show that

$$\{\nabla_\omega\}_3 (T_4 \text{rot} + T_4 \text{trans}) = \{H_{40}\}_{1/3}.$$

Solution. If

$$[I_4]_{4/n} = \begin{bmatrix} A & D & E \\ D & B & F \\ E & F & C \end{bmatrix}$$

then

$$2T_4 \text{rot} = \begin{bmatrix} \omega_x & \omega_y & \omega_z \end{bmatrix} \begin{bmatrix} A & D & E \\ D & B & F \\ E & F & C \end{bmatrix} \begin{bmatrix} \omega_x \\ \omega_y \\ \omega_z \end{bmatrix}$$

$$= A\omega_x^2 + B\omega_y^2 + C\omega_z^2 + 2(D\omega_x\omega_y + E\omega_x\omega_z + F\omega_y\omega_z)$$

and

Solution of Dynamics Problems

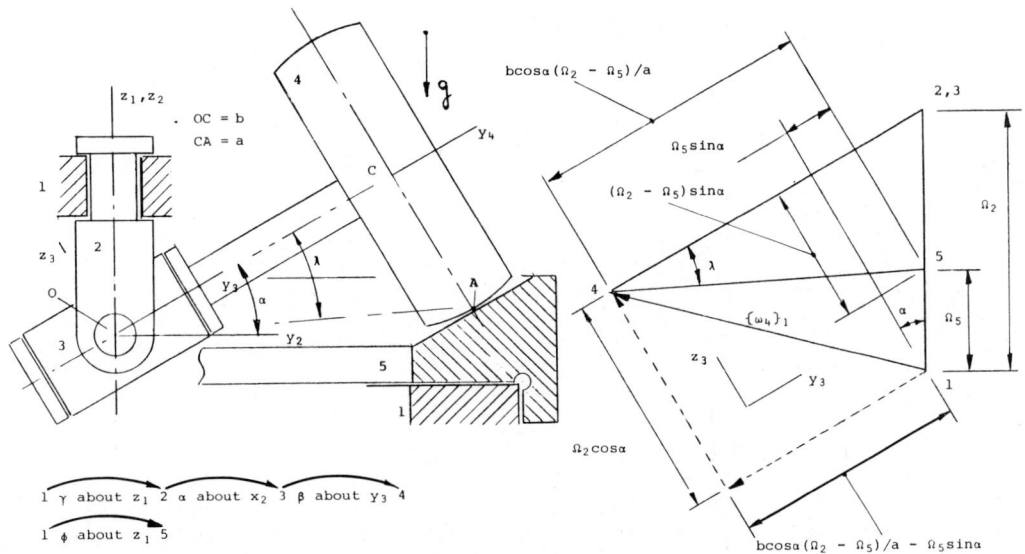

Fig. 4.35.

$$\begin{bmatrix} \partial T_4 rot/\partial \omega_x \\ \partial T_4 rot/\partial \omega_y \\ \partial T_4 rot/\partial \omega_z \end{bmatrix} = \begin{bmatrix} A\omega_x + D\omega_y + E\omega_z \\ D\omega_x + B\omega_y + F\omega_z \\ E\omega_x + F\omega_y + C\omega_z \end{bmatrix} = \{H_4\}_{1/n} \ .$$

Also

$$2T_4 trans = m_4 \begin{bmatrix} v_x & + v_y & + v_z \end{bmatrix} \begin{bmatrix} v_x \\ v_y \\ v_z \end{bmatrix} = m_4(v_x^2 + v_y^2 + v_z^2)$$

and

$$\begin{bmatrix} \partial T_4 trans/\partial v_x \\ \partial T_4 trans/\partial v_y \\ \partial T_4 trans/\partial v_z \end{bmatrix} = m_4 \begin{bmatrix} v_x \\ v_y \\ v_z \end{bmatrix} = \{G_4\}_{1/n} \ .$$

From Fig. 4.35

$$\{\omega_4\}_{1/3} = \begin{bmatrix} 0 \\ -\{b\cos\alpha(\Omega_2 - \Omega_5)/a - \Omega_5 \sin\alpha\} \\ \Omega_2 \cos\alpha \end{bmatrix} = \begin{bmatrix} 0 \\ \omega_y \\ \omega_z \end{bmatrix}$$

and

$$\{\omega_4\}_{1/2} = [\ell_3]_2 \{\omega_4\}_{1/3} = \begin{bmatrix} 0 \\ \omega_y \cos\alpha - \omega_z \sin\alpha \\ \omega_y \sin\alpha + \omega_z \cos\alpha \end{bmatrix} = \begin{bmatrix} 0 \\ \Omega_y \\ \Omega_z \end{bmatrix} \ .$$

Also
$$\{V_4\}_{1/3} = b\Omega_2\cos\alpha \begin{bmatrix} -1 \\ 0 \\ 0 \end{bmatrix} = \begin{bmatrix} v_x \\ 0 \\ 0 \end{bmatrix} = \begin{bmatrix} -b\omega_z \\ 0 \\ 0 \end{bmatrix}$$

Hence
$$2T_{4\,rot} = \{\omega_4\}_{1/3}^T [I_4]_{4/3} \{\omega_4\}_{1/3}$$

$$= \begin{bmatrix} 0 & \omega_y & \omega_z \end{bmatrix} \begin{bmatrix} I & 0 & 0 \\ 0 & J & 0 \\ 0 & 0 & I \end{bmatrix} \begin{bmatrix} 0 \\ \omega_y \\ \omega_z \end{bmatrix}$$

$$= J\omega_y^2 + I\omega_z^2$$

and
$$\{\nabla_\omega\}_3 T_{4\,rot} = \{H_4\}_{1/3} \ .$$

Also
$$2T_{4\,trans} = m_4 \{V_4\}_{1/3}^T \{V_4\}_{1/3} = m_4 v_x^2$$

and
$$\{\nabla_v\}_3 T_{4\,trans} = \{G_4\}_{1/3} \ .$$

Now
$$2T_{4\,rot} = \{\omega_4\}_{1/2} [I_4]_{4/2} \{\omega_4\}_{1/2}$$

where
$$[I_4]_{4/2} = [\ell_3]_2 [I_4]_{4/3} [\ell_3]_2^T = \begin{bmatrix} I & 0 & 0 \\ 0 & B & F \\ 0 & F & C \end{bmatrix} \ ,$$

$$B = J\cos^2\alpha + I\sin^2\alpha \ , \quad F = (J - I)\sin\alpha\cos\alpha$$

and
$$C = J\sin^2\alpha + I\cos^2\alpha \ .$$

Hence
$$2T_{4\,rot} = B\Omega_y^2 + C\Omega_z^2 + 2F\Omega_y\Omega_z$$

and
$$\{\nabla_\omega\}_2 T_{4\,rot} = \begin{bmatrix} 0 \\ B\Omega_y + F\Omega_z \\ F\Omega_y + C\Omega_z \end{bmatrix} = \begin{bmatrix} 0 \\ h_y \\ h_z \end{bmatrix} = \{H_4\}_{1/2}$$

Solution of Dynamics Problems

where
$$h_y = J\omega_y \cos\alpha - I\omega_z \sin\alpha$$
and
$$h_z = J\omega_y \sin\alpha + I\omega_z \cos\alpha \ .$$

The total kinetic energy of body 4 is given by

$$2(T_{4rot} + T_{4trans}) = \{\omega_4\}_{1/3}^T \lfloor I_4 \rfloor_{4/3} \{\omega_4\}_{1/3}$$
$$+ m_4 \{V_4\}_{1/3}^T \{V_4\}_{1/3}$$
$$= J\omega_y^2 + I\omega_z^2 + m_4 b^2 \omega_z^2$$

and

$$\{V_\omega\}_3 T_4 = \begin{bmatrix} 0 \\ J\omega_y \\ (I + m_4 b^2)\omega_z \end{bmatrix} = \{H_{40}\}_{1/3} \ .$$

Problem 4.36. Determine, for the system of Fig. 4.35,

$$\{F_{43}\}_3 \ , \ \{F_{45}\}_3 \ \text{ and } \ \{L_{43}\}_3$$

Assume that $\{F_{45}^y\}_3$, $\{L_{43}^x\}_3$ and $\{L_{43}^y\}_3$ are zero and the centre of mass of body 4 is at C.

Solution. Now

$$\{H_{40}\}_{1/3} = \{H_4\}_{1/3} + m_4 [R_{CO}]_{3/3} \{V_4\}_{1/3}$$
$$= \{H_4\}_{1/3} + m_4 [R_{CO}]_{3/3} \lfloor \omega_4 \rfloor_{1/3} \{R_{CO}\}_{3/3}$$
$$= \left[[I_4]_{4/3} - m_4 [R_{CO}]_{3/3}^2 \right] \{\omega_4\}_{1/3}$$
$$= [I_4]_{3/3} \{\omega_4\}_{1/3}$$
$$= \begin{bmatrix} I & 0 & 0 \\ 0 & J & 0 \\ 0 & 0 & I + m_4 b^2 \end{bmatrix} \begin{bmatrix} 0 \\ \omega_y \\ \omega_z \end{bmatrix} = \begin{bmatrix} 0 \\ J\omega_y \\ (I + m_4 b^2)\omega_z \end{bmatrix} = \begin{bmatrix} 0 \\ h_y \\ h_z \end{bmatrix}$$

which confirms the result of Problem 4.35. Hence

$$\{\dot{H}_{40}\}_{1/3} = \lfloor \omega_3 \rfloor_{1/3} \{H_{40}\}_{1/3}$$
$$= \Omega_2 \begin{bmatrix} 0 & -c\alpha & s\alpha \\ c\alpha & 0 & 0 \\ -s\alpha & 0 & 0 \end{bmatrix} \begin{bmatrix} 0 \\ h_y \\ h_z \end{bmatrix} = \Omega_2 \begin{bmatrix} h_z \sin\alpha - h_y \cos\alpha \\ 0 \\ 0 \end{bmatrix}$$

since $d\{H_{40}\}_{1/3}/dt$ is a null matrix.

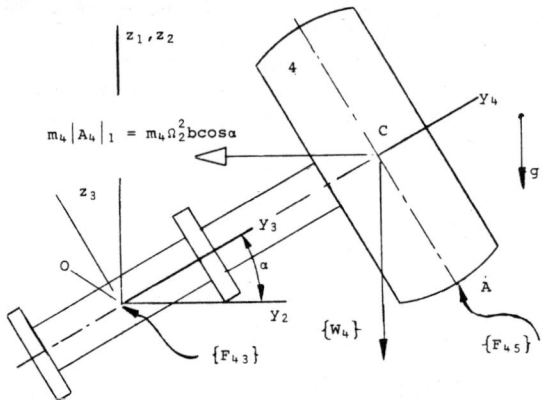

Fig. 4.36.

Taking moments about O for body 4, by reference to Fig. 4.36,

$$\{L_{43}\}_3 + [R_{AO}]_{3/3}\{F_{45}\}_3 + [R_{CO}]_{3/3}\{W_4\}_3 = \{\dot{H}_{40}\}_{1/3}$$

$$\begin{bmatrix} 0 \\ 0 \\ L_2 \end{bmatrix} + \begin{bmatrix} 0 & a & b \\ -a & 0 & 0 \\ -b & 0 & 0 \end{bmatrix} \begin{bmatrix} F_1 \\ 0 \\ F_2 \end{bmatrix} + m_4 g \begin{bmatrix} 0 & 0 & b \\ 0 & 0 & 0 \\ -b & 0 & 0 \end{bmatrix} \begin{bmatrix} 0 \\ -s\alpha \\ -c\alpha \end{bmatrix}$$

$$= \Omega_2 \begin{bmatrix} h_z \sin\alpha - h_y \cos\alpha \\ 0 \\ 0 \end{bmatrix} .$$

From the x component equation

$$F_2 = m_4 g \cos\alpha + \Omega_2 \{J\omega_y \sin\alpha - (I + m_4 b^2)\omega_z \cos\alpha\}/b$$

and from the y component equation $F_1 = 0$. L_2 is thus also zero. Equating applied and effective forces for body 4

$$\{F_{43}\}_3 + \{F_{45}\}_3 = m_4 \{A_4\}_{1/3}$$

$$\begin{bmatrix} F_3 \\ F_4 \\ F_5 \end{bmatrix} + \begin{bmatrix} 0 \\ 0 \\ F_2 \end{bmatrix} = m_4 \Omega_2^2 b \cos\alpha \begin{bmatrix} 0 \\ -\cos\alpha \\ \sin\alpha \end{bmatrix}$$

and therefore $F_3 = 0$, $F_4 = m_4 \Omega_2^2 b \cos^2\alpha$ and

$$F_5 = m_4 \Omega_2^2 b \sin\alpha \cos\alpha - F_2 .$$

Problem 4.37. A thin uniform straight rod, body 3, of length a and mass m, rests with ine end on a smooth vertical surface while the other end is retained in a smooth spherical bearing which is a fixed distance d (< a) from the vertical surface as shown in Fig. 4.37a. The rod is released from rest in the postion where $\alpha = \theta$. Find the value of α at which contact between the vertical surface and the rod ceases.

Solution.

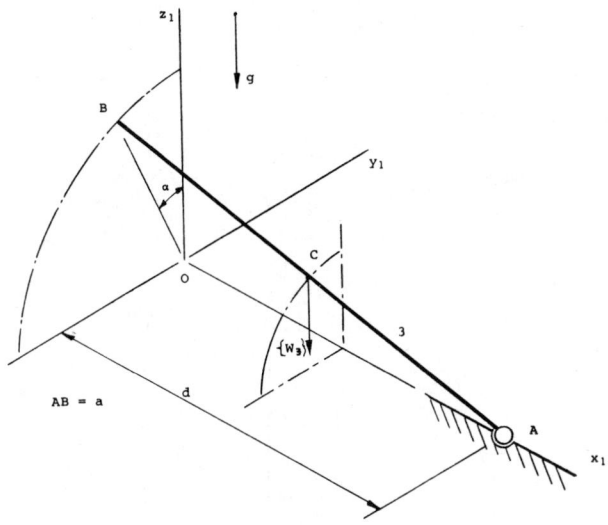

Fig. 4.37a.

Moments about A, by reference to Fig. 4.37b,

$$[R_{CA}]_{2/2}\{W_3\}_2 + [R_{BA}]_{2/2}\{F\}_2 = m[R_{CA}]_{2/2}\{A_3\}_{1/2}$$
$$+ [\ell_3]_2\{\dot{H}_3\}_{1/3} . \quad (1)$$

Evaluation of terms in Eq. 1, noting that β is constant,

$$\{R_{CA}\}_{1/1} = [\omega_2]_{1/2}\{R_{CA}\}_{2/2} ,$$

$$\{V_3\}_{1/2} = [\omega_2]_{1/2}\{R_{CA}\}_{2/2} = \begin{bmatrix} 0 & 0 & 0 \\ 0 & 0 & -\omega \\ 0 & \omega & 0 \end{bmatrix} \begin{bmatrix} -d/2 \\ 0 \\ r/2 \end{bmatrix} = -\frac{\omega r}{2} \begin{bmatrix} 0 \\ 1 \\ 0 \end{bmatrix}$$

where $\omega = \dot{\alpha}$,

$$\{A_3\}_{1/2} = [\omega_2]_{1/2}\{V_3\}_{1/2} + \frac{d}{dt}\{V_3\}_{1/2}$$

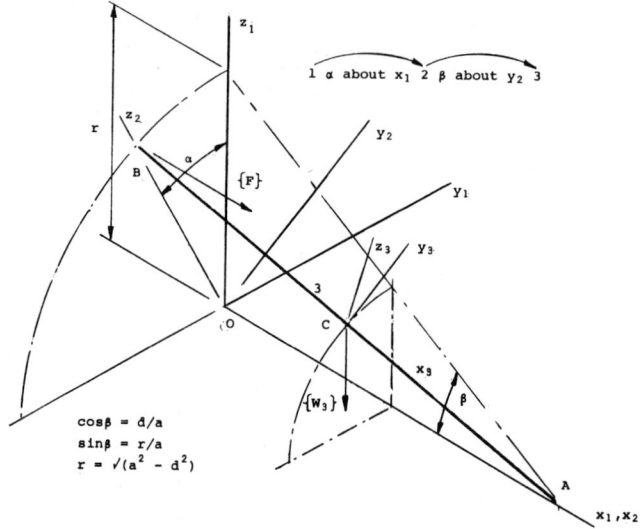

Fig. 4.37b.

$$= -\frac{\omega^2 r}{2}\begin{bmatrix} 0 & 0 & 0 \\ 0 & 0 & -1 \\ 0 & 1 & 0 \end{bmatrix}\begin{bmatrix} 0 \\ 1 \\ 0 \end{bmatrix} - \frac{\dot\omega r}{2}\begin{bmatrix} 0 \\ 1 \\ 0 \end{bmatrix} = -\frac{r}{2}\begin{bmatrix} 0 \\ \dot\omega \\ \omega^2 \end{bmatrix},$$

$$\{\omega_3\}_{1/3} = \{\omega_2\}_{1/2} = [\ell_2]_3\{\omega_2\}_{1/2}$$

$$= \omega\begin{bmatrix} d/a & 0 & -r/a \\ 0 & 1 & 0 \\ r/a & 0 & d/a \end{bmatrix}\begin{bmatrix} 1 \\ 0 \\ 0 \end{bmatrix} = \frac{\omega}{a}\begin{bmatrix} d \\ 0 \\ r \end{bmatrix},$$

$$\{\dot\omega_3\}_{1/3} = \frac{d}{dt}\{\omega_3\}_{1/3} = \frac{\dot\omega}{a}\begin{bmatrix} d \\ 0 \\ r \end{bmatrix},$$

$$[R_{CA}]_{2/2}\{W_3\}_2 = \frac{mg}{2}\begin{bmatrix} 0 & -r & 0 \\ r & 0 & d \\ 0 & -d & 0 \end{bmatrix}\begin{bmatrix} 1 & 0 & 0 \\ 0 & c\alpha & s\alpha \\ 0 & -s\alpha & c\alpha \end{bmatrix}\begin{bmatrix} 0 \\ 0 \\ -1 \end{bmatrix}$$

$$= \frac{mg}{2}\begin{bmatrix} r\sin\alpha \\ -d\cos\alpha \\ d\sin\alpha \end{bmatrix},$$

Solution of Dynamics Problems

$$[R_{BA}]_{2/2}\{F\}_2 = \begin{bmatrix} 0 & -r & 0 \\ r & 0 & d \\ 0 & -d & 0 \end{bmatrix} \begin{bmatrix} F \\ 0 \\ 0 \end{bmatrix} = Fr \begin{bmatrix} 0 \\ 1 \\ 0 \end{bmatrix}$$

$$m[R_{CA}]_{2/2}\{A_3\}_{1/2} = -\frac{mr}{4} \begin{bmatrix} 0 & -r & 0 \\ r & 0 & d \\ 0 & -d & 0 \end{bmatrix} \begin{bmatrix} 0 \\ \dot{\omega} \\ \omega^2 \end{bmatrix} = \frac{mr}{4} \begin{bmatrix} \dot{\omega}r \\ -\omega^2 d \\ -\dot{\omega}d \end{bmatrix},$$

$$\{\dot{H}_3\}_{1/3} = [\omega_3]_{1/3}[I_3]_{3/3}\{\omega_3\}_{1/3} + [I_3]_{3/3}\{\dot{\omega}_3\}_{1/3}$$

$$= \frac{m\omega^2}{12} \begin{bmatrix} 0 & -r & 0 \\ r & 0 & -d \\ 0 & d & 0 \end{bmatrix} \begin{bmatrix} 0 & 0 & 0 \\ 0 & 1 & 0 \\ 0 & 0 & 1 \end{bmatrix} \begin{bmatrix} d \\ 0 \\ r \end{bmatrix}$$

$$+ \frac{ma\dot{\omega}}{12} \begin{bmatrix} 0 & 0 & 0 \\ 0 & 1 & 0 \\ 0 & 0 & 1 \end{bmatrix} \begin{bmatrix} d \\ 0 \\ r \end{bmatrix} = \frac{mr}{12} \begin{bmatrix} 0 \\ -\omega^2 d \\ \dot{\omega}a \end{bmatrix}$$

and

$$\{\dot{H}_3\}_{1/2} = [\ell_3]_2\{\dot{H}_3\}_{1/3} = \frac{mr}{12} \begin{bmatrix} d/a & 0 & r/a \\ 0 & 1 & 0 \\ -r/a & 0 & d/a \end{bmatrix} \begin{bmatrix} 0 \\ -\omega^2 d \\ \dot{\omega}a \end{bmatrix}$$

$$= \frac{mr}{2} \begin{bmatrix} \dot{\omega}r \\ -\omega^2 d \\ \dot{\omega}d \end{bmatrix} .$$

The moments equation can thus be written

$$\frac{mg}{2} \begin{bmatrix} r\sin\alpha \\ -d\cos\alpha \\ d\sin\alpha \end{bmatrix} + Fr \begin{bmatrix} 0 \\ 1 \\ 0 \end{bmatrix} = \frac{mr}{4} \begin{bmatrix} \dot{\omega}r \\ -\omega^2 d \\ -\dot{\omega}d \end{bmatrix} + \frac{mr}{12} \begin{bmatrix} \dot{\omega}r \\ -\omega^2 d \\ \dot{\omega}d \end{bmatrix} .$$

From the x component equation

$$\dot{\omega} = \frac{3g}{2r}\sin\alpha$$

and since $\omega = \dot{\alpha}$,

$$\ddot{\alpha} = \frac{3g}{2r}\sin\alpha \quad \text{or} \quad \dot{\alpha}\frac{d\dot{\alpha}}{d\alpha} = \frac{3g}{2r}\sin\alpha .$$

This equation integrates to give

$$\omega^2 = \frac{3g}{r}(\cos\theta - \cos\alpha)$$

since the initial conditions are $\alpha = \theta$ and $\dot{\alpha} = 0$.

From the y component equation

$$F = \frac{mgd\cos\alpha}{2r} - \frac{m\omega^2 d}{3}$$

$$= \frac{mgd}{2r}(3\cos\alpha - 2\cos\theta)$$

and therefore $F = 0$ when $\cos\alpha = (2\cos\theta)/3$.

Alternatively, the angular velocity of body 3 can be determined from considerations of energy. However, force determination requires additional effort. Now

$$T_3\big|_\alpha = \frac{m}{2}\{V_3\}_{1/2}^T\{V_3\}_{1/2} + \frac{1}{2}\{\omega_3\}_{1/3}^T[I_3]_{3/3}\{\omega_3\}_{1/3}$$

$$= \frac{m\omega^2 r^2}{8} + \frac{m\omega^2 r^2}{24} = \frac{m\omega^2 r^2}{6} \;.$$

Since energy is conserved

$$T_3\big|_{\alpha=\theta} + V_3\big|_{\alpha=\theta} = T_3\big|_\alpha + V_3\big|_\alpha$$

$$0 + \frac{mgr\cos\theta}{2} = \frac{m\omega^2 r^2}{6} + \frac{mgr\cos\alpha}{2}$$

or

$$\omega^2 = \frac{3g}{r}(\cos\theta - \cos\alpha) \;.$$

Problem 4.38. A massive uniform rod, body 2, which is a solid of revolution, is constrained to move such that one end of its axis of generation B traces out a straight vertical path, while the other end A traces out a straight horizontal path as shown in Fig. 4.38. The constraints are conservative and exert no moment about the axis of generation of the rod.

Find the initial motion of the rod if it falls from rest in the position shown. Neglect the mass of the constraining bodies 3 and 4.

Solution. Since the constraints are conservative

$$\{F_{31}^y\}_1 \quad \text{and} \quad \{F_{41}^x\}_1$$

are zero so that $\{F_{31}\}$ and $\{F_{41}\}$ do no work.

Using the principle that the activity (rate of working) of the external forces is equal to the rate of change of kinetic energy, in this case,

Solution of Dynamics Problems 261

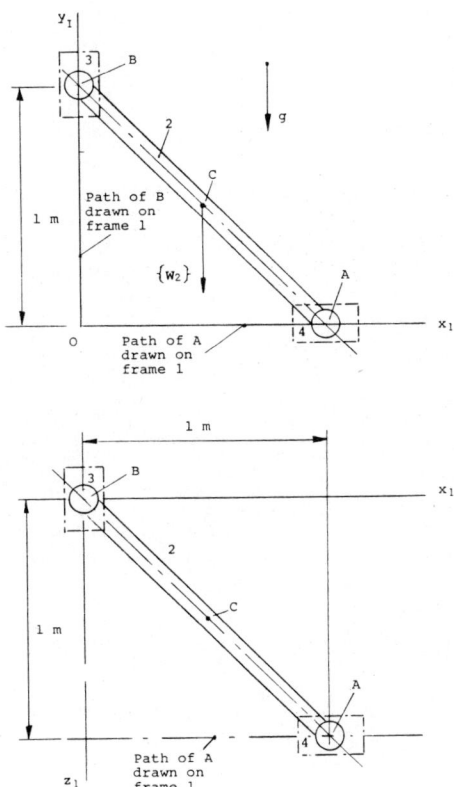

Fig. 4.38.

$$\{V_2\}^T_{1/1}\{W_2\}_1 = m_2\{V_2\}^T_{1/1}\{A_2\}_{1/1} + I\{\omega_2\}^T_{1/1}\{\dot{\omega}_2\}_{1/1} \qquad (1)$$

where I is the moment of inertia of body 2 about an axis perpendicular to the axis of generation which passes through C, the centre of mass. In the solution of this problem all vectors will be measures in and referred to frame 1 so that the 1 and 1/1 suffixes can, from this point, safely be omitted.

Now
$$\{V_2\} = \{V_A\} + [\omega_2]\{R_{CA}\}$$

and since $\{V_A\}$ is a null vector
$$\{V_2\} = -0.5[R_{BA}]\{\omega_2\}$$

and
$$\{V_2\}^T = -0.5\{[R_{BA}]\{\omega_2\}\}^T = 0.5\{\omega_2\}^T[R_{BA}] \; .$$

This result reduces Eq. 1 to

$$0.5[R_{BA}]\{W_2\} = 0.5m_2[R_{BA}]\{A_2\} + I\{\dot{\omega}_2\} \ . \tag{2}$$

Now
$$\{A_B\} = \{A_A\} + \{A_{BA}\}$$

and since $\{\omega_2\}_1$ is a null vector, $\{A_{BA}^p\}_1$ is also null. Hence

$$\{A_B\} = \{A_A\} + \{A_{BA}^n\}$$

$$A_B \begin{bmatrix} 0 \\ -1 \\ 0 \end{bmatrix} = A_A \begin{bmatrix} 1 \\ 0 \\ 0 \end{bmatrix} + \begin{bmatrix} r_2 \\ s_2 \\ t_2 \end{bmatrix}$$

and
$$\{A_{BA}\}^T\{R_{BA}\} = 0 \quad \text{or} \quad \begin{bmatrix} r_2 & s_2 & t_2 \end{bmatrix} \begin{bmatrix} -1 \\ 1 \\ -1 \end{bmatrix} = 0 \ .$$

These equations combine to give

$$\begin{bmatrix} 1 & 0 & 0 \\ 0 & 1 & 1 \\ -1 & 1 & 0 \end{bmatrix} \begin{bmatrix} r_2 \\ s_2 \\ A_B \end{bmatrix} = \begin{bmatrix} -A_A \\ 0 \\ 0 \end{bmatrix}$$

since $t_2 = 0$. This set of equations has the solution

$$\begin{bmatrix} r_2 \\ s_2 \\ A_B \end{bmatrix} = A_A \begin{bmatrix} -1 \\ -1 \\ 1 \end{bmatrix} \ .$$

Now
$$\{\dot{\omega}_2\} = \frac{[R_{BA}]\{A_{BA}^n\}}{|R_{BA}|^2} \tag{3}$$

and
$$\{A_2\} = \{A_A\} - 0.5[R_{BA}]\{\dot{\omega}_2\} = \{A_A\} - \frac{0.5[R_{BA}]^2\{A_{BA}^n\}}{|R_{BA}|^2} \tag{4}$$

Substituting Eqs. 3 and 4 in Eq. 2

$$0.5[R_{BA}]\{W_2\} = 0.5m_2[R_{BA}] \left\{ \{A_A\} - \frac{0.5[R_{BA}]^2\{A_{BA}^n\}}{|R_{BA}|^2} \right\}$$

$$+ \frac{m_2|R_{BA}|^2}{12} \frac{[R_{BA}]\{A_{BA}^n\}}{|R_{BA}|^2}$$

and this reduces to

$$0.5\{W_2\} = 0.5m_2\{A_A\} - \frac{0.25m_2[R_{BA}]^2\{A^n_{BA}\}}{|R_{BA}|^2} + \frac{m_2}{12}\{A^n_{BA}\}$$

$$0.5g\begin{bmatrix}0\\-1\\0\end{bmatrix} = 0.5A_A\begin{bmatrix}1\\0\\0\end{bmatrix} - \frac{A_A}{12}\begin{bmatrix}0 & 1 & 1\\-1 & 0 & 1\\-1 & -1 & 0\end{bmatrix}^2\begin{bmatrix}-1\\-1\\0\end{bmatrix} + \frac{A_A}{12}\begin{bmatrix}-1\\-1\\0\end{bmatrix}$$

$$6g\begin{bmatrix}0\\-1\\0\end{bmatrix} = 6A_A\begin{bmatrix}1\\0\\0\end{bmatrix} - A_A\begin{bmatrix}3\\3\\0\end{bmatrix} + A_A\begin{bmatrix}-1\\-1\\0\end{bmatrix}$$

giving

$$A_A = 1.5g \quad \text{and} \quad A_B = 1.5g.$$

Hence

$$\{A^n_{BA}\} = 1.5g\begin{bmatrix}-1\\-1\\0\end{bmatrix},$$

$$\{\dot{\omega}_2\} = 0.5g\begin{bmatrix}0 & 1 & 1\\-1 & 0 & 1\\-1 & -1 & 0\end{bmatrix}\begin{bmatrix}-1\\-1\\0\end{bmatrix} = 0.5g\begin{bmatrix}-1\\1\\2\end{bmatrix}$$

and

$$\{A_2\} = 1.5g\begin{bmatrix}1\\0\\0\end{bmatrix} - 0.25g\begin{bmatrix}0 & 1 & 1\\-1 & 0 & 1\\-1 & -1 & 1\end{bmatrix}\begin{bmatrix}-1\\1\\2\end{bmatrix} = 0.75g\begin{bmatrix}1\\-1\\0\end{bmatrix}.$$

Problem 4.39. A uniform rod, body 2, which is a solid of revolution, is constrained to move such that one end of its axis of generation B traces out a straight vertical path while the other end A traces out a straight horizontal path as shown in Fig. 4.39. The constraints are conservative and exert no moment about the axis of generation of the rod. The rod moves from rest under the action of gravity from the position in which B is y_i vertically above O. Obtain expressions for the velocity of B and the angular velocity of the rod.

Solution. For a conservative system

$$(T_2 + V_2)_{initially} = (T_2 + V_2)_{finally}$$

$$0 + mgh_i = \frac{m}{2}\{V_2\}^T_{1/1}\{V_2\}_{1/1} + \frac{I}{2}\{\omega_2\}^T_{1/1}\{\omega_2\}_{1/1} + mgh_f$$

where I is the moment of inertia of the rod about an axis perpendicular to the axis of generation through the centre of mass C.

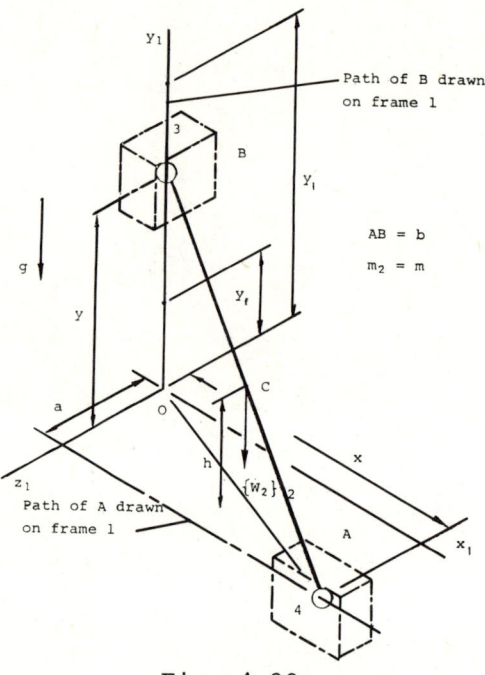

Fig. 4.39.

In the solution of this problem all vectors will be measured in and referred to frame 1 and therefore the 1/1 suffixes can be safely omitted.

It is necessary to obtain expressions for $\{V_2\}$ and $\{\omega_2\}$ in terms of $\{V_B\}$. Let

$$\{\omega_2\} = \begin{bmatrix} \omega_x \\ \omega_y \\ \omega_z \end{bmatrix}$$

and therefore

$$\{V_A\} = \{V_B\} + [\omega_2]\{R_{AB}\}$$

$$V_A \begin{bmatrix} 1 \\ 0 \\ 0 \end{bmatrix} = V_B \begin{bmatrix} 0 \\ -1 \\ 0 \end{bmatrix} + \begin{bmatrix} 0 & -\omega_z & \omega_y \\ \omega_z & 0 & -\omega_x \\ -\omega_y & \omega_x & 0 \end{bmatrix} \begin{bmatrix} x \\ -y \\ a \end{bmatrix}$$

where $x = |\sqrt{(b^2 - a^2 - y^2)}|$.

Also
$$\{\omega_2\}^T\{R_{BA}\} = x\omega_x - y\omega_y + a\omega_z = 0.$$

Hence

$$\begin{bmatrix} x & -y & a & 0 \\ y & x & 0 & 0 \\ a & 0 & -x & 0 \\ 0 & a & y & -1 \end{bmatrix} \begin{bmatrix} \omega_x \\ \omega_y \\ \omega_z \\ V_A \end{bmatrix} = \begin{bmatrix} 0 \\ 0 \\ -V_B \\ 0 \end{bmatrix}$$

and by row reduction

$$\begin{bmatrix} b^2 & 0 & 0 & 0 \\ y & x & 0 & 0 \\ a & 0 & -x & 0 \\ 0 & a & y & -1 \end{bmatrix} \begin{bmatrix} \omega_x \\ \omega_y \\ \omega_z \\ V_A \end{bmatrix} = \begin{bmatrix} -aV_B \\ 0 \\ V_B \\ 0 \end{bmatrix}.$$

This set of equations has the solution

$$\begin{bmatrix} \omega_x \\ \omega_y \\ \omega_z \\ V_A \end{bmatrix} = V_B \begin{bmatrix} -a/b^2 \\ ay/b^2x \\ (b^2 - a^2)/b^2x \\ y/x \end{bmatrix}.$$

Now

$$\{V_2\} = \{V_B\} + \{V_{CB}\}$$

$$= \{V_B\} - 0.5[R_{AB}]\{\omega_2\}$$

$$= \frac{V_B}{2}\begin{bmatrix} y/x \\ -1 \\ 0 \end{bmatrix},$$

$$2T_{2_{trans}} = \frac{mV_B^2}{4}(1 + y^2/x^2) = \frac{mV_B^2}{4x^2}(b^2 - a^2)$$

and

$$2T_{2_{rot}} = \frac{mV_B^2}{12b^2}\left\{a^2 + \frac{a^2y^2}{x^2} + \frac{(b^2 - a^2)^2}{x^2}\right\}$$

$$= \frac{mV_B^2}{12x^2}(b^2 - a^2).$$

The energy equation thus reduces to

$$0 + \frac{mgy_i}{2} = \frac{mV_B^2(b^2 - a^2)}{6x^2} + \frac{mgy_f}{2}$$

and therefore

$$V_B^2 = \frac{3g(y_i - y_f)(b^2 - a^2 - y_f^2)}{b^2 - a^2}$$

and

$$\{\omega_2\} = V_B \begin{bmatrix} -a/b^2 \\ ay/b^2x \\ (b^2 - a^2)/b^2x \end{bmatrix}$$

where

$$y/x = y_f/\sqrt{(b^2 - a^2 - y_f^2)} .$$

Problem 4.40. Figure 4.40 shows the schematic arrangement of a mechanism which is driven through the given position by an external couple $\{L_2\}$ applied to body 2. Determine that pert of the couple which is necessary to overcome the inertia of body 3. Treat body 3, which has a mass of 0.5 kg, as a solid of revolution with its axis of generation along AB. The centre of mass of body 3 is at C and its moment of inertia about an axis perpendicular to AB through C is 40×10^{-4} kg m^2. Assume that the angular velocity of body 3 about AB is zero.

Also determine the forces on the bearings 5, 6, 7 and 8. Bearings 5 and 7 carry only radial loads while bearings 6 and 8 carry both radial and axial loads.

The following data applies to the mechanism in the given position:

$$\{R_{AB}\}_{1/1} = \begin{bmatrix} 10 \\ 20 \\ 20 \end{bmatrix} \text{cm}, \quad \{R_{CB}\}_{1/1} = \frac{1}{3}\begin{bmatrix} 10 \\ 20 \\ 20 \end{bmatrix} \text{cm},$$

$$\{\omega_2\}_{1/1} = \begin{bmatrix} 0 \\ -6 \\ 0 \end{bmatrix} \text{rad/s}, \quad \{\omega_3^n\}_{1/1} = \frac{2}{3}\begin{bmatrix} -2 \\ -4 \\ 5 \end{bmatrix} \text{rad/s}$$

$$\{\omega_4\}_{1/1} = \begin{bmatrix} 0 \\ 0 \\ -6 \end{bmatrix} \text{rad/s}, \quad \{V_B\}_{1/1} = \begin{bmatrix} 120 \\ 0 \\ 0 \end{bmatrix} \text{cm/s},$$

$$\{V_{AH}\}_{1/1} = \begin{bmatrix} 0 \\ 60 \\ 0 \end{bmatrix} \text{cm/s}, \quad \{V_{AB}\}_{1/1} = \begin{bmatrix} -120 \\ -60 \\ 0 \end{bmatrix} \text{cm/s}$$

$$\{A_{BO}\}_{1/1} = \{A^n_{BO}\}_{1/1} = \begin{bmatrix} 0 \\ 0 \\ 720 \end{bmatrix} \text{cm/s}^2$$

and

$$\{\dot{\omega}^n_3\}_{1/1} = \begin{bmatrix} -8 \\ 16 \\ -12 \end{bmatrix} \text{rad/s}^2 .$$

Solution.

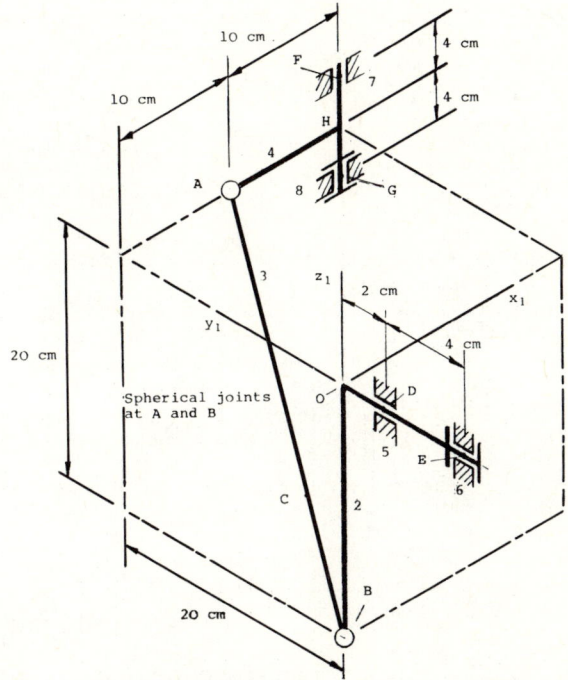

Fig. 4.40.

The velocity and acceleration of the centre of mass, C, of body 3 remain to be determined. Since all vectors will be measured in and referred to frame 1, the 1 and 1/1 suffixes can be omitted throughout the solution.

$$\{V_3\} = \{V_B\} + [\omega_3]\{R_{CB}\}$$

$$= \begin{bmatrix} 120 \\ 0 \\ 0 \end{bmatrix} + \frac{2}{9} \begin{bmatrix} 0 & -5 & -4 \\ 5 & 0 & 2 \\ 4 & -2 & 0 \end{bmatrix} \begin{bmatrix} 10 \\ 20 \\ 20 \end{bmatrix} = \begin{bmatrix} 80 \\ 20 \\ 0 \end{bmatrix} \text{cm/s}$$

$$\{A_{CB}^P\} = \frac{|V_{CB}|^2}{|R_{CB}|^2}\{R_{BC}\} = \frac{2000}{3\times100}\begin{bmatrix}-10\\-20\\-20\end{bmatrix} = \frac{1}{3}\begin{bmatrix}-200\\-400\\-400\end{bmatrix} \text{ cm/s}^2$$

$$\{A_{CB}^n\} = [\dot\omega_3^n]\{R_{CB}\}$$

$$= \frac{1}{3}\begin{bmatrix}0 & 12 & 16\\-12 & 0 & 8\\-16 & -8 & 0\end{bmatrix}\begin{bmatrix}10\\20\\20\end{bmatrix} = \frac{1}{3}\begin{bmatrix}560\\40\\0\end{bmatrix} \text{ cm/s}^2$$

and

$$\{A_3\} = \{A_B^P\} + \{A_{CB}^P\} + \{A_{CB}^n\}$$

$$= \frac{1}{3}\begin{bmatrix}360\\-360\\1760\end{bmatrix} \text{ cm/s}^2 .$$

Taking moments about B for body 3

$$[R_{AB}]\{F_{34}\} = m_3[R_{CB}]\{A_3\} + I\{\dot\omega_3^n\}$$

$$10^{-2}\begin{bmatrix}0 & -20 & 20\\20 & 0 & -10\\-20 & 10 & 0\end{bmatrix}\begin{bmatrix}F_1\\F_2\\F_3\end{bmatrix} = \frac{0.5\times10^{-4}}{9}\begin{bmatrix}0 & -20 & 20\\20 & 0 & -10\\-20 & 10 & 0\end{bmatrix}\begin{bmatrix}360\\-360\\1760\end{bmatrix}$$

$$+ 40\times10^{-4}\begin{bmatrix}-8\\16\\-12\end{bmatrix}$$

and applying the condition that the rate of supply of energy to body 4 is zero

$$\{F_{34}\}^T\{V_{AH}\} = 0$$

which, in this case, simply requires that $F_2 = 0$. Hence

$$\{F_{34}\} = \begin{bmatrix}0.54\\0\\1.0177\end{bmatrix} \text{ N} .$$

Equating applied and effective forces for body 3

$$\{F_{34}\} + \{F_{32}\} = m_3\{A_3\}$$

Solution of Dynamics Problems

$$\begin{bmatrix} F_1 \\ 0 \\ F_2 \end{bmatrix} + \begin{bmatrix} F_4 \\ F_5 \\ F_6 \end{bmatrix} = \frac{0.5 \times 10^{-2}}{3} \begin{bmatrix} 360 \\ -360 \\ 1760 \end{bmatrix} .$$

Hence

$$\{F_{32}\} = \begin{bmatrix} 0.06 \\ -0.6 \\ 1.916 \end{bmatrix} \text{ N} .$$

Taking moments about D for body 2

$$\{L_2\} + [R_{ED}]\{F_{26}\} + [R_{BD}]\{F_{23}\} = \{0\}$$

$$\begin{bmatrix} 0 \\ L \\ 0 \end{bmatrix} + 10^{-2} \begin{bmatrix} 0 & 0 & -4 \\ 0 & 0 & 0 \\ 4 & 0 & 0 \end{bmatrix} \begin{bmatrix} F_7 \\ F_8 \\ F_9 \end{bmatrix} + 10^{-2} \begin{bmatrix} 0 & 20 & 2 \\ -20 & 0 & 0 \\ 2 & 0 & 0 \end{bmatrix} \begin{bmatrix} -0.06 \\ 0.6 \\ -1.916 \end{bmatrix} = \begin{bmatrix} 0 \\ 0 \\ 0 \end{bmatrix}$$

giving

$F_7 = 0.032$ N, $F_9 = 2.042$ N and $L = -1.2 \times 10^{-2}$ N m.

Equating the forces on body 2 to zero

$$\{F_{23}\} + \{F_{25}\} + \{F_{26}\} = \{0\}$$

$$\begin{bmatrix} -0.06 \\ 0.6 \\ -1.916 \end{bmatrix} + \begin{bmatrix} F_{10} \\ 0 \\ F_{12} \end{bmatrix} + \begin{bmatrix} 0.03 \\ F_8 \\ 2.042 \end{bmatrix} = \begin{bmatrix} 0 \\ 0 \\ 0 \end{bmatrix}$$

and therefore

$$\{L_2\} = 1.2 \times 10^{-2} \begin{bmatrix} 0 \\ 1 \\ 0 \end{bmatrix} \text{ N m} , \quad \{F_{26}\} = \begin{bmatrix} 0.03 \\ -0.06 \\ 2.042 \end{bmatrix} \text{ N}$$

and

$$\{F_{25}\} = \begin{bmatrix} 0.03 \\ 0 \\ -0.126 \end{bmatrix} \text{ N}.$$

Taking moments about G for body 4

$$[R_{AG}]\{F_{43}\} + [R_{FG}]\{F_{47}\} = \{0\}$$

$$\begin{bmatrix} 0 & -4 & 0 \\ 4 & 0 & 10 \\ 0 & -10 & 0 \end{bmatrix} \begin{bmatrix} -0.54 \\ 0 \\ -1.0177 \end{bmatrix} + \begin{bmatrix} 0 & -8 & 0 \\ 8 & 0 & 0 \\ 0 & 0 & 0 \end{bmatrix} \begin{bmatrix} F_{13} \\ F_{14} \\ 0 \end{bmatrix} = \begin{bmatrix} 0 \\ 0 \\ 0 \end{bmatrix}$$

giving

$$\{F_{47}\} = \begin{bmatrix} 1.542 \\ 0 \\ 0 \end{bmatrix} \text{ N.}$$

Equating the external forces on body 4 to zero

$$\{F_{43}\} + \{F_{47}\} + \{F_{48}\} = \{0\}$$

$$\begin{bmatrix} -0.54 \\ 0 \\ -1.0177 \end{bmatrix} + \begin{bmatrix} 1.542 \\ 0 \\ 0 \end{bmatrix} + \begin{bmatrix} F_{15} \\ F_{16} \\ F_{17} \end{bmatrix} = \begin{bmatrix} 0 \\ 0 \\ 0 \end{bmatrix}$$

giving

$$\{F_{48}\} = \begin{bmatrix} -1.002 \\ 0 \\ 1.0177 \end{bmatrix} \text{ N.}$$

The value of L obtained earlier can be confirmed by equating the activity of $\{L_2\}$ to the rate of change of kinetic energy of body 3.

$$\{\omega_2\}^T \{L_2\} = m_3 \{V_3\}^T \{A_3\} + I\{\omega_3^n\}^T \{\dot{\omega}_3^n\}$$

$$\begin{bmatrix} 0 & -6 & 0 \end{bmatrix} \begin{bmatrix} 0 \\ L \\ 0 \end{bmatrix} = \frac{0.5 \times 10^{-4}}{3} \begin{bmatrix} 80 & 20 & 0 \end{bmatrix} \begin{bmatrix} 360 \\ -360 \\ 1760 \end{bmatrix}$$

$$+ \frac{2 \times 40 \times 10^{-4}}{3} \begin{bmatrix} -2 & -4 & 5 \end{bmatrix} \begin{bmatrix} -8 \\ 16 \\ -12 \end{bmatrix}$$

giving

$$L = -1.2 \ 10^{-2} \text{ N m.}$$

Problem 4.41. Determine, for the uniform hemisphere, body 2, of Fig. 4.41a

$$[I_2]_{3/3} \ , \ [I_2]_{2/2} \ , \ [I_2]_{4/4} \ \text{ and } \ [I_2]_{5/5}$$

where frame 5 is the set of principal axes which has its origin at A.

Solution of Dynamics Problems

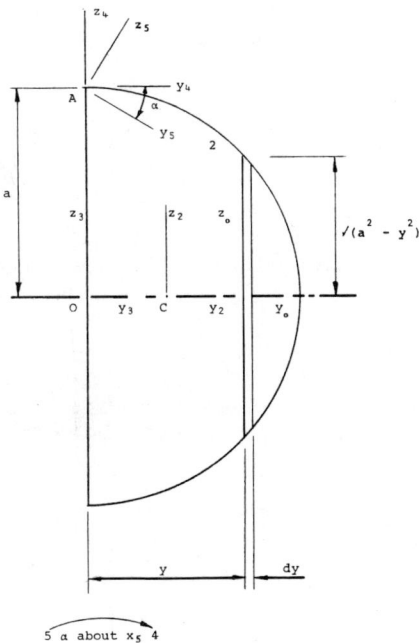

Fig. 4.41a.

Solution. The mass of the hemisphere is given by

$$m = \int_0^a \rho\pi(a^2 - y^2)\,dy = \rho\pi\left[a^2 y - \frac{y^3}{3}\right]_0^a = \frac{2}{3}\rho\pi a^3$$

and if the centre of mass is at C

$$m|R_{co}| = \int_0^a \rho\pi(a^2 - y^2)\,dy\,y = \rho\pi\left[\frac{a^2 y^2}{2} - \frac{y^4}{4}\right]_0^a = \frac{1}{4}\rho\pi a^4$$

giving

$$|R_{co}| = \frac{3}{8}a.$$

Terms in the inertia matrix are determined as follows. The moment of inertia of the elemental disc about the y_3 axis is given by

$$\frac{\rho\pi(a^2 - y^2)\,dy}{2}(a^2 - y^2)$$

the product of its (mass/2) times the radius squared. Hence

$$[I_2^y]_{3/3} = \int_0^a \frac{\rho\pi(a^2 - y^2)dy}{2}(a^2 - y^2)$$

$$= \frac{\rho\pi}{2}\left[a^4y - \frac{2}{3}a^2y^2 - \frac{y^5}{5}\right]_0^a = \frac{4}{15}\rho\pi a^5 = \frac{2}{5}ma^2$$

The moment of inertia of the elemental disc about the x_3 or z_3 axis is given by

$$\frac{\rho\pi}{4}(a^2 - y^2)^2 dy + \rho\pi(a^2 - y^2)dy\, y^2$$

the moment of inertia of the elemental disc about the x_o axis plus the mass of the disc times the distance between the x_o and x_3 axes squared. Hence

$$[I_2^x]_{3/3} = [I_2^z]_{3/3} = \int_0^a \frac{\rho\pi}{4}(a^2 - y^2)^2 dy + \rho\pi(a^2 - y^2)dy\, y^2$$

$$= \frac{4}{15}\rho\pi a^5 = \frac{2}{5}ma^2 \;.$$

By symmetry, the product of inertia terms are zero and therefore

$$[I_2]_{3/3} = \frac{2}{5}ma^2 \begin{bmatrix} 1 & 0 & 0 \\ 0 & 1 & 0 \\ 0 & 0 & 1 \end{bmatrix} \;.$$

Since frames 2 and 3 are aligned

$$[I_2]_{3/3} = [I_2]_{2/2} - m[R_{co}]^2_{3/3}$$

and, since

$$\{R_{co}\}_{3/3} = \frac{3}{8}a\begin{bmatrix} 0 \\ 1 \\ 0 \end{bmatrix}$$

$$[I_2]_{2/2} = [I_2]_{3/3} + m[R_{co}]^2_{3/3}$$

$$= \frac{2}{5}ma^2 \begin{bmatrix} 1 & 0 & 0 \\ 0 & 1 & 0 \\ 0 & 0 & 0 \end{bmatrix} + \frac{9}{64}ma^2 \begin{bmatrix} -1 & 0 & 0 \\ 0 & 0 & 0 \\ 0 & 0 & -1 \end{bmatrix}$$

$$= \frac{ma^2}{320}\begin{bmatrix} 83 & 0 & 0 \\ 0 & 128 & 0 \\ 0 & 0 & 83 \end{bmatrix}.$$

Also, since frame 4 is aligned with frame 2 and

$$\{R_{CA}\}_{4/4} = a\begin{bmatrix} 0 \\ 3/8 \\ -1 \end{bmatrix}$$

$$[I_2]_{4/4} = [I_2]_{2/2} - m[R_{CA}]^2_{4/4}$$

$$= \frac{ma^2}{320}\begin{bmatrix} 83 & 0 & 0 \\ 0 & 128 & 0 \\ 0 & 0 & 83 \end{bmatrix} + \frac{ma^2}{64}\begin{bmatrix} 73 & 0 & 0 \\ 0 & 64 & 24 \\ 0 & 24 & 9 \end{bmatrix}$$

$$= \frac{ma^2}{40}\begin{bmatrix} 56 & 0 & 0 \\ 0 & 56 & 15 \\ 0 & 15 & 16 \end{bmatrix}.$$

Let $[\ell_4]_5$ be the transformation matrix which relates the position of frame 4 to the set of principal axes, frame 5, at A. Then

$$[I_2]_{5/5} = [I_2]_{4/5} = [\ell_4]_5[I_2]_{4/4}[\ell_4]_5^T, \qquad (1)$$

where $[I_2]_{4/5}$ is a diagonal matrix, or

$$[\ell_4]_5^T[I_2]_{4/5} = [I_2]_{4/4}[\ell_4]_5^T$$

$$\begin{bmatrix} 1 & 0 & 0 \\ 0 & c\alpha & s\alpha \\ 0 & -s\alpha & c\alpha \end{bmatrix}\begin{bmatrix} A_5 & 0 & 0 \\ 0 & B_5 & 0 \\ 0 & 0 & C_5 \end{bmatrix} = \begin{bmatrix} A_4 & 0 & 0 \\ 0 & B_4 & F_4 \\ 0 & F_4 & C_4 \end{bmatrix}\begin{bmatrix} 1 & 0 & 0 \\ 0 & c\alpha & s\alpha \\ 0 & -s\alpha & c\alpha \end{bmatrix}$$

$$\begin{bmatrix} A_5 & 0 & 0 \\ 0 & B_5c\alpha & C_5s\alpha \\ 0 & -B_5s\alpha & C_5c\alpha \end{bmatrix} = \begin{bmatrix} A_4 & 0 & 0 \\ 0 & B_4c\alpha - F_4s\alpha & B_4s\alpha + F_4c\alpha \\ 0 & F_4c\alpha - C_4s\alpha & F_4s\alpha + C_4c\alpha \end{bmatrix}.$$

Hence
$$B_5\cos\alpha = B_4\cos\alpha - F_4\sin\alpha,$$

$$-B_5\sin\alpha + F_4\cos\alpha - C_4\sin\alpha$$

and therefore, eliminating B_5,

$$\frac{\tan\alpha}{1 - \tan^2\alpha} = \frac{\tan 2\alpha}{2} = \frac{F_4}{C_4 - B_4}$$

or

$$\tan 2\alpha = \frac{2F_4}{C_4 - B_4} \quad . \tag{2}$$

Alternatively

$$\tan \alpha = \frac{F_4}{C_4 - B_5} = \frac{B_4 - B_5}{F_4} \tag{3}$$

giving, on eliminating α,

$$B_5^2 - (B_4 + C_4)B_5 + B_4 C_4 - F_4^2 = 0 \tag{4}$$

the equation to Mohr's circle.

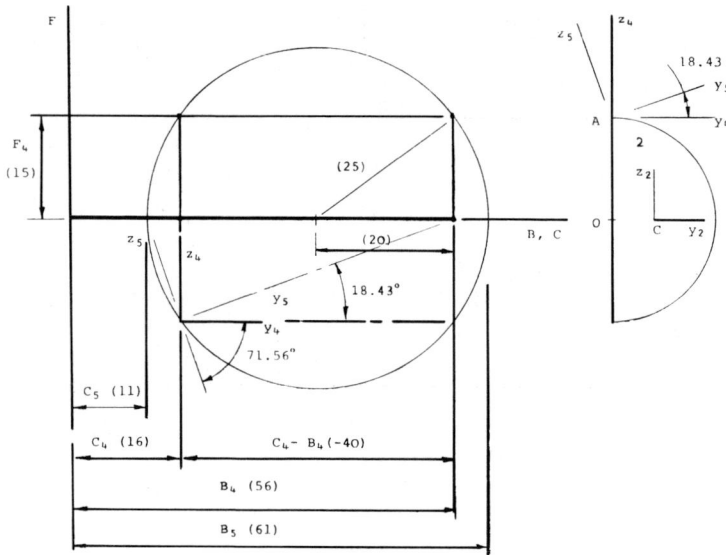

Fig. 4.41b.

By reference to the Mohr's circle of Fig. 41b, B_5 is readily seen to be

$$61ma^2/40 \quad \text{or} \quad 11ma^2/40.$$

From Eq. 2

$$\tan 2\alpha = \frac{2 \times 15}{15 - 56} = -0.75$$

giving

$$\alpha = -18.43° \quad \text{or} \quad 71.56° \quad .$$

Alternatively, from Eq. 3

$$\tan \alpha = \frac{15}{16 - 61} = -\frac{1}{3} \quad \text{and} \quad \alpha = -18.43°$$

or

Solution of Dynamics Problems

$$\tan\alpha = \frac{15}{16-11} = 3 \text{ and } \alpha = 71.56°\ .$$

As a check on the work, since $\sin\alpha = -1/\sqrt{10}$ and $\cos\alpha = 3/\sqrt{10}$, by substitution of the results in Eq. 1

$$[I_2]_{4/5} = \frac{ma^2}{400} \begin{bmatrix} \sqrt{10} & 0 & 0 \\ 0 & 3 & -1 \\ 0 & 1 & 0 \end{bmatrix} \begin{bmatrix} 56 & 0 & 0 \\ 0 & 56 & 15 \\ 0 & 15 & 0 \end{bmatrix} \begin{bmatrix} \sqrt{10} & 0 & 0 \\ 0 & 3 & 1 \\ 0 & -1 & 0 \end{bmatrix}$$

$$= \frac{ma^2}{40} \begin{bmatrix} 56 & 0 & 0 \\ 0 & 61 & 0 \\ 0 & 0 & 11 \end{bmatrix}\ .$$

Problem 4.42. A uniform thin straight rod, body 2, has a mass m and length $2b$. Find, for the axis system shown in Fig. 4.42,

$$[I_2]_{2/2}\ ,\ [I_2]_{2/4}\ ,\ [I_2]_{3/3}\ ,\ [I_2]_{4/4}\ ,\ [I_2]_{5/5}$$

and

$$[I_2]_{5/6}\ .$$

Solution. It is easy to show that

$$[I_2^x]_{2/2} = [I_2^z]_{2/2} = \frac{mb^2}{3}$$

and since body 2 is a thin rod

$$[I_2^y]_{2/2} \simeq 0\ .$$

Also, by symmetry, the products of inertia are zero. Hence

$$[I_2]_{2/2} = \frac{mb^2}{3} \begin{bmatrix} 1 & 0 & 0 \\ 0 & 0 & 0 \\ 0 & 0 & 1 \end{bmatrix}\ .$$

The inertia matrix for body 2, measured in frame 2 and referred to frame 4, is given by

$$[I_2]_{2/2} = [\ell_2]_4 [I_2]_{2/2} [\ell_2]_4^T$$

$$= \frac{mb^2}{3} \begin{bmatrix} 1 & 0 & 0 \\ 0 & c\alpha & s\alpha \\ 0 & -s\alpha & c\alpha \end{bmatrix} \begin{bmatrix} 1 & 0 & 0 \\ 0 & 0 & 0 \\ 0 & 0 & 1 \end{bmatrix} \begin{bmatrix} 1 & 0 & 0 \\ 0 & c\alpha & -s\alpha \\ 0 & s\alpha & c\alpha \end{bmatrix}$$

$$= \frac{mb^2}{3} \begin{bmatrix} 1 & 0 & 0 \\ 0 & \sin^2\alpha & \sin\alpha\cos\alpha \\ 0 & \sin\alpha\cos\alpha & \cos^2\alpha \end{bmatrix} .$$

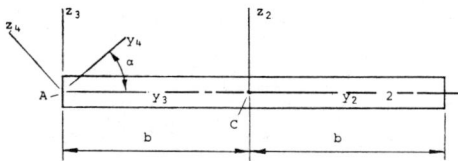

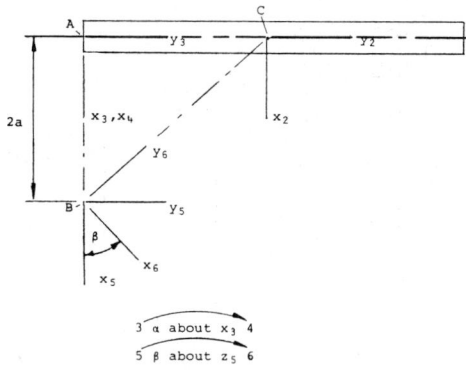

Fig. 4.42.

Since frame 3 is aligned with frame 2 the inertia matrix for body 2, measured in frame 3 and referred to frame 3, is given by

$$[I_2]_{3/3} = [I_2]_{2/2} - m[R_{CA}]^2_{3/3}$$

$$= \frac{mb^2}{3} \begin{bmatrix} 1 & 0 & 0 \\ 0 & 0 & 0 \\ 0 & 0 & 1 \end{bmatrix} - m \begin{bmatrix} -b^2 & 0 & 0 \\ 0 & 0 & 0 \\ 0 & 0 & -b^2 \end{bmatrix}$$

$$= \frac{4mb^2}{3} \begin{bmatrix} 1 & 0 & 0 \\ 0 & 0 & 0 \\ 0 & 0 & 1 \end{bmatrix}$$

Since the origins of frames 3 and 4 are coinicident the inertia matrix for body 2, measured in frame 4 and referred to frame 4, is given by

$$[I_2]_{4/4} = [\ell_3]_4 [I_2]_{3/3} [\ell_3]_4^T$$

$$= \frac{4mb^2}{3} \begin{bmatrix} 1 & 0 & 0 \\ 0 & \sin^2\alpha & \sin\alpha\cos\alpha \\ 0 & \sin\alpha\cos\alpha & \cos^2\alpha \end{bmatrix}.$$

Since frame 5 is aligned with frame 2 the inertia matrix for body 2, measured in frame 5 and referred to frame 5, is given by

$$[I_2]_{5/5} = [I_2]_{2/2} - m[R_{CB}]^2_{5/5}$$

$$= \frac{mb^2}{3}\begin{bmatrix} 1 & 0 & 0 \\ 0 & 0 & 0 \\ 0 & 0 & 1 \end{bmatrix} - m\begin{bmatrix} -b^2 & -ab & 0 \\ -ab & -a^2 & 0 \\ 0 & 0 & -(a^2+b^2) \end{bmatrix}$$

$$= \frac{m}{3}\begin{bmatrix} 4b^2 & 3ab & 0 \\ 3ab & 3a^2 & 0 \\ 0 & 0 & 3a^2+4b^2 \end{bmatrix}$$

The inertia matrix for body 2, measured in frame 5 and referred to frame 6 is given by

$$[I_2]_{5/6} = [\ell_5]_6 [I_2]_{5/5} [\ell_5]_6^T$$

$$= \begin{bmatrix} c\beta & -s\beta & 0 \\ s\beta & c\beta & 0 \\ 0 & 0 & 1 \end{bmatrix} \begin{bmatrix} A & D & 0 \\ D & B & 0 \\ 0 & 0 & C \end{bmatrix} \begin{bmatrix} c\beta & s\beta & 0 \\ -s\beta & c\beta & 0 \\ 0 & 0 & 1 \end{bmatrix}$$

$$= \frac{m}{3(a^2+b^2)} \begin{bmatrix} 4b^4+3a^4-6a^2b^2 & 7ab^3-6a^3b & 0 \\ 7ab^3-6a^3b & 13a^2b^2 & 0 \\ 0 & 0 & 4b^4+3a^4+7a^2b^2 \end{bmatrix}.$$

Problem 4.43. Body 5 is made up from three uniform thin rods, bodies 2, 3 and 4, as shown in Fig.4.43a. Determine the position of the centre of mass, C, of body 5 relative to the centre of mass, A, of body 2 and hence determine

$$[I_5]_{5/5}.$$

Also find the position of the set of principal axes, frame 6, at C and the principal moments of inertia associated with this frame.

Solution. The position of the centre of mass of body 5 relative to A is given by (the 2/2 suffix being understood)

278   Matrix Methods in Engineering Mechanics

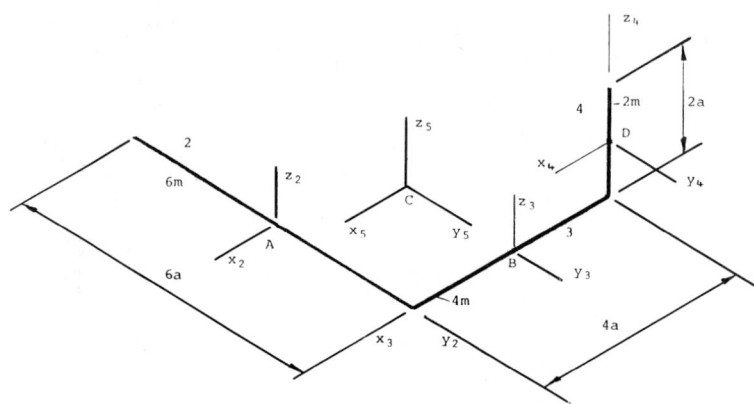

Fig. 4.43a

$$\{R_{CA}\} = \frac{4m\{R_{BA}\} + 2m\{R_{DA}\}}{12m}$$

$$= \frac{1}{12}\left[4\begin{bmatrix}-2a\\3a\\0\end{bmatrix} + 2\begin{bmatrix}-4a\\3a\\a\end{bmatrix}\right] = \frac{a}{6}\begin{bmatrix}-8\\9\\1\end{bmatrix}.$$

Hence

$$\{R_{BC}\} = \{R_{BA}\} - \{R_{CA}\}$$

$$= \frac{a}{6}\begin{bmatrix}-12\\18\\0\end{bmatrix} - \frac{a}{6}\begin{bmatrix}-8\\9\\1\end{bmatrix} = \frac{a}{6}\begin{bmatrix}-4\\9\\-1\end{bmatrix}$$

and

$$\{R_{DC}\} = \{R_{DA}\} - \{R_{CA}\}$$

$$= \frac{a}{6}\begin{bmatrix}-24\\18\\6\end{bmatrix} - \frac{a}{6}\begin{bmatrix}-8\\9\\1\end{bmatrix} = \frac{a}{6}\begin{bmatrix}-16\\9\\5\end{bmatrix}.$$

Since frames 2, 3 and 4 are aligned with frame 5, the inertia matrix for body 5, measured in frame 5 and referred to frame 5, is given by

$$[I_5]_{5/5} = [I_2]_{2/2} - 6m[R_{AC}]^2_{5/5} + [I_3]_{3/3} - 4m[R_{BC}]^2_{5/5}$$
$$+ [I_4]_{4/4} - 2m[R_{DC}]^2_{5/5}$$

and since

Solution of Dynamics Problems

$$[I_2]_{2/2} = \frac{ma^2}{18}\begin{bmatrix} 324 & 0 & 0 \\ 0 & 0 & 0 \\ 0 & 0 & 324 \end{bmatrix}, \quad [I_3]_{3/3} = \frac{ma^2}{18}\begin{bmatrix} 0 & 0 & 0 \\ 0 & 96 & 0 \\ 0 & 0 & 96 \end{bmatrix},$$

$$[I_4]_{4/4} = \frac{ma^2}{18}\begin{bmatrix} 12 & 0 & 0 \\ 0 & 12 & 0 \\ 0 & 0 & 0 \end{bmatrix},$$

$$6m[R_{AC}]^2_{5/5} = \frac{ma^2}{18}\begin{bmatrix} -246 & -216 & -24 \\ -216 & -195 & 27 \\ -24 & 27 & -435 \end{bmatrix},$$

$$4m[R_{BC}]^2_{5/5} = \frac{ma^2}{18}\begin{bmatrix} -164 & -72 & 8 \\ -72 & -34 & -18 \\ 8 & -18 & -194 \end{bmatrix}$$

and

$$2m[R_{DC}]^2_{5/5} = \frac{ma^2}{18}\begin{bmatrix} -106 & -144 & -80 \\ -144 & -281 & 45 \\ -80 & 45 & -337 \end{bmatrix}$$

$$[I_5]_{5/5} = \frac{ma^2}{18}\begin{bmatrix} 852 & 432 & 96 \\ 432 & 618 & -54 \\ 96 & -54 & 1\,386 \end{bmatrix}.$$

If frame 6 is a set of principal axes at C, then

$$[I_5]_{5/6} = [\ell_5]_6 [I_5]_{5/5} [\ell_5]_6^T$$

$$\begin{bmatrix} \lambda_1 & 0 & 0 \\ 0 & \lambda_2 & 0 \\ 0 & 0 & \lambda_3 \end{bmatrix} = \begin{bmatrix} \{a\}^T \\ \{b\}^T \\ \{c\}^T \end{bmatrix} \begin{bmatrix} A & D & E \\ D & B & F \\ E & F & C \end{bmatrix} \begin{bmatrix} \{a\} & \{b\} & \{c\} \end{bmatrix}$$

where λ_1, λ_2 and λ_3 are the principal moments of inertia which are found from

$$\lambda^3 - (A + B + C)\lambda^2 + (AB + BC + AC - F^2 - E^2 - D^2)\lambda$$
$$+ (AF^2 + BE^2 + CD^2 - ABC - 2DEF) = 0.$$

In this case the above equation reduces to

$$\lambda^3 - 2\,856\lambda^2 + 2\,365\,200\lambda - 458\,459\,136 = 0$$

which is of the form

$$\lambda^3 - \alpha\lambda^2 + \beta\lambda - \gamma = 0.$$

This equation must have three positive real roots which can be found using the root locus method by writing it in the form

$$\frac{\beta(\lambda - \gamma/\beta)}{\lambda^2(\lambda - \alpha)} = -1 = 1e^{j\pi}.$$

Taking the view that λ is the complex number $\sigma + j\omega$, the poles(x) and zeros(o) of the left hand side of the equation can be plotted as shown in Fig. 4.43b.

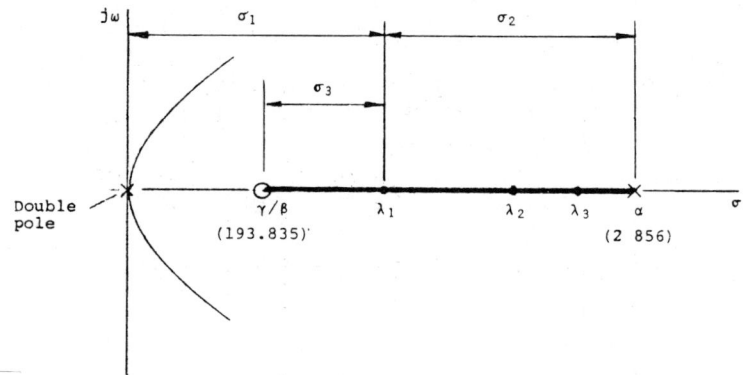

Fig. 4.43b.

On the line joining the pole at α and the zero at γ/β

$$\text{Arg}\frac{(\lambda - \gamma/\beta)}{\lambda^2(\lambda - \alpha)} = \pi.$$

Thus, three values of λ can be found which satisfy the condition

$$\frac{\beta|\lambda - \gamma/\beta|}{|\lambda^2||\lambda - \alpha|} = \frac{\beta|\lambda - 193.835|}{|\lambda^2||\lambda - 2\,856|} = \frac{\beta\sigma_3}{\sigma_1\sigma_2} = 1.$$

By trial and error, a value of $\lambda_1 = 278.35$ gives

$$\frac{2\,365\,200 \times 84.515}{278.35^2 \times 2\,577.65} = 1.000\,9.$$

By further trial and error

$\lambda_2 = 1\,170$ (1.000 3) and

$\lambda_3 = 1\,408$ (1.000 3).

Now $\lambda_1 + \lambda_2 + \lambda_3 \simeq 2\,856$, which is the magnitude of the coefficient of λ^2 in the λ equation as it should be. Also

$$\lambda_1 \times \lambda_2 \times \lambda_3 \simeq 458\,542\,656$$

which is acceptably close to the magnitude of the λ^0 term in the λ equation.

Solution of Dynamics Problems

The columns $\{a\}$, $\{b\}$, $\{c\}$ of

$$[\ell_5]_6^T$$

can be found by evaluating a column of

$$\text{Adjoint}\left[\lambda [1] - [I_5]_{5/5}\right] = \left[\{C\}_1 \ \{C\}_2 \ \{C\}_2\right],$$

say $\{C\}_1$, for the three values of λ. Hence

$$\{C_i\}_1 = \begin{bmatrix} \lambda_i^2 - (B+C)\lambda_i + BC - F^2 \\ D\lambda_i + EF - CD \\ E\lambda_i + DF - BE \end{bmatrix} = \begin{bmatrix} p_i \\ q_i \\ r_i \end{bmatrix}$$

giving

$$\{a\} = \frac{1}{\sqrt{(p_1^2 + q_1^2 + r_1^2)}} \begin{bmatrix} p_1 \\ q_1 \\ r_1 \end{bmatrix}, \quad \{b\} = \frac{1}{\sqrt{(p_2^2 + q_2^2 + r_2^2)}} \begin{bmatrix} p_2 \\ q_2 \\ r_2 \end{bmatrix}$$

and

$$\{c\} = \frac{1}{\sqrt{(p_3^2 + q_3^2 + r_3^2)}} \begin{bmatrix} p_3 \\ q_3 \\ r_3 \end{bmatrix}.$$

The columns of $[\ell_5]_6^T$ corresponding to $\lambda_1 = 278.35$, $\lambda_2 = 1\ 170$ and $\lambda_1 = 1\ 408$ are respectively

$$\{a\} = \begin{bmatrix} -0.608\ 4 \\ 0.788\ 3 \\ 0.091\ 2 \end{bmatrix}, \quad \{b\} = \begin{bmatrix} 0.764\ 9 \\ 0.616\ 8 \\ -0.185\ 7 \end{bmatrix} \text{ and } \{c\} = \begin{bmatrix} 0.264\ 7 \\ 0.079\ 1 \\ 0.961\ 1 \end{bmatrix}.$$

Transformation using these columns gives

$$[I_5]_{5/6} = \frac{ma^2}{18}\begin{bmatrix} 278.13 & 1.30 & -2.95 \\ 1.30 & 1174 & 86.59 \\ -2.95 & 86.59 & 1\ 402 \end{bmatrix}.$$

Better estimates of the principal moments of inertia are

$$\lambda_1 = 278.17, \quad \lambda_2 = 1\ 174.12 \quad \text{and} \quad \lambda_3 = 1\ 403.71$$

and the corresponding columns of the transformation matrix are

$$\{a\} = \begin{bmatrix} -0.608\ 6 \\ 0.788\ 2 \\ 0.091\ 2 \end{bmatrix}, \quad \{b\} = \begin{bmatrix} 0.766\ 2 \\ 0.613\ 7 \\ -0.190\ 7 \end{bmatrix} \text{ and } \{c\} = \begin{bmatrix} 0.206\ 3 \\ 0.046\ 2 \\ 0.977\ 4 \end{bmatrix}.$$

Transformation using these colums gives

$$[I_5]_{5/6} = \frac{ma^2}{18} \begin{bmatrix} 278.19 & -5.34\times10^{-4} & 0.01 \\ -5.34\times10^{-4} & 1\,174.12 & 0.08 \\ -0.01 & 0.08 & 1\,403.74 \end{bmatrix}.$$

4.2 Problems For Solution

Problem 4.44. At a particular instant of time, a particle has the velocity

$$\{V\}_{1/1} = \begin{bmatrix} 10 \\ -6 \\ 12 \end{bmatrix} \text{ m/s}$$

and the force acting on it is

$$\{F\}_1 = \begin{bmatrix} 40 \\ 20 \\ 30 \end{bmatrix} \text{ N}.$$

What is the power, rate of working or activity of the force?

Problem 4.45. Show that the force

$$\{F\} = \begin{bmatrix} 2xy + z^3 \\ x^2 \\ 3xz^2 \end{bmatrix}$$

is conservative and hence obtain an expression for the work it does when its point of application moves from point A to point B.

Problem 4.46. In the system shown in Fig. 4.46, a small block of mass m, body 2, is constrained to move in a circular path, drawn on an inertial frame 1, which lies in the vertical plane. Motion of body 2 is induced by a light spring, body 3, which has an unstretched length r and stiffness k. Find, for the case in which the block moves from rest at A to B, when the length of the spring is r,

$$\{V_D\}_{1/2}, \quad \{F_{21}\}_2, \quad \{F_{23}\}_2 \quad \text{and} \quad \{A_D\}_{1/2}.$$

Also find the above vectors when the y coordinate of B is 1.452 times it former value. What is the y coordinate of B when the block first comes to rest after its release from A?

Neglect the effects of the mass of the spring and assume that the system is conservative. Also, assume that the centre of mass of the block is at D and the force due to the spring is at all times along CD.

Take a = 0.05 m, b = 0.1 m, r = 0.5 m, k = 357 kN/m and m_2 = 0.5 kg.

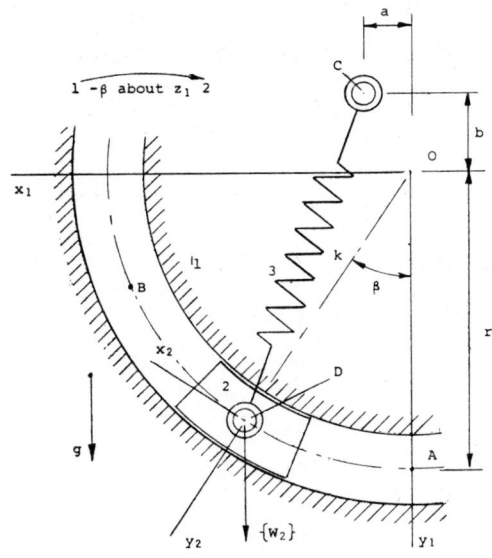

Fig. 4.46.

Problem 4.47. The position of a system can be specified by the position of a point P on a rotating z_3 axis, which is positioned relative to frame 1 as shown in Fig. 4.47. Show that $\{\nabla\}_3 V$, i.e. del V referred to the rotating frame 3, is given by

$$\{\nabla\}_3 V = \begin{bmatrix} \dfrac{1}{r} \dfrac{\partial V}{\partial \theta} \\ \dfrac{1}{r\sin\theta} \dfrac{\partial V}{\partial \phi} \\ \dfrac{\partial V}{\partial r} \end{bmatrix}.$$

Problem 4.48. A couple is formed by a pair of parallel forces which are equal in magnitude, but opposite in direction as shown in Fig. 4.48. Show that the moment of the couple is independent of the point A about which its turning effect is computed.

Problem 4.49. A given force $\{F\}$ passes through the point P as shown in Fig. 4.49. Obtain an expression for the component of the moment of $\{F\}$ about a given point A which is parallel to a line through A and a further point B.

284 Matrix Methods in Engineering Mechanics

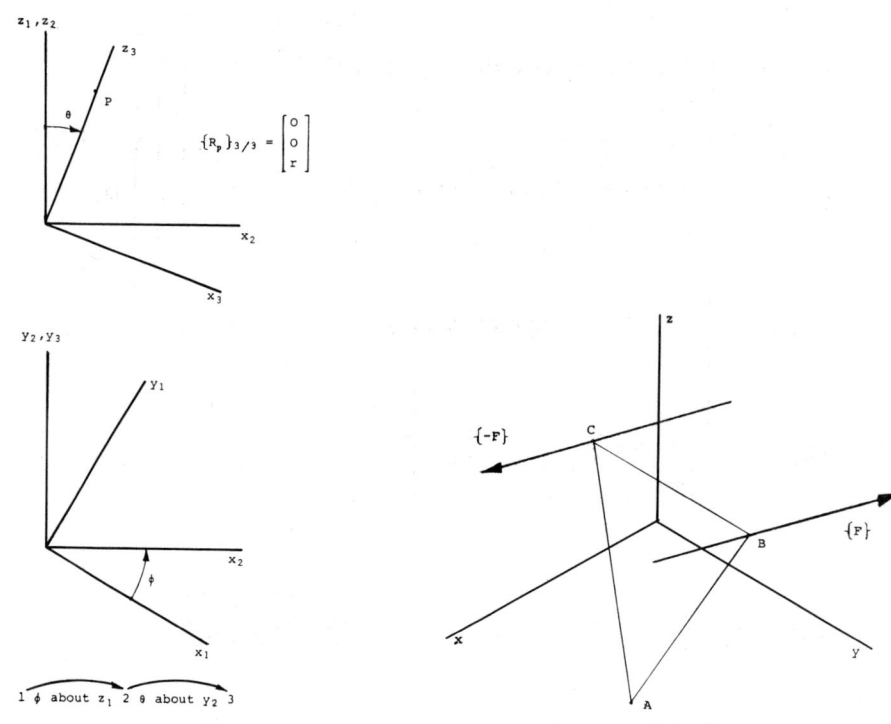

Fig. 4.47.   Fig. 4.48.

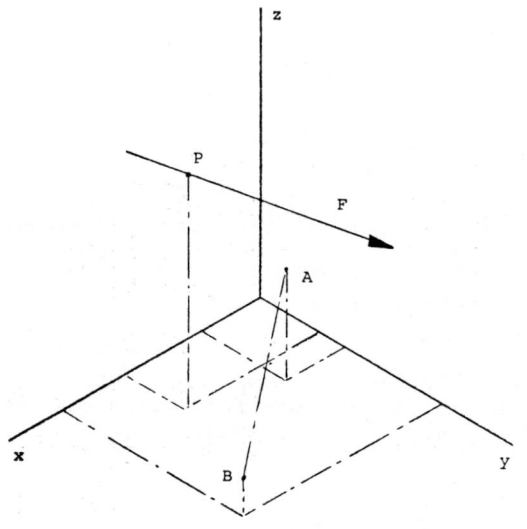

Fig. 4.49.

Problem 4.50. A system of forces and couples, all of which are referred to the same frame, act on body 2 and is specified as follows:

$$\{F_{23}\} = \begin{bmatrix} 4 \\ 3 \\ 0 \end{bmatrix} \text{ kN} \quad \text{through the point} \quad \{R_A\} = \begin{bmatrix} 1 \\ 0 \\ 1 \end{bmatrix} \text{ m,}$$

$$\{F_{24}\} = \begin{bmatrix} 5 \\ 5 \\ 5 \end{bmatrix} \text{ kN} \quad \text{through the point} \quad \{R_B\} = \begin{bmatrix} 1 \\ 1 \\ 1 \end{bmatrix} \text{ m,}$$

$$\{F_{25}\} = \begin{bmatrix} 0 \\ 2 \\ 4 \end{bmatrix} \text{ kN} \quad \text{through the point} \quad \{R_C\} = \begin{bmatrix} 0 \\ 1 \\ 2 \end{bmatrix} \text{ m,}$$

$$\{L_{23}\} = \begin{bmatrix} 6 \\ 6 \\ 6 \end{bmatrix} \text{ kN m} \quad \text{and} \quad \{L_{26}\} = \begin{bmatrix} 3 \\ 0 \\ -3 \end{bmatrix} \text{ kN m.}$$

Find the resultant force and couple of the system for the point

(a) $\{R_O\} = \begin{bmatrix} 0 \\ 0 \\ 0 \end{bmatrix}$ m and (b) $\{R_B\} = \begin{bmatrix} 1 \\ 1 \\ 1 \end{bmatrix}$ m.

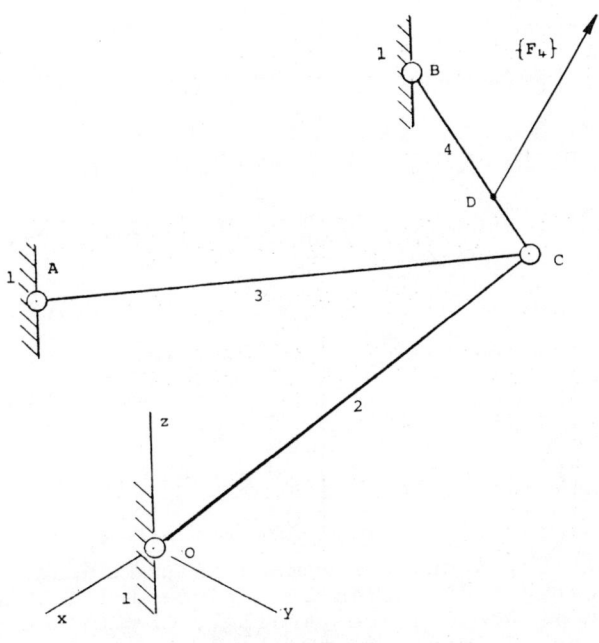

Fig. 4.51.

Problem 4.51. Figure 4.51 shows a structure subjected to an external force $\{F_4\}$. The members of the structure, bodies 2,3 and 4 are supported at their ends by fittings which are incapable of inducing bending effects. For the case in which

$$\{R_{AO}\} = \begin{bmatrix} 2 \\ 0 \\ 5 \end{bmatrix} m, \quad \{R_{BO}\} = \begin{bmatrix} -5 \\ 0 \\ 5 \end{bmatrix} m, \quad \{R_{CO}\} = \begin{bmatrix} -2 \\ 5 \\ 6 \end{bmatrix} m$$

$$\{F_4\} = \begin{bmatrix} -1 \\ 2 \\ 5 \end{bmatrix} \text{kN} \quad \text{and} \quad |R_{DB}| = 0.75|R_{CB}|$$

find

$\{F_{21}\}$, $\{F_{31}\}$ and $\{F_{41}\}$.

Problem 4.52. Figure 4.50 shows a mechanism which is maintained in the given position by an external couple

$$\{L_2\}_1 = \begin{bmatrix} 0 \\ L \\ 0 \end{bmatrix}$$

when an external force

$$\{F_3\}_1 = \begin{bmatrix} -200 \\ 200 \\ -100 \end{bmatrix} \text{N}$$

is applied to body 3 at C. Determine L and the forces due to the bearings at D, E, F and H. Obtain any necessary data from Problem 4.40.

Problem 4.53. Figure 4.53 specifies the positions of the lines of action of the following forces:

$$\{F_1\} = \begin{bmatrix} 0 \\ 0 \\ -8 \end{bmatrix} \text{N}, \quad \{F_2\} = \begin{bmatrix} 10 \\ 0 \\ 0 \end{bmatrix} \text{N}, \quad \{F_3\} = \begin{bmatrix} 0 \\ 6 \\ 0 \end{bmatrix} \text{N},$$

$$\{F_4\} = \begin{bmatrix} 0 \\ 0 \\ 2 \end{bmatrix} \text{N}, \quad \{F_5\} = \begin{bmatrix} 0 \\ -5 \\ 0 \end{bmatrix} \text{N} \quad \text{and} \quad \{F_6\} = \begin{bmatrix} 0 \\ -8 \\ 0 \end{bmatrix} \text{N}.$$

Find the resultant force and the moment of the forces about O. Reduce this force and moment combination to a force and a couple such that the force and couple vectors are parallel. Obtain the position of a point G on the line of action of the force relative to O.

Solution of Dynamics Problems

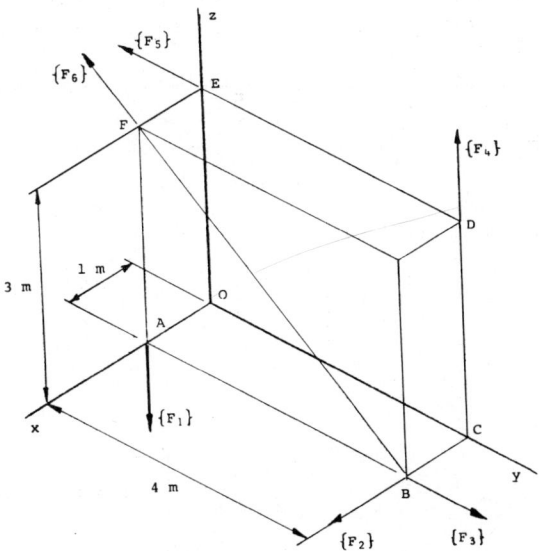

Fig. 4.53.

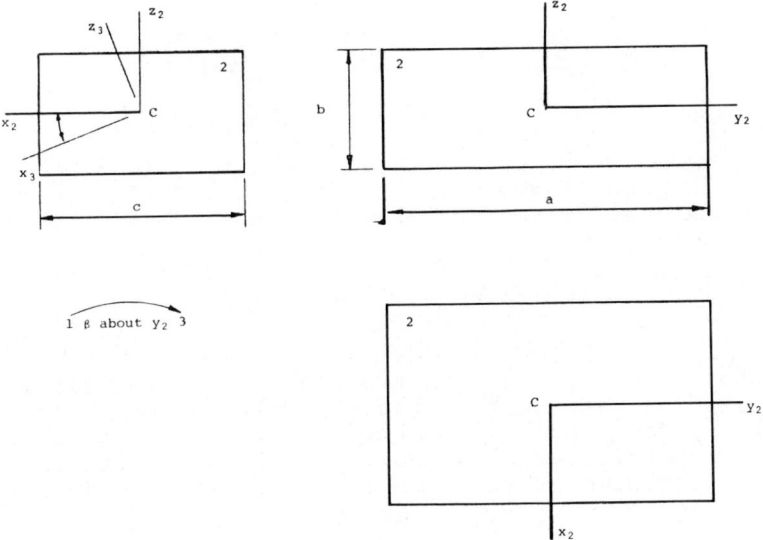

Fig. 4.54.

Problem 4.54. Determine

$$[I_2]_{2/2}$$

for the uniform rectangular parallelepiped shown in Fig. 4.54. Also determine

$$[I_2]_{2/3}$$

and hence show that when $b = c$, the inertia matrix is invariant for the $[\ell_2]_3$ transformation.

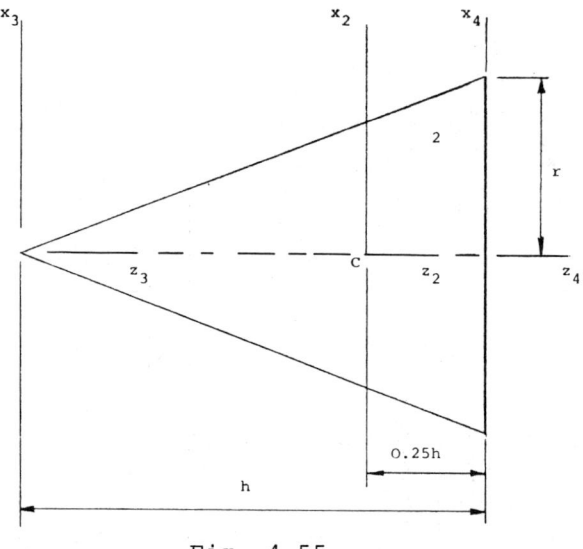

Fig. 4.55.

Problem 4.55. Figure 4.55 shows a uniform, solid right circular cone. Demonstrate that the centre of mass is in the position shown. Also, determine

$$[I_2]_{2/2}, \quad [I_2]_{3/3} \quad \text{and} \quad [I_2]_{4/4}.$$

Problem 4.56. An aircraft with a two bladed airscrew flies at a steady rate in a circular path such that the angular velocity of the fuselage about an axis perpendicular to the plane of the wings has a magnitude Ω relative to inertial axes as shown in Fig. 4.56. Determine the couple which the engine must exert on the airscrew to maintain a constant speed of rotation ω.

Problem 4.57. An aircraft, body 2, is rolling at a constant rate

$$\{\omega_2\}_{1/2} = \begin{bmatrix} \omega \\ 0 \\ 0 \end{bmatrix}$$

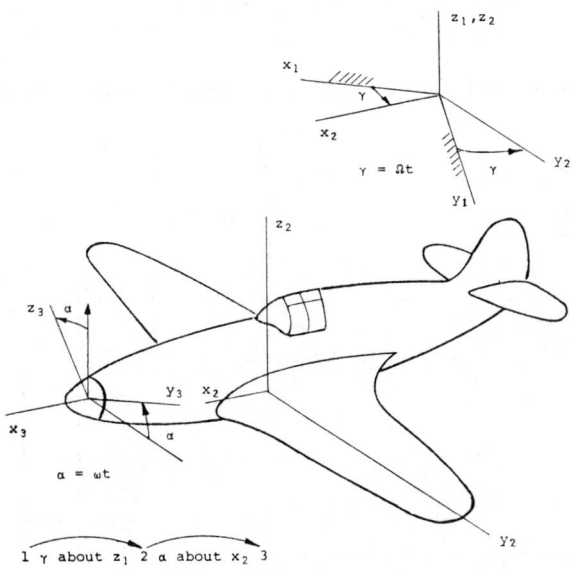

Fig. 4.56.

To induce this motion, the control surfaces, collectively designated body 3, must exert a couple

$$\{L_{23}\}_2 = \begin{bmatrix} L_1 \\ L_2 \\ L_3 \end{bmatrix}$$

while the aerodynamic couple opposing the motion is

$$\{L_2\}_2 = \begin{bmatrix} -c\omega \\ 0 \\ 0 \end{bmatrix}.$$

The rotational inertia of the aircraft can be described by

$$[I_2]_{2/2} = \begin{bmatrix} A & 0 & -E \\ 0 & B & 0 \\ -E & 0 & C \end{bmatrix}$$

Determine

$$\{L_{23}\}_2 .$$

Problem 4.58. Figure 4.58 shows a circular saw arranged to cut grooves wider than the saw blade thickness by changing the plane of rotation of the blade. This is achieved by clamping the saw blade to the motor shaft between washers which have surfaces mating with the saw blade inclined to the motor shaft axis. Obtain expressions for

$$\{L_{23}\}_2 \quad \text{and} \quad \{L_{23}\}_1$$

when the saw blade is driven at the constant rate ω. Take

$$[I_3]_{3/3} = \begin{bmatrix} I & 0 & 0 \\ 0 & I & 0 \\ 0 & 0 & J \end{bmatrix}.$$

Evaluate the expression for $\{L_{23}\}_2$ for the case in which

$$I = 6 \times 10^{-6} \text{ kg m}^2, \quad J = 12 \times 10^{-6} \text{ kg m}^2, \quad \omega \equiv 3\,600 \text{ rev/min}$$

and

$$\alpha = 10°.$$

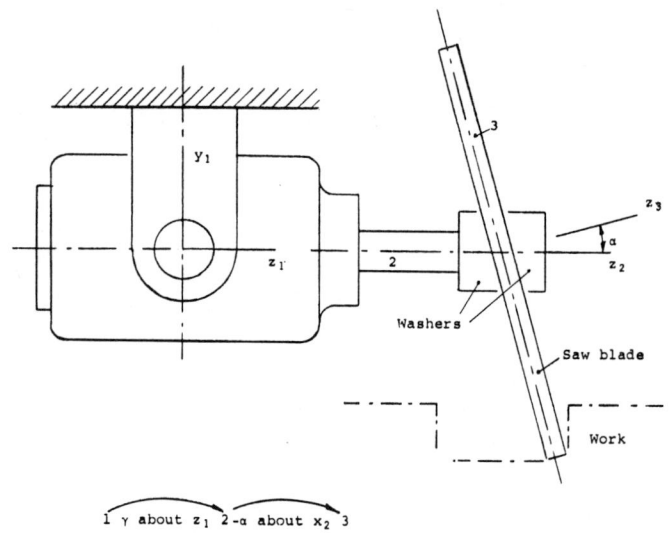

Fig. 4.58.

Problem 4.59. Figure 4.59 shows a diagrammatic arrangement of a device used to measure rate of rotation relative to an inertial frame. The rotor, body 4, is driven at a constant high speed Ω (16 000 rev/min) relative to the gimbal, body 3. The gimbal is mounted in bearings in the case of the device, body 2. The motion of the gimbal relative to the case is controlled by a torsion spring, of rate k, and a viscous damper which exerts a couple on the gimbal proportional to the relative angular velocity between gimbal and case, the constant of proportionality being c. Show that the angular displacement of the gimbal relative to the case about the x_2 output axis is proportional to the rate at which the case turns about the y_2 input axis relative to an inertial frame. Neglect the effects due to the inertia of the gimbal.

Problem 4.60. A rotor, body 4, of Fig. 4.60a runs at a constant rate ω in bearings, bodies 5 and 6. The bearings are mounted on a disc, body 2, which turns at a constant rate $\dot{\gamma}$, relative to an inertial reference,

Solution of Dynamics Problems

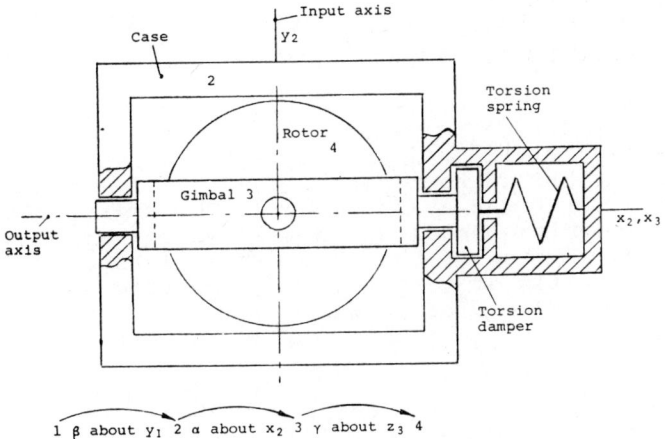

1 β about y_1 2 α about x_2 3 γ about z_3 4

Fig. 4.59.

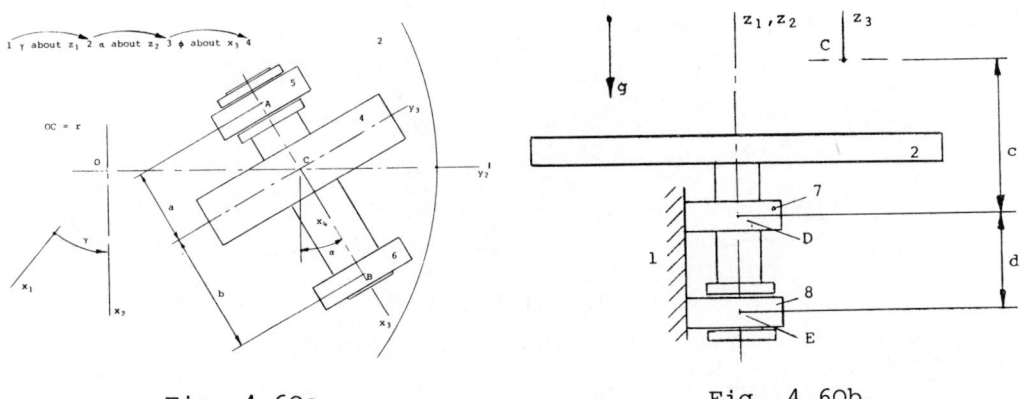

Fig. 4.60a. Fig. 4.60b.

in bearings, bodies 7 and 8, as shown in Fig. 4.60b. Determine the forces

$$\{F_{45}\}_3 = \begin{bmatrix} F_1 \\ F_2 \\ F_3 \end{bmatrix}, \quad \{F_{46}\}_3 = \begin{bmatrix} 0 \\ F_4 \\ F_5 \end{bmatrix}, \quad \{F_{28}\}_2 = \begin{bmatrix} F_6 \\ F_7 \\ F_8 \end{bmatrix}$$

and

$$\{F_{27}\}_2 = \begin{bmatrix} F_9 \\ F_{10} \\ 0 \end{bmatrix}$$

due to the motion of the rotor.

Problem 4.61. A solid uniform rotor, body 3, of length b and diameter 2a, is mounted on a light shaft, body 2. The shaft rotates at a constant rate Ω in bearings, bodies 4 and 5, which can be considered inertial. The centre of mass of the rotor is a distance e from the axis of the shaft and the axis of generation of the rotor is inclined to the shaft axis at the angle α as shown in Fig. 4.61. Determine

$$\{F_{24}\}_2 \quad \text{and} \quad \{F_{25}\}_2.$$

Neglect the effects due to the weight of the rotor.

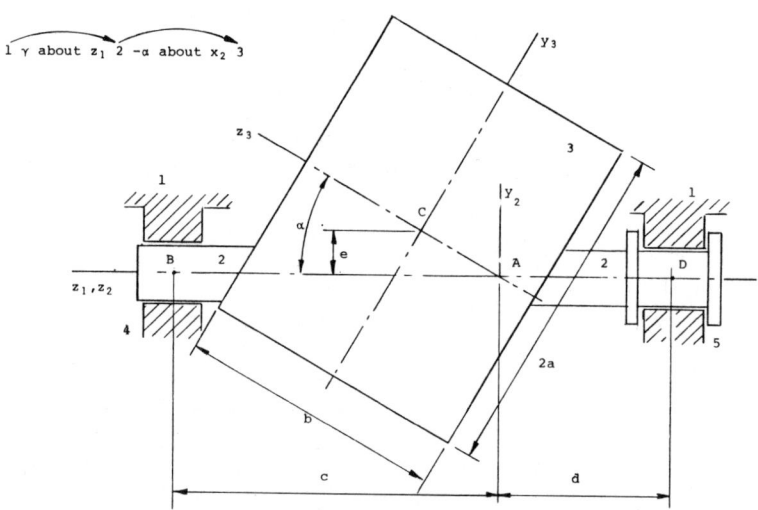

Fig. 4.61.

Problem 4.62. A uniform rectangular parallelepiped of cross-sectional dimensions 2a by 2b and length 2c, is supported by a smooth pivot as shown in Fig. 4.62. Determine the attitude of the parallelepiped when the system rotates about the vertical x_1 axis at the constant rate Ω. Calculate the particular value of γ when $\Omega = 5$ rad/s, e = 0.3 m and
(a) 2a = 0.08 m, 2b = 0.16 m and
(b) 2a = 0.16 m, 2b = 0.08 m.

Problem 4.63. A straight uniform rod, body 3, of length 2b and mass m, is connected by a smooth pin joint at one end to a vertical shaft, body 2, as shown in Fig. 4.63. The shaft is driven at a constant rate Ω relative to an inertial bearing, body 1. Obtain an expression for the steady state attitude of the rod. Deduce expressions for the natu-

Solution of Dynamics Problems

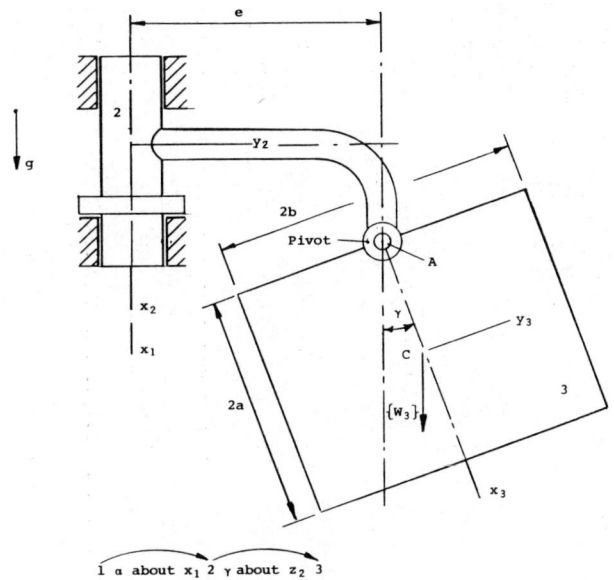

Fig. 4.62.

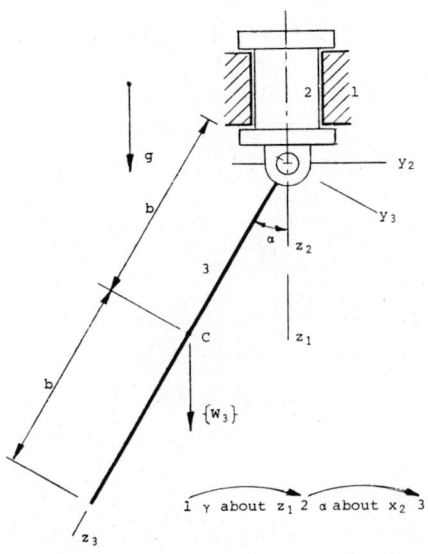

Fig. 4.63.

ral frequency of small vibrations of the rod (a) in the vertical position and (b) the inclined equilibrium position α_o. Also obtain an expression for the couple which the vertical shaft exerts on the rod.

Examine the motion of the rod for the case in which b = 0.92 m and (i) Ω = 2 rad/s and (ii) Ω = 4 rad/s.

Problem 4.64. A uniform rod, body 3, having a circular cross section of radius a and of length b, is mounted in a frame, body 2, as shown in Fig. 4.64. The motion of body 3 relative to body 2 is controlled by a torsion spring and torsion damper. The spring has a torsional stiffness k and the torsion damper exerts a viscous couple equal to $c\dot{\alpha}$. Frame 2 is driven at a constant rate Ω relative to bearings fixed in an inertial body 1.

Obtain the equation of motion for the rod when the motion relative to the equilibrium position $\alpha = \alpha_o$ is small. Consider the cases for which the spring couple is zero when (a) $\alpha_o = 0$, (b) $\alpha_o = \pi/6$, (c) $\alpha_o = \pi/4$ and (d) $\alpha_o = \pi/2$.

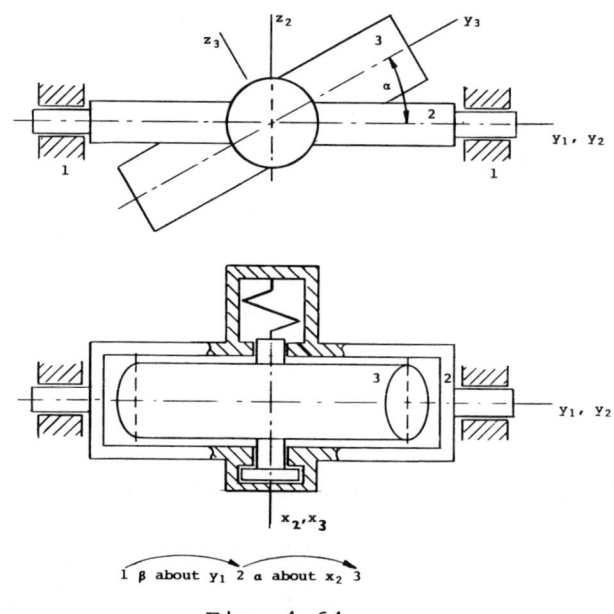

Fig. 4.64.

Problem 4.65. A uniform solid rotor, body 4, is constrained as shown in Fig. 4.65. Body 4 turns freely relative to the pivoted axle body 4. Body 3 can turn freely relative to body 2 about a pivot at A. Find

$$\{F_{45}\}_3 = \begin{bmatrix} F_1 \\ 0 \\ F_2 \end{bmatrix} \quad \text{and} \quad \{F_{32}\}_3 = \begin{bmatrix} F_3 \\ F_4 \\ F_5 \end{bmatrix}$$

for the case in which bodies 2 and 5 are driven at the constant resp-

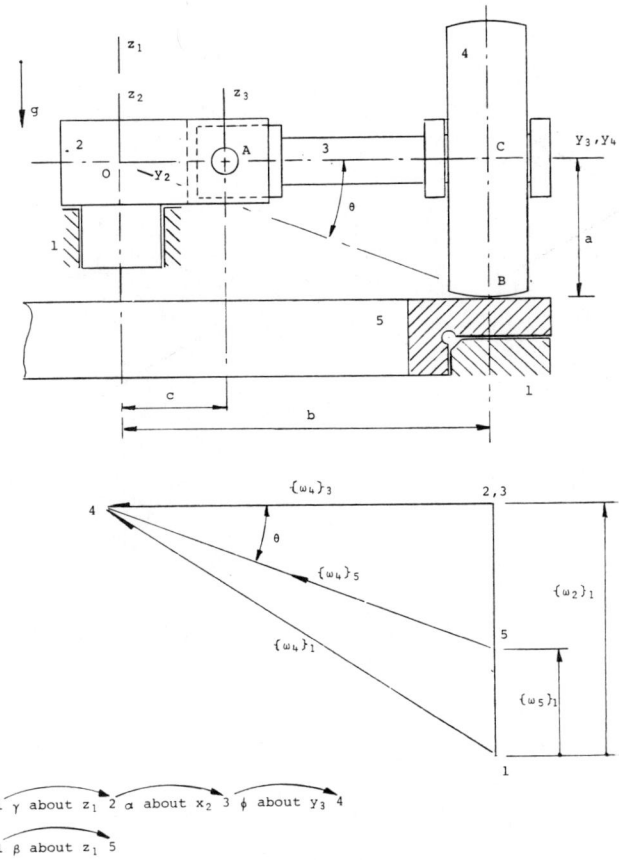

Fig. 4.65.

ective rates ω and Ω relative to the inertial body 1 and body 4 rolls without slip on body 5. Also, determine

$$T_4 = T_{4rot} + T_{4trans}.$$

Neglect the weight of body 3 and take

$$[I_4]_{4/4} = \begin{bmatrix} I & 0 & 0 \\ 0 & J & 0 \\ 0 & 0 & I \end{bmatrix}.$$

Problem 4.66. A uniform rotor, body 4, is constrained as shown in Fig. 4.66. Body 4 turns freely relative to the pivoted axle body 3. Body 3 can turn freely relative to body 2 about a pivot at A. Find

$$\{F_{41}\}_3 = \begin{bmatrix} F_1 \\ O \\ F_2 \end{bmatrix} \quad \text{and} \quad \{F_{32}\}_3 = \begin{bmatrix} F_3 \\ F_4 \\ F_5 \end{bmatrix}$$

for the case in which body 2 is driven at the constant rate ω relative to the inertial body 1 and body 4 rolls without slip on body 1. Neglect the weight of body 3 and take

$$[I_4]_{4/4} = \begin{bmatrix} I & O & O \\ O & J & O \\ O & O & I \end{bmatrix}.$$

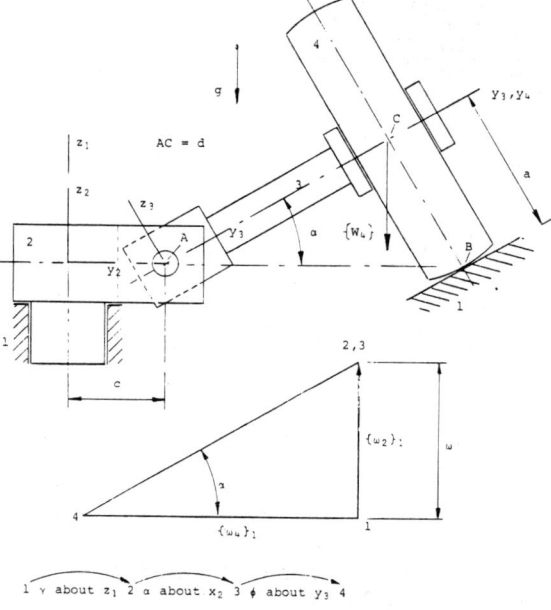

Fig. 4.66.

Problem 4.67. A solid uniform right circular cone, body 4, rolls at a constant rate and without slipping when supported as shown in Fig. 4.67. Determine the contact forces

$$\{F_{45}\}_2 = \begin{bmatrix} F_1 \\ F_2 \\ F_3 \end{bmatrix} \quad \text{and} \quad \{F_{46}\}_2 = \begin{bmatrix} F_4 \\ O \\ F_5 \end{bmatrix}$$

for the case in which $|\omega_4|_1 = \omega$. Take

$$[I_4]_{4/4} = \begin{bmatrix} I & 0 & 0 \\ 0 & J & 0 \\ 0 & 0 & I \end{bmatrix}.$$

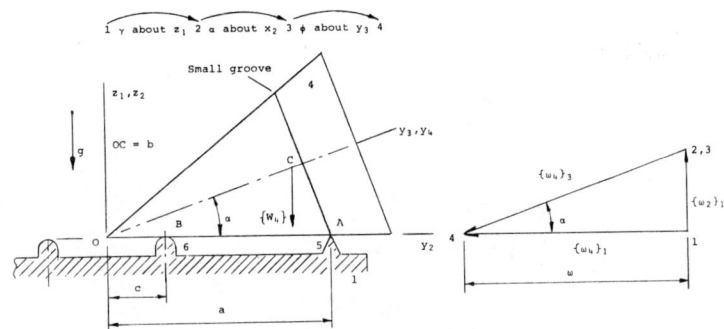

Fig. 4.67.

Problem 4.68. A uniform solid rotor, body 4, is constrained as shown in Fig. 4.68. Body 4 turns freely relative to the pivoted axle body 3. Body 3 can turn freely relative to body 2 about the pivot at A. Find

$$\{F_{45}\}_2 = \begin{bmatrix} F_1 \\ 0 \\ F_2 \end{bmatrix}$$

when the given angular velocity vector diagram applies and   and   are constants. Neglect the weight of body 3 and take

$$[I_4]_{4/4} = \begin{bmatrix} I & 0 & 0 \\ 0 & J & 0 \\ 0 & 0 & I \end{bmatrix}.$$

Problem 4.69. Body 4, which is a uniform solid of revolution, is constrained as shown in Fig. 4.69. Find the constant value of

$$|\omega_2|_1 = \dot{\beta}$$

for the case in which

$$|\omega_3|_2 = \dot{\theta} = 0 \quad \text{and} \quad |\omega_4|_3 = \omega, \text{ a constant. Also find}$$

$$\{F_{41}\}_1 = \begin{bmatrix} F_1 \\ F_2 \\ F_3 \end{bmatrix}$$

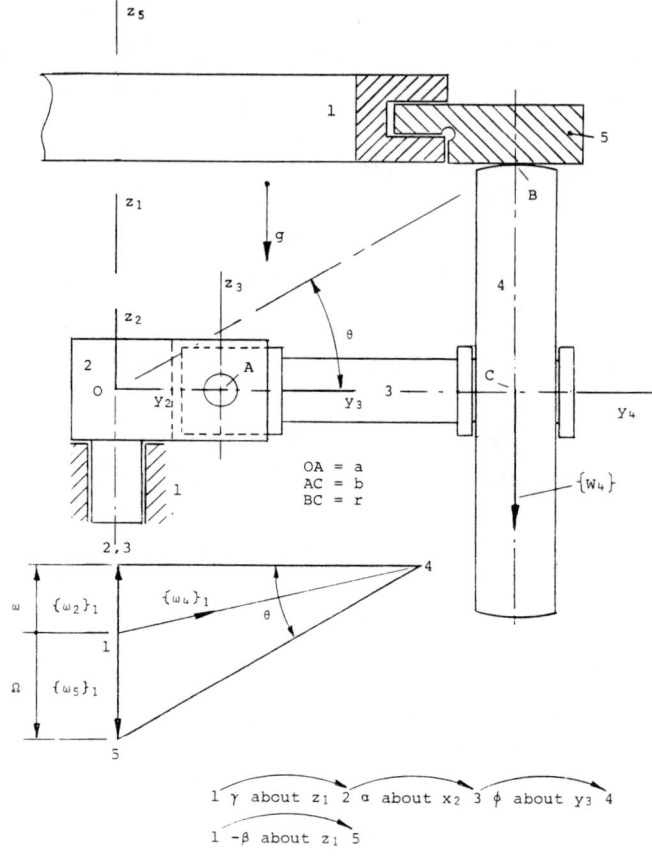

Fig. 4.68.

Take

$$[I_4]_{4/4} = \begin{bmatrix} I & 0 & 0 \\ 0 & J & 0 \\ 0 & 0 & I \end{bmatrix}.$$

Problem 4.70. In the system of Fig.4.70 the uniform heavy rotor, body 5, rotates relative to the light pivoted axle, body 3, at the constant high speed ω. Body 2 is driven at the constant rate $\dot{\alpha}$ relative to an inertial body 1. Show that, for small values of β,

$$\ddot{\beta} + \frac{\{I + mb(a + b) - J\}\dot{\alpha}^2}{I + mb^2}\beta = \frac{mgb - J\dot{\alpha}\omega}{I + mb^2}$$

where

$$[I_4]_{4/4} = \begin{bmatrix} J & 0 & 0 \\ 0 & I & 0 \\ 0 & 0 & I \end{bmatrix}.$$

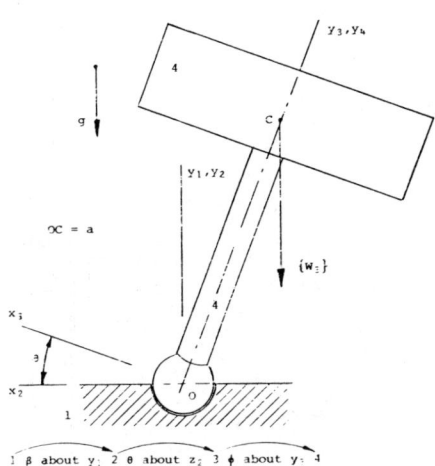

Fig. 4.69.

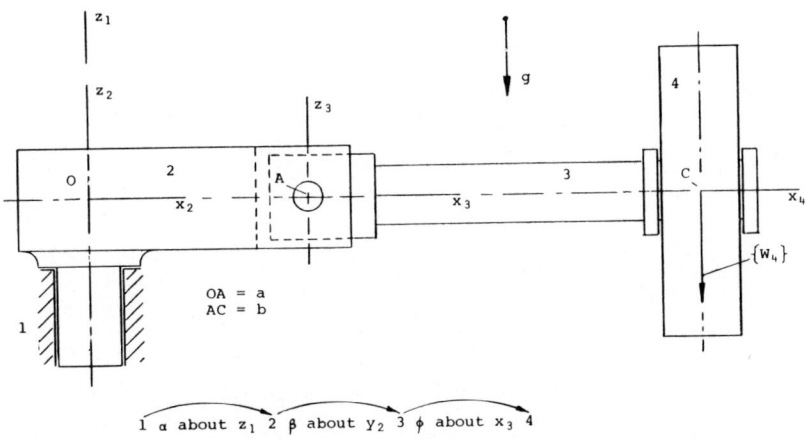

Fig. 4.70.

Problem 4.71. The uniform rotor and attached light shaft, body 4, shown in Fig. 4.71, turns at a constant high speed ω relative to body 3. The motion of body 3 relative to body 2 is constrained by torsionally elastic supports of combined stiffness k. Body 2 is driven at a constant rate $\dot{\gamma} = \Omega(<<\omega)$ about the vertical z_1 axis of an inertial frame 1. Show that, when β is small and the inertia of body 3 is neglected,

$$\ddot{\beta} + \frac{(ma^2\Omega^2 + k)}{I + ma^2}\beta = \frac{mga - J\omega\Omega}{I + ma^2}$$

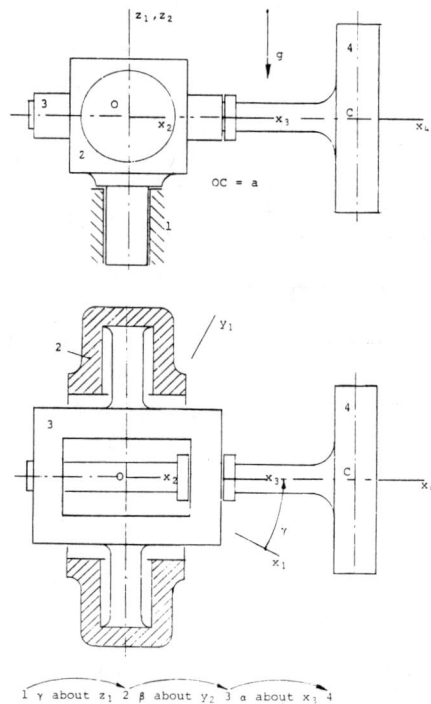

Fig. 4.71.

Take

$$[I_4]_{4/4} = \begin{bmatrix} J & 0 & 0 \\ 0 & I & 0 \\ 0 & 0 & 0 \end{bmatrix}.$$

Problem 4.72. A high speed rotor, body 4, is supported pendulously in gimbals as shown in Fig. 4.72. The outer gimbal, body 2, is free to rotate about the x_1 axis fixed in an inertial reference, while the inner gimbal, body 3, which supports the rotor directly, is free to rotate about the y_2 axis fixed in the outer gimbal. Obtain the equations of motion for the system making the following assumptions:

(i) the angular velocity of the rotor relative to the inner gimbal is constant,

(ii) the mass of bodies 2 and 3 is negligible,

(iii) the angular motion of bodies 2 and 3 relative to the given position is small and

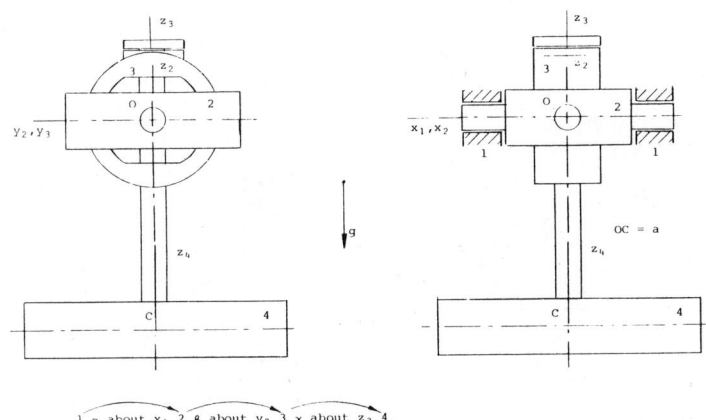

Fig. 4.72.

(iv) the $\dot{\alpha}\dot{\beta}$ products can be neglected.

Take

$$[I_4]_{4/4} = \begin{bmatrix} I & 0 & 0 \\ 0 & I & 0 \\ 0 & 0 & J \end{bmatrix}.$$

Problem 4.73. A high speed rotor, body 4, is mounted in gimbals, bodies 3 and 2, as shown in Fig. 4.73. The motion of the inner gimbal relative to the outer gimbal is controlled by a spring, body 5, of torsional stiffness k. The outer gimbal, body 2, is free to rotate about a horizontal axis through O. Obtain the equations of motion for the system appropriate to small displacements from the equilibrium position shown. Take

$$[I_4]_{4/4} = \begin{bmatrix} I & 0 & 0 \\ 0 & J & 0 \\ 0 & 0 & 0 \end{bmatrix}, \quad [I_3]_{3/3} = \begin{bmatrix} A_3 & 0 & 0 \\ 0 & B_3 & 0 \\ 0 & 0 & C_3 \end{bmatrix},$$

$$[I_2]_{2/2} = \begin{bmatrix} A_2 & 0 & 0 \\ 0 & B_2 & 0 \\ 0 & 0 & C_2 \end{bmatrix}$$

and assume that the centres of mass of bodies 2, 3 and 4 are at C.

Problem 4.74. A high speed rotor, body 4, is mounted in a light frame, body 3, as shown in Fig. 4.74. The frame can turn freely about a horizontal axis through A which is fixed in body 2. Body 2 is driven at the constant rate $\dot{\beta} = \Omega$ relative to an inertial frame 1. Obtain the γ

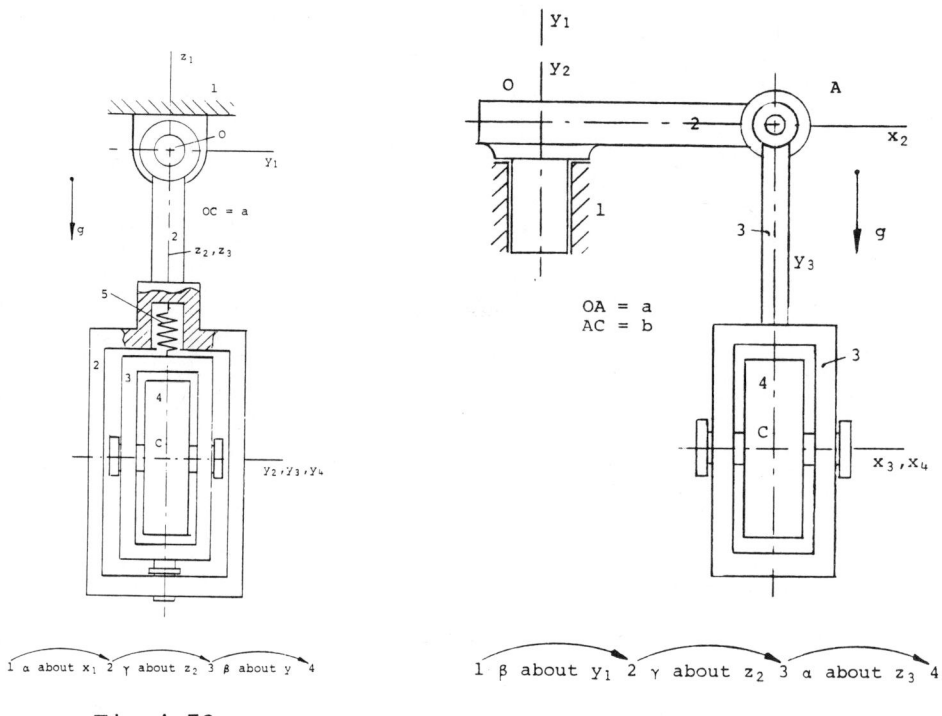

Fig.4.73.

Fig. 4.74.

equation of motion for the case in which γ is small. Also obtain the γ equation for small values of γ about the $\gamma = \gamma_o$ position.

Problem 4.75. A uniform rotor, body 4, is mounted on a light shaft, body 3, as shown in Fig. 4.75, the position for which $\beta = 0$. The rotor turns at a constant rate ω relative to the shaft. The shaft is mounted in bearings, bodies 5 and 6, which are fixed in an inertial body 1. Motion of the system is induced by applying an external couple

$$\{L_3\}_2 = \begin{bmatrix} 0 \\ L \\ 0 \end{bmatrix}$$

to the shaft, the motion of the shaft being opposed by a torsion spring, body 7, of stiffness k. Determine the y equation of motion,

$$\{F_{35}\}_2 = \begin{bmatrix} F_1 \\ F_2 \\ F_3 \end{bmatrix} \quad \text{and} \quad \{F_{36}\}_2 = \begin{bmatrix} F_4 \\ 0 \\ F_5 \end{bmatrix}.$$

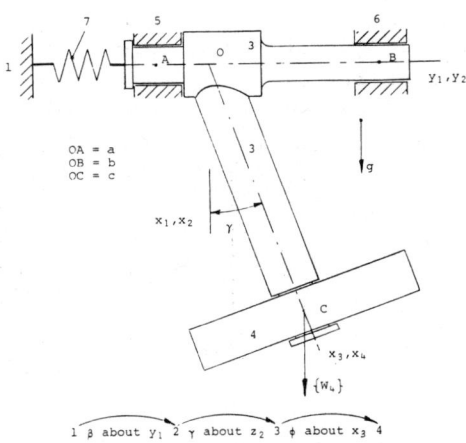

Fig. 4.75.

Problem 4.76. A uniform rotor, body 3, is mounted in bearings in body 2 and turns at a constant rate ω relative to it. Body 2 is mounted in bearings, bodies 4 and 5, which are fixed in an inertial body 1. The motion of body 2 relative to body 1 is opposed by a torsion spring, body 6, of stiffness k. Determine the natural frequency of the system by an energy method. Neglect the effects due to the inertia of body 2. btain expressions for the bearing forces

$$\{F_{24}\}_2 = \begin{bmatrix} F_1 \\ F_2 \\ F_3 \end{bmatrix} \quad \text{and} \quad \{F_{25}\}_2 = \begin{bmatrix} F_4 \\ F_5 \\ 0 \end{bmatrix}$$

for the case in which the system is excited by an external couple

$$\{L_2\}_2 = \begin{bmatrix} 0 \\ 0 \\ L \end{bmatrix}$$

which is adjusted such as to make

$\gamma = A \sin pt$.

Problem 4.77. Figure 4.77 shows, diagrammatically, the essential features of a vibration absorber which employs a gimbal mounted high speed rotor, torsion spring and damper. The vibrating system, the vibration characteristics of which are to be modified, can be represented by a torsionally elastic shaft and a rotor, body 2. The shaft has a torsional stiffness k_o and the rotor an inertia J_o about the z_2 axis. The high speed rotor, body 4, is mounted in a gimbal, body 3. The motion of the gimbal relative to body 2 is controlled by a damper and a spring.

304   Matrix Methods in Engineering Mechanics

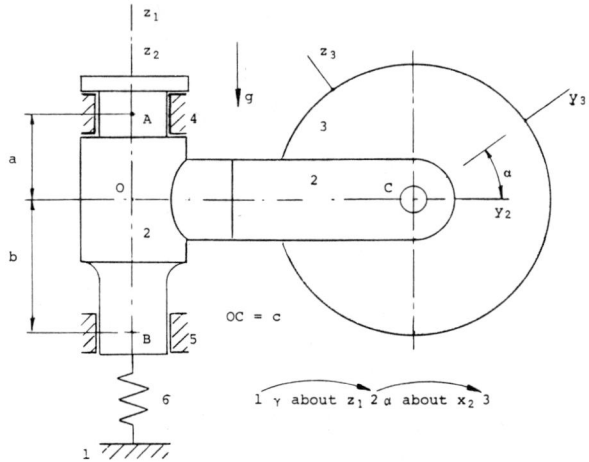

Fig. 4.76.

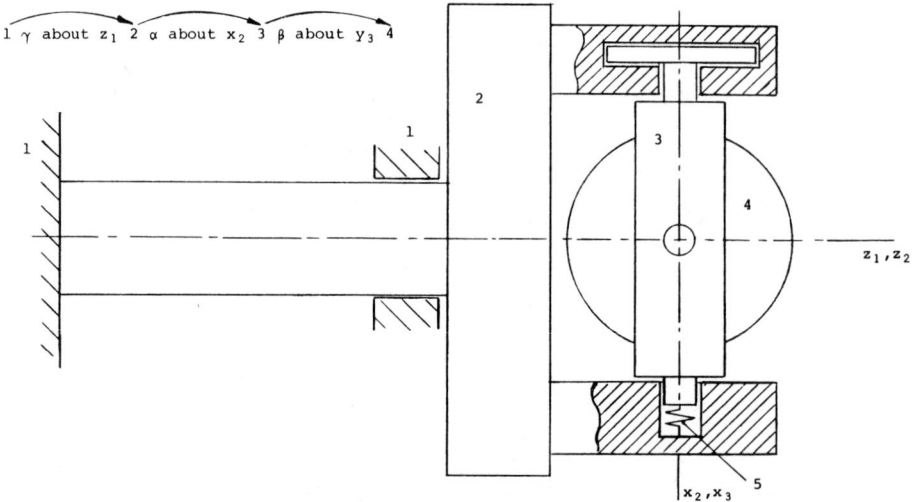

Fig. 4.77.

The damper exerts a couple $c|\omega_3|_2$ such as to oppose the relative motion. The spring has a torsional stiffness k. The system is excited by an external couple

$$\{L_2\}_2 = \begin{bmatrix} 0 \\ 0 \\ \ell \end{bmatrix}.$$

Show that, when $\ddot{\alpha}\gamma$ products are neglected and α is small, the equations

Solution of Dynamics Problems

the equations of motion can be written

$$J_o \ddot{\gamma} + h\dot{\alpha} + k_o \gamma = \ell$$

and

$$I\ddot{\alpha} - h\dot{\gamma} + c\dot{\alpha} + k\alpha = 0,$$

where

$$[I_4]_{4/4} = \begin{bmatrix} I & 0 & 0 \\ 0 & J & 0 \\ 0 & 0 & 0 \end{bmatrix}, \quad |\omega_4|_3 = \omega \quad \text{and} \quad h = J\omega.$$

Further show that, on taking Laplace transforms of the above equations of motion and writing the initial conditions equal to zero, they can be written

$$\frac{\Gamma}{L} = \frac{Is^2 + cs + k}{J_o I s^4 + cJ_o s^3 + (kJ_o + k_o I + h^2)s^2 + ck_o s + kk_o}$$

and

$$\frac{A}{L} = \frac{hs}{J_o I s^4 + cJ_o s^3 + (kJ_o + k_o I + h^2)s^2 + ck_o s + kk_o}$$

where Γ, A and L are the transforms of γ, α and ℓ respectively.

The reader is referred to Green, W. (1954) *Theory of Machines*, Blackie, London for a numerical analysis of this device. The reader is also referred to Inglis, C. (1951) *Applied Mechanics for Engineers*, Cambridge University Press, Cambridge for a discussion on gyroscope principles and applications.

Problem 4.78. A vehicle, body 3, travels due north with a velocity $v = \dot{a}R$ relative to the earth as shown in Fig. 4.78. A rotor, body 5, which is driven at a constant high speed ω relative to a gimbal, body 4, is mounted in the vehicle. The gimbal is free to turn relative to the vehicle about the vertical y_3 axis. Show that, when products of $\dot{\gamma} = \Omega$, $\dot{\alpha}$ and $\dot{\beta}$ are neglected, the equation of motion for the rotor reduces to

$$\ddot{\beta} + \frac{h\Omega}{I}\cos\alpha\sin\beta = \frac{\dot{\alpha}}{I}\cos\beta$$

where

$$[I_5]_{5/5} = \begin{bmatrix} I & 0 & 0 \\ 0 & I & 0 \\ 0 & 0 & J \end{bmatrix} \quad \text{and} \quad h = J\omega$$

and hence that the axle of the rotor has a north seeking property, but there is a steady state deflection

$$\tan\beta = v/(\Omega R\cos\alpha)$$

from true north.

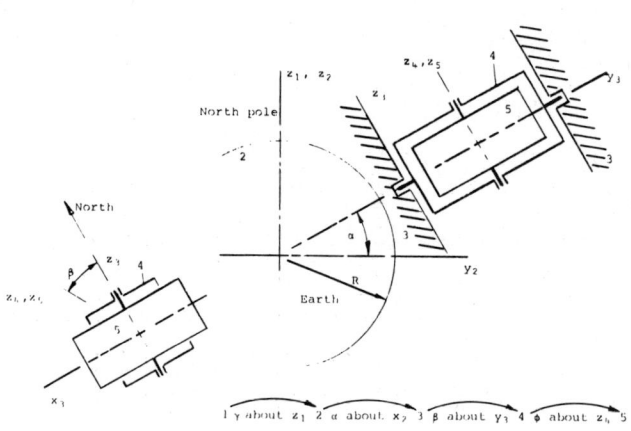

Fig. 4.78.

Problem 4.79. A thin uniform circular disc, body 4, of radius r rolls without slipping on an inertial horizontal surface, body 1, as shown in Fig. 4.79. The path of the centre of mass of body 4, drawn on body 1 is a circle of radius a. Show that the velocity of the centre of mass of the disc is given by

$$v^2 = \frac{4a^2 g \tan\alpha}{6a + r\sin\alpha} \; .$$

Also show that

$$\{F_{41}\}_2 = \begin{bmatrix} 0 \\ mv^2/a \\ mg \end{bmatrix} \; .$$

Problem 4.80. A thin uniform rod, body 2, is retained in a smooth spherical bearing fixed in an inertial body 1 at one end and is supported at a point B part way along its length on a smooth horizontal ridge on body 3, fixed relative to body 1, as shown in Fig. 4.80. Given the relative positions of A, B, C and D, derive a set of equations from which the following could be determined at the instant the rod is released from rest:
    (i) the angular acceleration of the rod,
    (ii) the force on the rod at A and
    (iii) the force on the rod at B.
Organise the equations into the matrix form

$$[a]\{b\} = \{c\}$$

where $\{b\}$ is a column of unknown quantities and $[a]$ is an array of known quantities.

Solution of Dynamics Problems

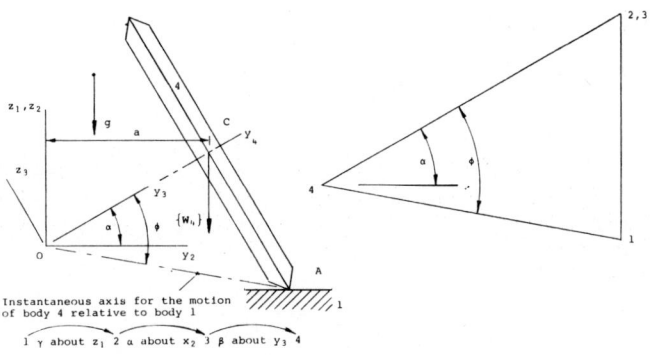

Fig. 4.79.

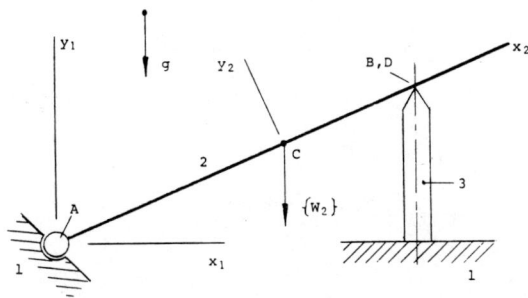

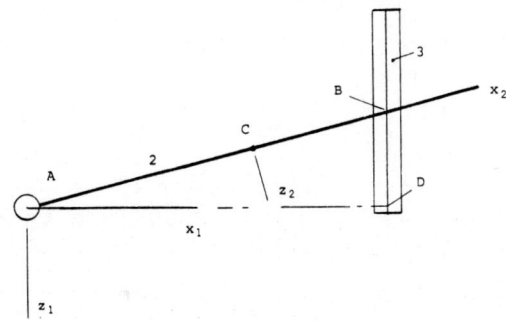

Fig. 4.80.

Problem 4.81. A projectile, body 2, which can be treated as a homogeneous solid of revolution, has the angular velocity

$$\{\omega_2\}_{1/2} = \begin{bmatrix} 0 \\ \Omega \\ 0 \end{bmatrix}$$

as shown in Fig. 4.81. The y_2 axis is the axis of generation of the projectile and frame is is inertial. Show that the frequency of the oscillations resulting from an externally applied impulsive couple which is perpendicular to the axis of generation is, in the absence of damping effects, given by

$$\frac{|A - B|}{B}\Omega$$

where

$$[I_2]_{2/2} = \begin{bmatrix} B & 0 & 0 \\ 0 & A & 0 \\ 0 & 0 & B \end{bmatrix}$$

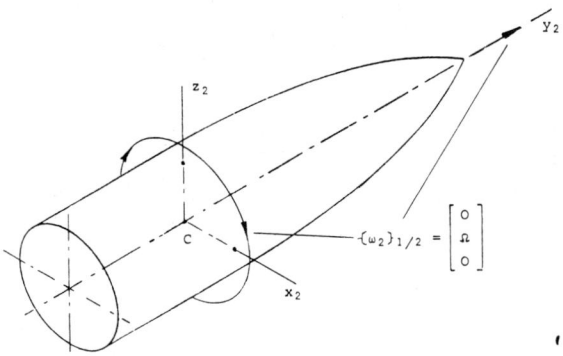

Fig. 4.81.

Problem 4.82. A thin uniform disc, body 4, of radius a and mass m is mounted on a light shaft. The shaft runs in a bearing, body 3 as shown in Fig. 4.82. Body is freely pivoted to body 2 and body 2 is free to turn about a vertical axis. The disc is released from rest in the position shown with body 4 rotating at a high speed $\dot{\psi}_o = \omega$ relative to body 3. Show that when the mass of bodies 2 and 3 is neglected

$$\dot{\omega} = \dot{\psi} - \dot{\phi}\sin\beta, \quad \ddot{\phi} = (2\omega\sin\beta)/(5\cos^2\beta)$$

and

$$\dot{\psi} = \omega(1 + 2\tan^2\beta).$$

Note that since there are no external couples on the disc about the y_3 and z_2 axes there can be no change of angular momentum about them. Use the principle of energy conservation to determine ω for a given maximum value of β.

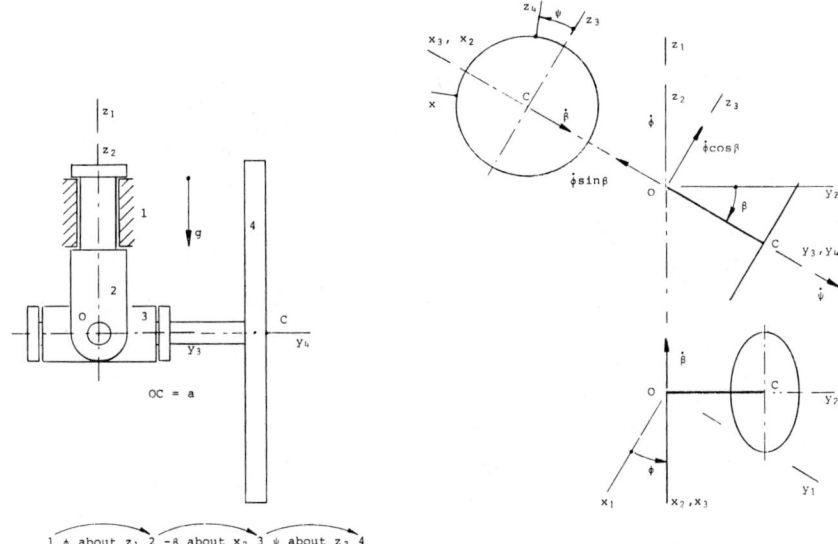

Fig. 4.82.

Problem 4.83. If a rigid body, body 2, is not subjected to external couples, then the equation for its rotational motion can be written

$$[I_2]_{2/2}\{\dot{\omega}_2\}_{1/2} + \lfloor \omega_2 \rfloor_{1/2}[I_2]_{2/2}\{\omega_2\}_{1/2} = \{0\}$$

where

$$[I_2]_{2/2} = \begin{bmatrix} A & 0 & 0 \\ 0 & B & 0 \\ 0 & 0 & C \end{bmatrix}.$$

Show that, on premultiplying this equation by $\{\omega_2\}^T_{1/2}$, the rotational kinetic energy of body 2 is constant.

Also show that, on premultiplying the equation by

$$\{\omega_2\}^T_{1/2}[I_2]_{2/2} .$$

the angular momentum of body 2 is constant.

Problem 4.84. Figure 4.84 shows a mechanism which is driven through the position for which $\theta = 30°$ by an external couple applied to body 2

$$\{L_2\}_1 = \begin{bmatrix} 0 \\ 0 \\ L_z \end{bmatrix}$$

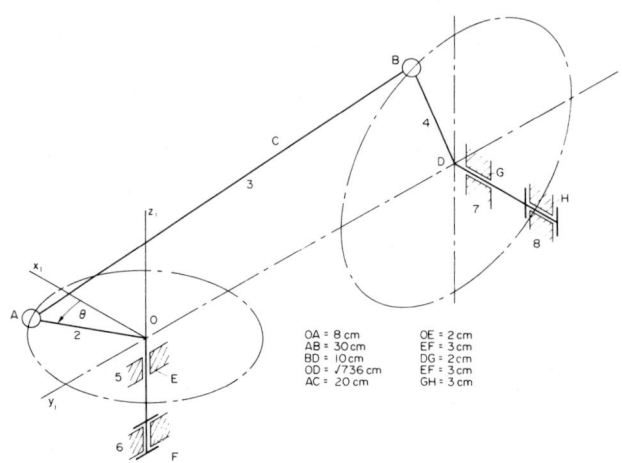

Fig. 4.84.

when

$$\{\omega_2\}_{1/1} = \begin{bmatrix} 0 \\ 0 \\ 100 \end{bmatrix} \text{ rad/s} \quad \text{and} \quad \{\dot{\omega}_2\}_{1/1} = \begin{bmatrix} 0 \\ 0 \\ -10^4 \end{bmatrix} \text{ rad/s}^2$$

Body 3 is a solid of revolution with its axis of generation along AB, the line joining the frictionless ball joints at its ends. The mass of body 3 is 0.6 kg and its moment of inertia about an axis through C, the centre of mass, and perpendicular to AB is 0.005 kg m^2. Determine that part of L_z which is due to the mass of body 3. Neglect the effect of the weight of body 3.

Also determine the forces on bodies 2 and 4 due to the short bearings 5, 6, 7 and 8. Bearings 5 and 7 can sustain only radial forces while bearings 6 and 8 can sustain both radial and axial forces.

Problem 4.85. Figure 4.85 shows a solid uniform wedge, body 2. Determine

$$[I_2]_{3/3} = \begin{bmatrix} A_3 & 0 & 0 \\ 0 & B_3 & F_3 \\ 0 & F_3 & C_3 \end{bmatrix} \quad \text{and} \quad [I_2]_{2/2} = \begin{bmatrix} A_2 & 0 & 0 \\ 0 & B_2 & F_2 \\ 0 & F_2 & C_2 \end{bmatrix}$$

in terms of the dimensions of the wedge.

Also determine, for the case in which a, b and c are in the ratio 5:3:2,

$$[I_2]_{3/4} = \begin{bmatrix} A_4 & 0 & 0 \\ 0 & B_4 & 0 \\ 0 & 0 & C_4 \end{bmatrix}, \quad [I_2]_{2/5} = \begin{bmatrix} A_5 & 0 & 0 \\ 0 & B_5 & 0 \\ 0 & 0 & C_5 \end{bmatrix},$$

$[\ell_3]_4$ and $[\ell_2]_5$,

where frames 4 and 5 are sets of principal axes corresponding to the points A and C respectively.

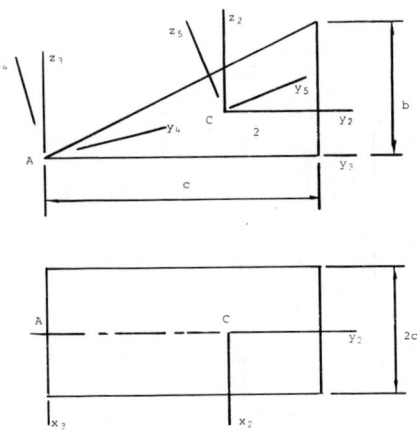

Fig. 4.85.

Problem 4.86. Body 4 is a composite of the uniform steel bodies 2 and 3 as shown in Fig. 4.86. Determine

$$[I_4]_{4/4} \quad \text{and} \quad [I_4]_{4/5}$$

where frame 5 is the set of principal axes through C, the centre of mass of body 4. Also find

$[\ell_4]_5$.

Take the density of steel as 7.8 g/cm^3.

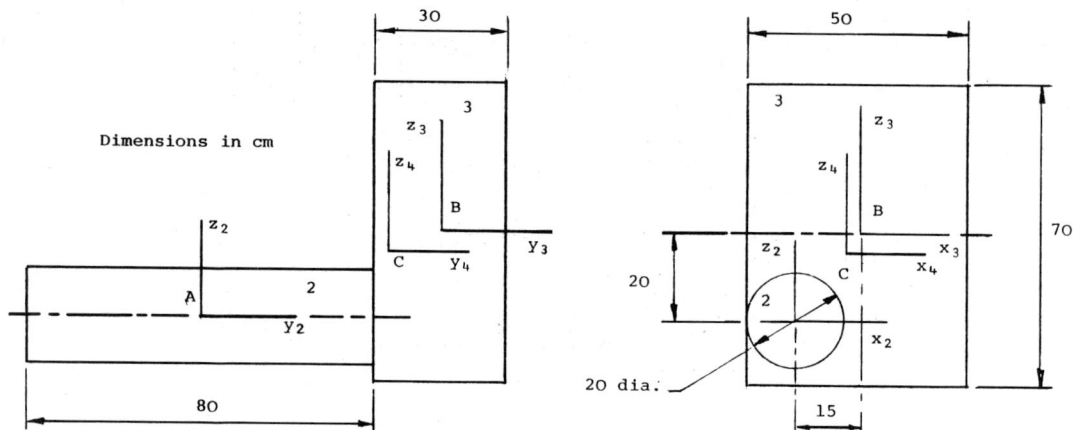

Fig. 4.86.

Answers to Problems for Solution

Chapter 3

3.26.
$$\{V_{BA}\}_{2/2}\big|_{t=0} = \{v_B - (v_A b/a)\}\begin{bmatrix}0\\1\\0\end{bmatrix}$$

$$\{A_{BA}\}_{2/2}\big|_{t=0} = \{(2v_A v_B/a) - (2a+b)v_A^2/a^2\}\begin{bmatrix}1\\0\\0\end{bmatrix}$$

3.27.
$$\{V_{BA}\}_{2/2}\big|_{t=0} = \{v_B - (v_A c/a)\}\begin{bmatrix}0\\1\\0\end{bmatrix}$$

$$\{A_{BA}\}_{2/2}\big|_{t=0} = \begin{bmatrix}(2v_A v_B/a) - v_A^2 c/a^2\\ 0\\ v_B^2/b\end{bmatrix}$$

3.28.

$$1 \overset{\theta \text{ about } z_1}{\longrightarrow} 2 \overset{-\phi \text{ about } y_2}{\longrightarrow} 3$$

$$[\ell_2]_1 = \begin{bmatrix}\cos\theta & -\sin\theta & 0\\ \sin\theta & \cos\theta & 0\\ 0 & 0 & 1\end{bmatrix} \quad [\ell_3]_2 = \begin{bmatrix}\cos\phi & 0 & -\sin\phi\\ 0 & 1 & 0\\ \sin\phi & 0 & \cos\phi\end{bmatrix}$$

3.29.
$$[\ell_2]_1 = \begin{bmatrix}0.7926 & 0.457 & 0.4036\\ -0.6097 & 0.594 & 0.5247\\ 0 & 0.662 & 0.7495\end{bmatrix}$$

(ctd.)

$$= \begin{bmatrix} c\gamma c\beta & -s\gamma c\beta & s\beta \\ s\gamma c\alpha + s\alpha c\gamma s\beta & c\alpha c\gamma - s\alpha s\gamma s\beta & -s\alpha c\beta \\ s\gamma s\alpha - c\alpha c\gamma s\beta & s\alpha c\gamma + c\alpha s\gamma s\beta & c\alpha c\beta \end{bmatrix}$$

$$= \begin{bmatrix} c\phi & s\phi c\psi & -s\phi s\psi \\ -s\phi & c\phi c\psi & -c\phi s\psi \\ 0 & s\psi & c\psi \end{bmatrix}$$

3.30. Start with

$$\{R_{AO}\}_{3/3} = [\ell_2]_3[\ell_1]_2\{R_{AO}\}_{1/1}$$

3.32.

$$\{R_{QA}\}_{1/1} = \begin{bmatrix} 243.6 \\ 140.3 \\ 46.51 \end{bmatrix} \text{ mm} \quad \{R_{SB}\}_{1/1} = \begin{bmatrix} 0 \\ 165.46 \\ -3.44 \end{bmatrix} \text{ mm}$$

$$\{V_Q\}_{1/1} = 1.19 \begin{bmatrix} 9.75 \\ -1 \\ -48 \end{bmatrix} \text{ mm/s}$$

Direction cosines of $\{\omega_3\}_{1/1}$

$$\cos\alpha = \frac{0.346}{\sqrt{3.031}} = 0.1987 \quad \cos\beta = \frac{-0.0355}{\sqrt{3.031}} = -0.0204$$

$$\cos\gamma = \frac{-1.706}{\sqrt{3.031}} = -0.9798$$

Equation to central axis, $X(x,y,z)$

$$|R_{XQ}| = \frac{x - 286.9}{0.1987} = \frac{y - 165}{-0.0204} = \frac{z - 46.5}{-0.9789}$$

$z = 0$, $x = 287$ mm, $y = 164.5$ mm.

3.36.

$$\{R_{BO}\}_{1/1} = [\ell_2]_1\{R_{AO}\}_{2/2} + [\ell_2]_1\{R_{BA}\}_{2/2}$$

$$\{R_{BA}\}_{2/2} = na \begin{bmatrix} \cos\beta \\ \sin\beta \\ b \end{bmatrix}$$

$$\{R_{BO}\}_{1/1} = [\ell_2]_1\{R_{AO}\}_{2/2} + [\ell_2]_1[\ell_3]_2\{R_{BA}\}_{3/3}$$

$$\{R_{BA}\}_{3/3} = \begin{bmatrix} na \\ 0 \\ b \end{bmatrix}$$

(ctd.)

Answers

$$\{V_{BO}\}_{1/2} = \begin{bmatrix} -a\dot{\alpha} - na(\dot{\alpha} + \dot{\beta})\sin\beta \\ na(\dot{\alpha} + \dot{\beta})\cos\beta \\ 0 \end{bmatrix}$$

$$\{A_{BO}\}_{1/2} = \begin{bmatrix} -na(\dot{\alpha} + \dot{\beta})^2 \cos\beta \\ -a\dot{\alpha}^2 - na(\dot{\alpha} + \dot{\beta})^2 \sin\beta \\ 0 \end{bmatrix}$$

3.37.

$$\{R_{PO}\}_{1/1} = [\ell_2]_1 \{R_{AO}\}_{1/1} + [\ell_2]_1 [\ell_3]_2 \{R_{PA}\}_{3/3}$$

$$\{V_{PO}\}_{1/2} = \begin{bmatrix} -a\dot{\alpha} - x(\dot{\alpha} + \dot{\beta})\sin\beta + v\cos\beta \\ x(\dot{\alpha} + \dot{\beta})\cos\beta + v\sin\beta \\ 0 \end{bmatrix}$$

$$\{A_{PO}\}_{1/2} = \begin{bmatrix} -2v(\dot{\alpha} + \dot{\beta})\sin\beta - x(\dot{\alpha} + \dot{\beta})^2 \cos\beta \\ 2v(\dot{\alpha} + \dot{\beta})\cos - x(\dot{\alpha} + \dot{\beta})^2 \sin\beta - a\dot{\alpha}^2 \\ 0 \end{bmatrix}$$

3.38.

$$\{A_{BO}\}_{1/2} = \begin{bmatrix} 2\dot{\alpha}\dot{\gamma}b\sin\gamma - c\dot{\alpha}^2 \\ -\dot{\alpha}^2(a + b\cos\gamma) - b\dot{\gamma}^2\cos\gamma \\ -b\dot{\gamma}^2\sin\gamma \end{bmatrix}$$

3.39.

$$\{V_{QO}\}_{1/2} = \omega \begin{bmatrix} a\{(b\cos\beta/r) - 1\} \\ b\sin\beta \\ ab\sin\beta/r \end{bmatrix} \qquad \{V_{PO}\}_{1/2} = \omega \begin{bmatrix} a(\cos\beta - 1) \\ r\sin\beta \\ a\sin\beta \end{bmatrix}$$

$$\{A_{QO}\}_{1/2} = \omega^2 \begin{bmatrix} -b\sin\beta(1 + a^2/r^2) \\ a\{(2b\cos\beta/r) - 1\} \\ a^2 b\cos\beta/r^2 \end{bmatrix}$$

$$\{A_{PO}\}_{1/2} = \omega^2 \begin{bmatrix} -r\sin\beta(1 + a^2/r^2) \\ a(2\cos\beta - 1) \\ a^2 \cos\beta/r \end{bmatrix}$$

(ctd.)

Position of P	A $\beta = 0$	B $\beta = \pi/2$	E $\beta = \pi$	D $\beta = -\pi/2$
$\{V_{PO}\}_{1/2}$	$\begin{bmatrix} 0 \\ 0 \\ 0 \end{bmatrix}$	$\omega \begin{bmatrix} -a \\ r \\ a \end{bmatrix}$	$\omega \begin{bmatrix} -2a \\ 0 \\ 0 \end{bmatrix}$	$\omega \begin{bmatrix} -a \\ -r \\ -a \end{bmatrix}$
$\{A_{PO}\}_{1/2}$	$\omega^2 \begin{bmatrix} 0 \\ a \\ a^2/r \end{bmatrix}$	$\omega^2 \begin{bmatrix} r(1+a^2/r^2) \\ -a \\ 0 \end{bmatrix}$	$\omega^2 \begin{bmatrix} 0 \\ -3a \\ -a^2/r \end{bmatrix}$	$\omega^2 \begin{bmatrix} -r(1+a^2/r^2) \\ -a \\ 0 \end{bmatrix}$

3.40.

$$\lambda = \omega_4/\omega_2 \qquad \{\omega_3\}_{2/2} = \omega_2(\lambda - 1) \begin{bmatrix} 0 \\ a/b \\ 0 \end{bmatrix}$$

$$\{\omega_3\}_{1/2} = \omega_2 \begin{bmatrix} 0 \\ a(\lambda - 1) \\ 1 \end{bmatrix} \qquad \{\dot{\omega}_3\}_{1/2} = \omega_2^2 (\lambda - 1) \begin{bmatrix} -a/b \\ 0 \\ 0 \end{bmatrix}$$

$$\{V_{AO}\}_{1/2} = a\omega_2 \begin{bmatrix} -(\lambda - 1)\cos\beta - 1 \\ b\sin\beta/r \\ -(\lambda - 1)\sin\beta \end{bmatrix}$$

$$\{A_{AO}\}_{1/2} = a\omega_2^2 \begin{bmatrix} -\{b^2 + (\lambda - 1)^2\}\sin\beta/ab \\ -2(\lambda - 1)\cos\beta - 1 \\ a(\lambda - 1)^2\cos\beta/b \end{bmatrix}$$

3.41.

$$\{\omega_4\}_{1/3} = \dot{\gamma} \begin{bmatrix} 0 \\ -(R + a\cos\alpha)/r \\ \cos\alpha \end{bmatrix}$$

$$\{\dot{\omega}_4\}_{1/3} = \dot{\gamma}^2 \cos\alpha (R + a\cos\alpha + r\sin\alpha) \begin{bmatrix} 1/r \\ 0 \\ 0 \end{bmatrix}$$

$$\{\dot{\omega}_5\}_{1/3} = \dot{\gamma}^2 \cos\alpha (R - a\cos\alpha + r\sin\alpha) \begin{bmatrix} 1/r \\ 0 \\ 0 \end{bmatrix}$$

(ctd.)

$$\{V_{DO}\}_{1/3} = \{V_{DE}\}_{1/3} = 2a\dot{\gamma}\cos\alpha \begin{bmatrix} 1 \\ 0 \\ 0 \end{bmatrix}$$

3.42.

$$\{\omega_3\}_{1/2} = \begin{bmatrix} \dot{\alpha} \\ 0 \\ \dot{\beta} \end{bmatrix} \qquad \{\dot{\omega}_3\}_{1/2} = \begin{bmatrix} 0 \\ \dot{\alpha}\dot{\beta} \\ 0 \end{bmatrix}$$

$$\{V_{AO}\}_{1/2} = \begin{bmatrix} -z\dot{\beta}\sin\alpha \\ z\dot{\alpha}\cos\alpha + v\sin\alpha \\ -z\dot{\alpha}\sin\alpha + v\cos\alpha \end{bmatrix}$$

$$\{A_{AO}\}_{1/2} = \begin{bmatrix} -2\dot{\beta}(v\sin\alpha + z\dot{\alpha}\cos\alpha) \\ 2v\dot{\alpha}\cos\alpha - (\dot{\alpha}^2 + \dot{\beta}^2)z\sin\alpha \\ -2v\dot{\alpha}\sin\alpha - z\dot{\alpha}^2\cos\alpha \end{bmatrix}$$

3.43.

$$\{V_{EO}\}_{1/3} = a\dot{\alpha}\begin{bmatrix} -\sin\theta - \lambda_2 \\ \cos\theta + \lambda_1 + \lambda_3\sin\beta_1 \\ 0 \end{bmatrix} + a\dot{\theta}\begin{bmatrix} \lambda_2 \\ -\lambda_1 \\ 0 \end{bmatrix}$$

$$+ a\lambda_3\begin{bmatrix} \dot{\beta}_1\cos\beta_1 \\ -\dot{\theta}\sin\beta_1 \\ \dot{\beta}_1\sin\beta_1 \end{bmatrix}$$

$$\lambda_3\dot{\beta}_1 = \dot{\alpha}(\sin\theta + \lambda_2) \qquad \lambda_3\dot{\beta}_2 = \dot{\alpha}(\sin\theta - \lambda_2)$$

$$\{A_{EO}\}_{1/3} = \lfloor \omega_3 \rfloor_{1/3}\{V_{EO}\}_{1/3} + \frac{d}{dt}\{V_{EO}\}_{1/3}$$

For $\dot{\theta} = 0$ and $\dot{\alpha}$ and $\dot{\beta}_1$ constants, when E is at the point of contact $\{V_{EO}\}_1$ is a null vector. Hence

$$\left\{\frac{d}{dt}\{V_{EO}\}_{1/3}\right\}\bigg|_{\dot{\alpha},\dot{\theta},\dot{\beta}_1 \text{ constant}}$$

$$= a\dot{\alpha}\begin{bmatrix} 0 \\ \lambda_3\dot{\beta}_1\cos\beta_1 \\ 0 \end{bmatrix} + a\lambda_3\begin{bmatrix} -\dot{\beta}_1^2\sin\beta_1 \\ 0 \\ \dot{\beta}_1^2\cos\beta_1 \end{bmatrix}$$

When E is at the point of contact, $\beta_1 = 0$ and

$$\{A_{EA}\}_{1/3} = a\dot{\alpha}^2(\sin\theta + \lambda_2)\begin{bmatrix} 0 \\ 1 \\ (\sin\theta + \lambda_2)/\lambda_3 \end{bmatrix}$$

3.44.
$$\{\dot{\omega}_4\}_{1/3} = \begin{bmatrix} n\omega(\omega\cos\gamma - \omega_s) \\ -\omega\omega_s\sin\gamma \\ -n\omega^2\sin\gamma \end{bmatrix}$$

3.45. To ensure that rubbing does not occur at F arrange that F is on OE.

3.46.
$$\{\omega_4\}_{1/3} = \{\omega_3\}_{1/3} + \{\omega_4\}_{3/3}$$

$$\omega \begin{bmatrix} t_2/t_3 \\ 0 \\ 0 \end{bmatrix} + \{\Omega - (\omega t_2/t_1)\} \begin{bmatrix} 0 \\ 0 \\ t_5/t_4 \end{bmatrix}$$

$$\{\dot{\omega}_4\}_{1/3} = \omega\{\Omega - (\omega t_2/t_1)\} \begin{bmatrix} 0 \\ t_2 t_5/t_3 t_4 \\ 0 \end{bmatrix}$$

$$|\omega_6|_1 = (2\omega t_2/t_3) - \Omega \qquad |\omega_4|_3 = t_5\{\Omega - (\omega t_2/t_3)\}/t_4$$

3.47.
$$\{\omega_4\}_{1/2} = \begin{bmatrix} 0 \\ -t_3(\omega - \Omega)/t_4 \\ \Omega \end{bmatrix} \qquad \{\dot{\omega}_4\}_{1/2} = \omega(\omega - \Omega) \begin{bmatrix} t_3/t_4 \\ 0 \\ 0 \end{bmatrix}$$

$$\{V_{AO}\}_{1/2} = \begin{bmatrix} -\omega r - t_3 R\cos\beta(\omega - \Omega)/t_4 \\ \Omega R\sin\beta \\ t_3 R\sin\beta(\omega - \Omega)/t_4 \end{bmatrix}$$

$$\{A_{AO}\}_{1/2} = \begin{bmatrix} R\sin\beta\{t_3^2(\omega - \Omega)^2/t_4^2 - \omega\Omega\} \\ -\omega^2 r + R\cos\beta\{t_3^2(\omega - \Omega)^2/t_4^2 - t_3\omega(\omega - \Omega)/t_4\} \\ Rt_3^2\cos\beta(\omega - \Omega)^2/t_4^2 \end{bmatrix}$$

3.49. From the angular velocity vector diagram

$$|\omega_3|_2 = \Omega a/b = r_3\Omega \qquad |\omega_4|_3 = nr_3\Omega = r_4\Omega$$
$$|\omega_5|_2 = 2r_3\Omega = r_5\Omega$$

$$\{\omega_4\}_{1/3} = \Omega \begin{bmatrix} \sin\beta \\ -r_3 \\ r_4 + \cos\beta \end{bmatrix} \qquad \{\dot{\omega}_4\}_{1/3} = \Omega^2 \begin{bmatrix} r_3(\cos\beta - r_4) \\ -r_4\sin\beta \\ -r_3\sin\beta \end{bmatrix}$$

from which $\{\omega_4\}_{1/2}$ and $\{\dot{\omega}_4\}_{1/2}$ can be determined on premultiplying the above results by

(ctd.)

$$[\ell_3]_2 = \begin{bmatrix} \cos\beta & 0 & -\sin\beta \\ 0 & 1 & 0 \\ \sin\beta & 0 & \cos\beta \end{bmatrix}$$

$$\{\omega_5\}_{1/2} = \Omega \begin{bmatrix} 0 \\ -2r_3 \\ 1 \end{bmatrix} \qquad \{\dot{\omega}_5\}_{1/2} = 2r_3\Omega^2 \begin{bmatrix} 1 \\ 0 \\ 0 \end{bmatrix}$$

3.50.
$$\{V_{AC}\}_{2/1} = \begin{bmatrix} 0 \\ 0 \\ -1.1 \end{bmatrix} \text{m/s} \qquad \{A_{AC}\}_{2/1} = \begin{bmatrix} 0 \\ 0 \\ -8.775 \end{bmatrix} \text{m/s}^2$$

$$\{\omega_4^n\}_{1/1} = \begin{bmatrix} -1.768 \\ 1.989 \\ 1.9776 \end{bmatrix} \text{rad/s} \qquad \{\dot{\omega}_4^n\}_{1/1} = \begin{bmatrix} 5.738 \\ 5.842 \\ 15.584 \end{bmatrix} \text{rad/s}^2$$

$$\{A_{AC}\}_{2/1} = \begin{bmatrix} 0 \\ 0 \\ -8.775 \end{bmatrix} \text{m/s}^2 \qquad \{A_{CB}^n\}_{1/1} = \begin{bmatrix} 7.54 \\ -4.4 \\ -1.127 \end{bmatrix} \text{m/s}^2$$

$$\{A_{AO}^P\}_{1/1} = \begin{bmatrix} 0 \\ 0 \\ -12.36 \end{bmatrix} \text{m/s}^2 \qquad \{A_{BC}^P\}_{1/1} = \begin{bmatrix} 2.2 \\ 4.4 \\ 2.458 \end{bmatrix} \text{m/s}^2$$

$$2[\omega_2]_{1/1}\{V_{AC}\}_{2/1} = \begin{bmatrix} -22.1 \\ 0 \\ 0 \end{bmatrix} \text{m/s}^2$$

3.51.
$$\{\omega_4\}_{1/1} = \{\omega_2\}_{1/1} + [\ell_2]_1\{\omega_3\}_{2/2} + [\ell_2]_1[\ell_3]_2\{\omega_4\}_{3/3}$$

$$= \begin{bmatrix} \dot{\psi} \\ 0 \\ 0 \end{bmatrix} + \begin{bmatrix} 1 & 0 & 0 \\ 0 & c\psi & -s\psi \\ 0 & s\psi & c\psi \end{bmatrix} \begin{bmatrix} 0 \\ \dot{\theta} \\ 0 \end{bmatrix}$$

$$+ \begin{bmatrix} 1 & 0 & 0 \\ 0 & c\psi & -s\psi \\ 0 & s\psi & c\psi \end{bmatrix} \begin{bmatrix} c\theta & 0 & s\theta \\ 0 & 1 & 0 \\ -s\theta & 0 & c\theta \end{bmatrix} \begin{bmatrix} 0 \\ 0 \\ \dot{\phi} \end{bmatrix}$$

$$\{\dot{\omega}_4\}_{1/3} = -[\omega_4]_{3/3}\{\omega_3\}_{1/3} + \frac{d}{dt}\{\omega_4\}_{1/3}$$

(ctd.)

$$\{V_{CA}\}_{1/3} = \lfloor \omega_4 \rfloor_{1/3} \lfloor \ell_4 \rfloor_3 \{R_{CA}\}_{4/4}$$

$$= b \begin{bmatrix} -(\dot{\psi}\sin\theta + \dot{\phi})\sin\phi \\ (\dot{\psi}\sin\theta + \dot{\phi})\cos\phi \\ -\dot{\theta}\cos\phi + \psi\cos\theta\sin\phi \end{bmatrix}$$

$$\{V_{CP}\}_{1/3} = \{V_{CA}\}_{1/3}\Big|_{\phi=0} = \{V_4\}_{1/3}$$

$$\{A_4\}_{1/3} = \lfloor \omega_3 \rfloor_{1/3} \{V_4\}_{1/3} + \frac{d}{dt}\{V_4\}_{1/3}$$

Answers to Problems for Solution

Chapter 4

4.44. 640 W

4.45. $W_{A \to B} = (x^2 y^2 + z^3 x + c) \Big|_A^B$.

4.46. $y = 0.1722$ m

$$\{V_D\}_{1/2} = \begin{bmatrix} 0.998 \\ 0 \\ 0 \end{bmatrix} \text{ m/s} \quad \{F_{21}\}_2 = \begin{bmatrix} 0 \\ -2.686 \\ 0 \end{bmatrix} \text{ N}$$

$$\{F_{23}\}_2 = \begin{bmatrix} 0 \\ 0 \\ 0 \end{bmatrix} \text{ N} \quad \{A_D\}_{1/2} = \begin{bmatrix} -9.21 \\ -1.994 \\ 0 \end{bmatrix} \text{ m/s}^2$$

$y = 0.25$ m

$$\{V_D\}_{1/2} = \begin{bmatrix} 1.495 \\ 0 \\ 0 \end{bmatrix} \text{ m/s} \quad \{F_{21}\}_2 = \begin{bmatrix} 0 \\ 1.59 \\ 0 \end{bmatrix} \text{ N}$$

$$\{F_{23}\}_2 = \begin{bmatrix} 1.382 \\ -6.276 \\ 0 \end{bmatrix} \text{ N} \quad \{A_D\}_{1/2} = \begin{bmatrix} -5.72 \\ -4.46 \\ 0 \end{bmatrix} \text{ m/s}$$

$y = 0.121\ 4$ m

4.47. $\{R_p\} = \{R\}$

$$\{R\}_{1/1} = r \begin{bmatrix} \sin\theta\cos\phi \\ \sin\theta\sin\phi \\ \cos\theta \end{bmatrix} \quad \{R_\theta\}_{1/1} = r \begin{bmatrix} \cos\theta\cos\phi \\ \cos\theta\sin\phi \\ -\sin\theta \end{bmatrix}$$

$$\{R_\theta\}_{1/3} = r\begin{bmatrix} 1 \\ 0 \\ 0 \end{bmatrix} \quad \{R_\phi\}_{1/1} = r\begin{bmatrix} -\sin\theta\sin\phi \\ \sin\theta\cos\phi \\ 0 \end{bmatrix} \quad \{R_\phi\}_{1/3} = \begin{bmatrix} 0 \\ r\sin\theta \\ 0 \end{bmatrix}$$

$$\{R_r\}_{1/1} = \begin{bmatrix} \sin\theta\cos\phi \\ \sin\theta\sin\phi \\ \cos\theta \end{bmatrix} \quad \{R_r\}_{1/3} = \begin{bmatrix} 0 \\ 0 \\ 1 \end{bmatrix}$$

4.48.
$$\{M_A\} = [R_{BA}]\{F\} + [R_{CA}]\{-F\}$$
$$= [R_{BA}]\{F\} + [[R_{BA}] + [R_{CB}]]\{-F\}$$
$$= [R_{CB}]\{-F\} = [R_{BC}]\{F\}$$

4.49.
$$\{M_A\} = [R_{PA}]\{F\} \quad \{R_{BA}\}^T\{M_A\} = |R_{BA}||M_A|\cos\theta$$

Component parallel to AB

$$|M_A^P| = \{R_{BA}\}^T\{M_A\}/|R_{BA}|$$

$$\{M_A^P\} = (\{R_{BA}\}^T\{M_A\})\{R_{BA}\}/|R_{BA}|^2$$

4.50.
$$\{F_2\} = \begin{bmatrix} 9 \\ 10 \\ 9 \end{bmatrix} \text{kN} \quad \{M_O\} = \begin{bmatrix} 6 \\ 10 \\ 6 \end{bmatrix} \text{kN m} \quad \{M_B\} = \begin{bmatrix} 7 \\ 10 \\ 5 \end{bmatrix} \text{kN m}$$

4.51.
$$\{F_{21}\} = \begin{bmatrix} 1.218 \\ -3.218 \\ -3.86 \end{bmatrix} \text{kN} \quad \{F_{31}\} = \begin{bmatrix} -0.956 \\ 1.195 \\ 0.239 \end{bmatrix} \text{kN}$$

$$\{F_{41}\} = \begin{bmatrix} 0.669 \\ 0.023 \\ -1.38 \end{bmatrix} \text{kN}$$

4.52.
$$\{L_2\}_1 = \begin{bmatrix} 0 \\ -20 \\ 0 \end{bmatrix} \text{N m} \quad \{F_{26}\} = \begin{bmatrix} -50 \\ -200 \\ 1\,090 \end{bmatrix} \text{N} \quad \{F_{25}\} = \begin{bmatrix} 150 \\ 0 \\ -1\,270 \end{bmatrix} \text{N}$$

$$\{F_{47}\} = \begin{bmatrix} 50 \\ 0 \\ 0 \end{bmatrix} \text{N} \quad \{F_{48}\} = \begin{bmatrix} 50 \\ 0 \\ 280 \end{bmatrix} \text{N}$$

4.53.
$$\sum\{F\} = \begin{bmatrix} 10 \\ -7 \\ 0 \end{bmatrix} \text{N} \quad \{M_o\} = \begin{bmatrix} 47 \\ 2 \\ -42 \end{bmatrix} \text{N m}$$

(ctd.)

$$\{R_{GO}\} = \begin{bmatrix} 1.973 \\ 2.819 \\ 2.342 \end{bmatrix} \text{ m} \qquad \{L^P\} = \begin{bmatrix} 30.60 \\ -21.42 \\ 0 \end{bmatrix} \text{ N m}.$$

4.54.
$$[I_2]_{2/2} = \begin{bmatrix} A & 0 & 0 \\ 0 & B & 0 \\ 0 & 0 & C \end{bmatrix} \qquad \begin{aligned} A &= m(a^2 + b^2)/12 \\ B &= m(b^2 + c^2)/12 \\ C &= m(a^2 + c^2)/12 \end{aligned}$$

$$[I_2]_{2/3} = \begin{bmatrix} Ac^2\beta + Cs^2\beta & 0 & (C - A)s\beta c\beta \\ 0 & B & 0 \\ (C - A)s\beta c\beta & 0 & As^2\beta + Cc^2\beta \end{bmatrix}$$

If $b = c$, then $A = C$ and $[I_2]_{2/3}$ is invariant.

4.55.
$$[I_2]_{n/n} = \begin{bmatrix} I & 0 & 0 \\ 0 & I & 0 \\ 0 & 0 & J \end{bmatrix} \qquad J = 3mr^2/10, \text{ a constant}.$$

$n = 2 \qquad I = 3m(r^2 + h^2/4)/20$

$n = 3 \qquad I = 3m(r^2 + 4h^2)/20$

$n = 4 \qquad I = 3m(r^2 + 2h^2/3)/20$

4.56. Let the engine be body 2 and the airscrew body 3.

$$[I_3]_{3/3} = J\begin{bmatrix} 1 & 0 & 0 \\ 0 & 0 & 0 \\ 0 & 0 & 1 \end{bmatrix} \quad \{\omega_3\}_{1/2} = \begin{bmatrix} \omega \\ 0 \\ \Omega \end{bmatrix} \quad \{\dot{\omega}_3\}_{1/2} = \begin{bmatrix} 0 \\ \omega\Omega \\ 0 \end{bmatrix}$$

$$[I_3]_{3\ 2} = J\begin{bmatrix} 1 & 0 & 0 \\ 0 & s^2\alpha & -s\alpha c\alpha \\ 0 & -s\alpha c\alpha & c^2\alpha \end{bmatrix}$$

$$\{L_{32}\}_2 = J\begin{bmatrix} (\Omega^2 \sin 2\omega t)/2 \\ \omega\Omega(1 - \cos 2\omega t) \\ -\omega\Omega \sin 2\omega t \end{bmatrix}$$

4.57.
$$\{L_{32}\}_2 = \begin{bmatrix} c\omega \\ \omega^2 E \\ 0 \end{bmatrix}$$

4.58.
$$\{L_{32}\}_2 = \omega^2 (J - I) \sin\alpha \cos\alpha \begin{bmatrix} -1 \\ 0 \\ 0 \end{bmatrix}$$

(ctd.)

$$\{L_{32}\}_1 = \omega^2(J - I)\sin\alpha\cos\alpha \begin{bmatrix} -\cos\gamma \\ -\sin\gamma \\ 0 \end{bmatrix} \quad |L_{32}| = 525 \text{ N m}$$

4.59.
$$\{\omega_4\}_{1/3} = \begin{bmatrix} \dot\alpha \\ \dot\beta\cos\alpha \\ \Omega - \dot\beta\sin\alpha \end{bmatrix} \simeq \begin{bmatrix} \dot\alpha \\ \dot\beta \\ \Omega \end{bmatrix} \quad \dot\beta\sin\alpha \ll \Omega \text{ and } \alpha \text{ small}$$

$$\{\dot\omega_4\}_{1/3} \simeq \begin{bmatrix} \ddot\alpha + \Omega\dot\beta \\ \Omega\dot\alpha \\ -(\alpha\ddot\beta + \dot\alpha\dot\beta) \end{bmatrix} \quad \dot\beta \ll \dot\alpha\Omega \text{ and } \alpha \text{ small}$$

$$\{\dot H_4\}_{1/3} = \begin{bmatrix} J\Omega\dot\beta + I\ddot\alpha \\ I\Omega\dot\alpha \\ -J(\alpha\ddot\beta + \dot\alpha\dot\beta) \end{bmatrix} \quad \text{where } [I_4]_{4/4} = \begin{bmatrix} I & 0 & 0 \\ 0 & I & 0 \\ 0 & 0 & J \end{bmatrix}$$

$$\{L_{32}\}_3 = \begin{bmatrix} -k\alpha - c\dot\alpha \\ L_y \\ L_z \end{bmatrix}$$

$$\ddot\alpha + \frac{c}{I}\dot\alpha + \frac{k}{I}\alpha = \frac{J\Omega}{I}\dot\beta$$

For a sustained constant rate turn $\alpha = (J\Omega\dot\beta)/k$

4.60.
$$[I_4]_{4/4} = \begin{bmatrix} J & 0 & 0 \\ 0 & I & 0 \\ 0 & 0 & J \end{bmatrix} \quad \{H_4\}_{1/3} = J\dot\gamma\omega \begin{bmatrix} 0 \\ 1 \\ 0 \end{bmatrix}$$

$F_1 = mr\dot\gamma^2\sin\alpha \quad F_2 = -(mbr\dot\gamma^2\cos\alpha)/(a + b) \quad F_3 = J\dot\gamma\omega/(a + b)$

$F_4 = -(mar\dot\gamma^2\cos\alpha)/(a + b) \quad F_5 = -J\dot\gamma\omega/(a + b)$

$F_6 = -(J\dot\gamma\omega\cos\alpha)/d \quad F_7 = (mcr\dot\gamma^2 - J\dot\gamma\omega\sin\alpha)/d$

$F_8 = 0 \quad F_9 = (J\dot\gamma\omega\cos\alpha)/d$

$F_{10} = -mr\dot\gamma^2(1 + c/d) + (J\dot\gamma\omega\sin\alpha)/d$

4.61. $\dot h_x = m\Omega^2\{12e + (b^2 - 3a^2)\sin\alpha\}/(12\tan\alpha)$

$\{F^y_{24}\}_2 = (m\Omega^2 ed - \dot h_x)/(c + d)$

$\{F^y_{25}\}_2 = (m\Omega^2 ec + \dot h_x)/(c + d)$

4.62. $3g a \tan\gamma = \Omega^2\{3ae + 4(a^2 - b^2)\sin\gamma\}$

(a) $\gamma = 37.4°$ (b) $\gamma = 43.2°$ (by trial)

4.63.
$$\{\omega_3\}_{1/3} = \begin{bmatrix} \alpha \\ \Omega\sin\alpha \\ \Omega\cos\alpha \end{bmatrix} = \begin{bmatrix} \omega_x \\ \omega_y \\ \omega_z \end{bmatrix} \quad \{V_3\}_{1/3} = \begin{bmatrix} b\omega_y \\ -b\omega_x \\ 0 \end{bmatrix}$$

$$\{H_{3_0}\}_{1/3} = \frac{4}{3}mb^2 \begin{bmatrix} \omega_x \\ \omega_y \\ 0 \end{bmatrix} = \begin{bmatrix} h_x \\ h_y \\ 0 \end{bmatrix}$$

$$\{\dot{H}_{3_0}\}_{1/3} = \lfloor\omega_3\rfloor_{1/3}\{H_{3_0}\}_{1/3} + \frac{d}{dt}\{H_{3_0}\}_{1/3}$$

$$= \begin{bmatrix} \dot{h}_x - h_y\omega_z \\ \dot{h}_y + h_x\omega_z \\ \dot{h}_y\omega_x - h_x\omega_y \end{bmatrix}$$

$$\{L_{32}\}_3 + [R_{CO}]_{3/3}\{W_3\}_3 = \{\dot{H}_{3_0}\}_{1/3}$$

$$\ddot{\alpha} + \{(3g/4b) - \Omega^2\cos\alpha\}\sin\alpha = 0$$

Small vibrations about $\alpha = 0$

$$\ddot{\alpha} + \{(3g/4b) - \Omega^2\}\alpha = 0 \qquad \omega_n^2 = (3g/4b) - \Omega^2$$

Stable if $\Omega^2 < 3g/4b$

Small vibrations about α_o. Let Ω_o be the Ω which gives the α_o equilibrium position

$$\cos\alpha_o = 3g/4b\Omega_o^2 \qquad \ddot{\alpha} + \Omega_o^2\sin^2\alpha_o\,\alpha = 0$$

$$\omega_n^2 = \Omega_o^2(1 - 9g^2/16b^2\Omega_o^4)$$

Stable if $\Omega_o^2 > 3g/4b$.

$\Omega = 2$ rad/s. $\alpha_o = 0$. $\omega_n = 2$ rad/s(0.32 Hz)

$\cos\alpha_o = 2(>1)$. No stable vibration in the inclined position.

$\Omega = 4$ rad/s. $\alpha_o = 0$. $\Omega^2 \not< 8$. No stable vibration in the vertical position.

$\cos\alpha = 0.5$. $\alpha_o = 60°$. $\omega_n = \Omega_o\sin\alpha_o = 3.464$ rad/s(0.55 Hz)

$$\{L_{32}^y\}_3 = 8mb^2\Omega\dot{\alpha}\cos\alpha/3.$$

4.64.
$$[I_3]_{3/3} = \begin{bmatrix} I & 0 & 0 \\ 0 & J & 0 \\ 0 & 0 & I \end{bmatrix}$$

(ctd.)

$$I\ddot{\alpha} + c\dot{\alpha} + (J - I)\Omega^2 \sin\alpha\cos\alpha + k\alpha = 0$$

$$I\ddot{\alpha} + c\dot{\alpha} + 0.5(J - I)\Omega^2 \sin 2(\alpha_o + \alpha) + k\alpha = 0$$

$\alpha_o = 0 \quad I\ddot{\alpha} + c\dot{\alpha} + \{(J - I)\Omega^2 + k\}\alpha = 0$

$\alpha_o = \pi/6 \quad I\ddot{\alpha} + c\dot{\alpha} + \{0.5(J - I)\Omega^2 + k\}\alpha = \sqrt{3}(I - J)\Omega^2/4$

$\alpha_o = \pi/4 \quad I\ddot{\alpha} + c\dot{\alpha} + k\alpha = (I - J)\Omega^2$

$\alpha_o = \pi/2 \quad I\ddot{\alpha} + c\dot{\alpha} + \{(I - J)\Omega^2 + k\}\alpha = 0$

4.65.
$$F_1 = 0 \quad F_2 = \{Jab\omega(\omega - \Omega)\}/d(b - c) + mg$$
$$F_4 = -mb\omega^2 \quad F_5 = -F_2$$

4.66.
$$F_1 = 0 \quad F_2 = m\omega^2 \sin\alpha(c + d\cos\alpha) + \omega^2(J\cos^2\alpha + I\sin^2\alpha)/a + mg\cos\alpha$$

$$F_3 = 0 \quad F_4 = -m\omega^2 \cos\alpha(c + d\cos\alpha)$$

$$F_5 = -mg\cos\alpha - \omega^2(J\cos^2\alpha + I\sin^2\alpha)/a$$

4.67.
$$F_1 = F_4 = 0 \quad F_2 = -m\omega^2 b\sin^2\alpha$$

$$(a - c)F_5 = mg(a - b\cos\alpha) - \omega^2 \tan\alpha\{(I + mb^2)\sin^2\alpha + J\cos^2\alpha\}$$

$$(a - c)F_3 = mg(b\cos\alpha - c) + \omega^2 \tan\alpha\{(I + mb^2)\sin^2\alpha + J\cos^2\alpha\}$$

4.68.
$$F_1 = 0 \quad F_2 = -\{J\omega(a + b)(\Omega + \omega)\}/(rb) + mg$$

4.69.
$$\dot{\beta} = \frac{-h \pm \sqrt{\{h^2 + 4mga\cos\theta(J - I - ma^2)\}}}{2\cos\theta(J - I - ma^2)}$$

$$F_1 = m\omega^2 \sin\theta\cos\beta \quad F_2 = mg \quad F_3 = -m\omega^2 \sin\theta\sin\beta$$

4.70.
$$\{\omega_4\}_{1/3} = \begin{bmatrix} -\dot{\alpha}\sin\beta + \omega \\ \dot{\beta} \\ \dot{\alpha}\cos\beta \end{bmatrix} = \begin{bmatrix} \omega_x \\ \omega_y \\ \omega_z \end{bmatrix}$$

$$\{\dot{\omega}_4\}_{1/3} = \begin{bmatrix} -\dot{\alpha}\dot{\beta}\cos\beta \\ \dot{\alpha}\omega\cos\beta + \ddot{\beta} \\ \omega\dot{\beta} - \dot{\alpha}\dot{\beta}\sin\beta \end{bmatrix} = \begin{bmatrix} \dot{\omega}_x \\ \dot{\omega}_y \\ \dot{\omega}_z \end{bmatrix}$$

$$\{V_4\}_{1/3} = \begin{bmatrix} 0 \\ \dot{\alpha}(a + b\cos\beta) \\ -b\dot{\beta} \end{bmatrix}$$

(ctd.)

Answers

$$\{A_4\}_{1/3} = \begin{bmatrix} -\dot{\alpha}^2(a + b\cos\beta) - b\dot{\beta}^2 \\ -2\dot{\alpha}\dot{\beta}b\sin\beta \\ -\dot{\alpha}^2\sin\beta(a + b\cos\beta) - b\ddot{\beta} \end{bmatrix} = \begin{bmatrix} a_x \\ a_y \\ a_z \end{bmatrix}$$

$$mgb\cos\beta = (J - I)\omega_x\omega_y + I\dot{\omega}_y - mba_z$$

4.71.
$$\{\omega_4\}_{1/3} = \begin{bmatrix} \omega - \Omega\sin\beta \\ \dot{\beta} \\ \omega\cos\beta \end{bmatrix} = \begin{bmatrix} \omega_x \\ \omega_y \\ \omega_z \end{bmatrix} \quad \{V_4\}_{1/3} = \begin{bmatrix} 0 \\ \Omega\cos\beta \\ -a\dot{\beta} \end{bmatrix}$$

$$\{\dot{\omega}_4\}_{1/3} = \begin{bmatrix} \omega\dot{\beta}\cos\beta \\ \ddot{\beta} + \omega\Omega\cos\beta \\ \omega\dot{\beta} \end{bmatrix} = \begin{bmatrix} \dot{\omega}_x \\ \dot{\omega}_y \\ \dot{\omega}_z \end{bmatrix}$$

$$\{A_4\}_{1/3} = \begin{bmatrix} -(\omega^2 a\cos^2\beta + a\dot{\beta}^2) \\ -2\omega a\dot{\beta}\sin\beta \\ -(\omega^2 a\sin\beta\cos\beta + a\ddot{\beta}) \end{bmatrix} = \begin{bmatrix} a_x \\ a_y \\ a_z \end{bmatrix}$$

$$mga - k\beta = (J - I)\omega_x\omega_z + I\dot{\omega}_y - maa_z$$

4.72.
$$|I_4|_{4/4} = \begin{bmatrix} I & 0 & 0 \\ 0 & I & 0 \\ 0 & 0 & J \end{bmatrix} \quad \omega = \dot{\gamma} \quad h = J\omega$$

$$\{\omega_4\}_{1/3} = \begin{bmatrix} \dot{\alpha}\cos\beta \\ \dot{\beta} \\ \omega + \dot{\alpha}\sin\beta \end{bmatrix} \quad \{V_4\}_{1/3} = \begin{bmatrix} -a\dot{\beta} \\ a\dot{\alpha}\cos\beta \\ 0 \end{bmatrix}$$

$$\{\dot{\omega}_4\}_{1/3} = \begin{bmatrix} \omega\dot{\beta} + \ddot{\alpha}\cos\beta - \dot{\alpha}\dot{\beta}\sin\beta \\ \ddot{\beta} - \omega\dot{\alpha}\cos\beta \\ \ddot{\alpha}\sin\beta + \dot{\alpha}\dot{\beta}\cos\beta \end{bmatrix}$$

$$\{A_4\}_{1/3} = \begin{bmatrix} -a\dot{\alpha}^2\sin\beta\cos\beta - a\ddot{\beta} \\ -2a\,\dot{\alpha}\dot{\beta}\sin\beta + a\ddot{\alpha}\cos\beta \\ a\dot{\beta}^2 + a\dot{\alpha}^2\cos^2\beta \end{bmatrix}$$

$$mga\begin{bmatrix} -\alpha \\ -\beta \\ 0 \end{bmatrix} = \begin{bmatrix} h\dot{\beta} + (I + ma^2)\ddot{\alpha} \\ -h\dot{\alpha} + (I + ma^2)\ddot{\beta} \\ J\ddot{\alpha}\beta \end{bmatrix}$$

4.73. $\{W\} = \{W_1\} + \{W_2\} + \{W_3\} \quad m = m_1 + m_2 + m_3 \quad h = J\omega$

(ctd.)

$$\{\omega_3\}_{1/3} = \begin{bmatrix} \dot\alpha\cos\gamma \\ \dot\alpha\sin\gamma \\ \dot\gamma \end{bmatrix} = \begin{bmatrix} \omega_x \\ \omega_y \\ \omega_z \end{bmatrix} \quad \{\omega_4\}_{1/3} = \begin{bmatrix} \omega_x \\ \omega_y \\ \omega_z \end{bmatrix} + \omega$$

$$\{\dot\omega_4\}_{1/3} = \begin{bmatrix} \dot\omega_x - \omega\omega_z \\ \dot\omega_y \\ \dot\omega_z + \omega\omega_x \end{bmatrix} \quad \{L_{43}\}_3 = \{\dot H_4\}_{1/3}$$

$$\{L_{34}\}_3 + \{L_{32}\}_3 + \{L_{35}\}_3 = \{\dot H_3\}_{1/3}$$

$$\{L_{32}\}_3 + \{L_{35}\}_3 = \{\dot H_3\}_{1/3} + \{\dot H_4\}_{1/3}$$

$$\{L_{21}\}_2 + [R_{CO}]_{2/2}\{W\}_2 = m[R_{CO}]_{2/2}\{A_{CO}\}_{1/2} + \{\dot H_2\}_{1/2}$$
$$+ [\ell_3]_2 \left\{\{\dot H_3\}_{1/3} + \{\dot H_4\}_{1/3}\right\}$$

$$-k\gamma = (I + C_3)\ddot\gamma + h\dot\alpha$$

$$-mga = (I + A_2 + A_3 + ma^2)\ddot\alpha - h\dot\gamma$$

4.74.

$$\{\omega_3\}_{1/3} = \begin{bmatrix} \Omega\sin\gamma \\ \Omega\cos\gamma \\ \dot\gamma \end{bmatrix} = \begin{bmatrix} \omega_x \\ \omega_y \\ \omega_z \end{bmatrix} \quad \{\omega_4\}_{1/3} = \begin{bmatrix} \omega_x + \omega \\ \omega_y \\ \omega_z \end{bmatrix}$$

$$\{V_4\}_{1/3} = \begin{bmatrix} b\dot\gamma \\ 0 \\ -\Omega(a + b\sin\gamma) \end{bmatrix} = \begin{bmatrix} v_x \\ 0 \\ v_z \end{bmatrix}$$

$$\{H_{4A}\}_{1/3} = \{H_4\}_{1/3} + m|R_{CA}|_{3/3}\{V_4\}_{1/3}$$
$$= \begin{bmatrix} J(\omega + \omega_x) - mbv_z \\ I\omega_y \\ I\omega_z + mbv_x \end{bmatrix} = \begin{bmatrix} h_x \\ h_y \\ h_z \end{bmatrix}$$

$$\{\dot H_{4A}\}_{1/3} = [\omega_3]_{1/3}\{H_{4A}\}_{1/3} + \frac{d}{dt}\{H_{4A}\}_{1/3}$$
$$= \begin{bmatrix} h_z\omega_y - h_y\omega_z + \dot h_x \\ h_x\omega_z - h_z\omega_x + \dot h_y \\ h_y\omega_x - h_x\omega_y + \dot h_z \end{bmatrix}$$

$$\{L_{23}\}_3 + [R_{CA}]_{3/3}\{W_4\}_3 = \{\dot H_{4A}\}_{1/3}$$

$$-mgb\sin\gamma = I\omega_x\omega_y - \{J(\omega + \omega_x) - mbv_z\}\omega_y + I\dot\omega_z + mb\dot v_x$$

(ctd.)

Small γ $h = J\omega$

$$(I + mb^2)\ddot{\gamma} + \{(I - mb^2 - J)\Omega^2 + mgb\}\gamma = (h + mab\Omega)\Omega$$

γ_0 is given by

$$(I - mb^2 - J)\Omega^2 \sin\gamma_0 = (h + mab\Omega)\Omega - mgb\tan\gamma_0$$

$$(I + mb^2)\ddot{\gamma} + (I - mb^2 - J)\Omega^2 \sin\gamma\cos\gamma = (h + mab\Omega)\Omega\cos\gamma$$
$$- mgb\sin\gamma$$

$$A\ddot{\gamma} + B\sin\gamma\cos\gamma = C\cos\gamma - D\sin\gamma$$

$$A\ddot{\gamma} + \{B(\cos^2\gamma_0 - \sin^2\gamma_0) + C\sin\gamma_0 + D\cos\gamma_0\}\gamma$$
$$= C\cos\gamma_0 - D\sin\gamma_0 - B\sin\gamma_0\cos\gamma_0$$

4.75.

$\{\omega_4\}_{1/3} = \{\omega_2\}_{1/3} + \{\omega_4\}_{3/3}$, $\{\omega_3\}_{2/2}$ is null

$$= \dot{\beta}\begin{bmatrix}\sin\gamma\\\cos\gamma\\0\end{bmatrix} + \begin{bmatrix}\omega\\0\\0\end{bmatrix} = \begin{bmatrix}\omega_x + \omega\\\omega_y\\0\end{bmatrix} \qquad \{V_4\}_{1/3} = \begin{bmatrix}0\\0\\-c\omega_y\end{bmatrix}$$

$$\{A_4\}_{1/3} = \begin{bmatrix}c\omega_y^2\\-c\omega_x\omega_y\\c\dot{\omega}_y\end{bmatrix} \qquad \{H_{4_0}\}_{1/3} = \begin{bmatrix}J(\omega_x + \omega)\\(I + mc^2)\omega_y\\0\end{bmatrix} = \begin{bmatrix}J\omega_x + h\\I_0\omega_y\\0\end{bmatrix}$$

$$\{\dot{H}_{4_0}\}_{1/3} = \begin{bmatrix}\dot{h}_x\\\dot{h}_y\\h_y\omega_x - h_x\omega_z\end{bmatrix} \qquad \begin{array}{l}h_x = J\omega_x + h\\h_y = I_0\omega_y\end{array}$$

Moments about O

$$\{L_7\}_2 + \{L_2\}_2 + [R_{AO}]_{2/2}\{F_{35}\}_2 + [R_{BA}]_{2/2}\{F_{36}\}_2$$
$$+ [\ell_3]_2[R_{CO}]_{3/3}[\ell_1]_3\{W_4\}_1 = [\ell_3]_2\{\dot{H}_{4_0}\}_{1/3}$$

Equating applied to effective forces

$$\{F_{35}\}_2 + \{F_{36}\}_2 + [\ell_1]_2\{W_4\}_1 = m[\ell_3]_2\{A_4\}_{1/3}$$

$$(a + b)F_1 = (I_0 - J)\dot{\beta}^2\sin\gamma\cos\gamma - h\dot{\beta}\cos\gamma + mbc\dot{\beta}^2\cos\gamma$$
$$- mg\cos\beta(c\cos\gamma + b)$$

$$(a + b)F_4 = -(I_0 - J)\dot{\beta}^2\sin\gamma\cos\gamma + h\dot{\beta}\cos\gamma + mac\dot{\beta}^2\cos\gamma$$
$$- mg\cos\beta(c\cos\gamma - a)$$

$$(a + b)F_3 = -(J - I_o)\ddot{\beta}\sin\gamma\cos\gamma + mbc\ddot{\beta}\cos\gamma - mgb\sin\beta$$

$$(a + b)F_5 = (J - I_o)\ddot{\beta}\sin\gamma\cos\gamma + mac\ddot{\beta}\cos\gamma - mga\sin\beta$$

$$F_2 = 0$$

$$L = \ddot{\beta}(J\sin^2\gamma + I_o\cos^2\gamma) + mgc\cos\gamma\sin\beta + k\beta$$

4.76.

$$\{\omega_3\}_{1/2} = \{\omega_2\}_{1/2} + \{\omega_3\}_{2/2}$$

$$= \begin{bmatrix} 0 \\ 0 \\ \dot{\gamma} \end{bmatrix} + \begin{bmatrix} \omega \\ 0 \\ 0 \end{bmatrix} = \begin{bmatrix} \omega \\ 0 \\ \dot{\gamma} \end{bmatrix}$$

$$\{V_3\}_{1/2} = \lfloor \omega_2 \rfloor_{1/2} \{R_{co}\}_{2/2} = \begin{bmatrix} -c\dot{\gamma} \\ 0 \\ 0 \end{bmatrix}$$

$$\{A_3\}_{1/2} = \lfloor \omega_2 \rfloor_{1/2} \{V_3\}_{1/2} + \frac{d}{dt}\{V_3\}_{1/2} = \begin{bmatrix} -c\ddot{\gamma} \\ -c\dot{\gamma}^2 \\ 0 \end{bmatrix}$$

$$2T_3 = \{\omega_3\}_{1/2}^T \lfloor I_3 \rfloor_{3/2} \{\omega_3\}_{1/2} + \{V_3\}_{1/2}^T \{V_3\}_{1/2}$$

$$= J\omega^2 + (I + mc^2)\dot{\gamma}^2$$

$$V_6 = k\gamma^2/2$$

For a conservative system $T_3 + V_6 = $ constant

$$\frac{d}{dt}(T_3 + V_6) = 0 \qquad (I + mc^2)\ddot{\gamma} + k\gamma = 0$$

$$\omega_n^2 = k/(I + mc^2) = k/I_o$$

$$\{H_{3o}\}_{1/2} = \{H_3\}_{1/2} + m \lfloor R_{co} \rfloor_{2/2} \{V_3\}_{1/2}$$

$$= \begin{bmatrix} J\omega \\ 0 \\ I_o\dot{\gamma} \end{bmatrix} = \begin{bmatrix} \partial T_3/\partial \omega \\ 0 \\ \partial T_3/\partial \dot{\gamma} \end{bmatrix}$$

$$\{\dot{H}_{3o}\}_{1/2} = \lfloor \omega_2 \rfloor_{1/2} \{H_{3o}\}_{1/2} + \frac{d}{dt}\{H_{3o}\}_{1/2}$$

$$= \begin{bmatrix} 0 \\ J\omega\dot{\gamma} \\ I_o\ddot{\gamma} \end{bmatrix}$$

(ctd.)

Moments about O

$$\{L_2\}_2 + \{L_{26}\}_2 + [R_{AO}]_{2/2}\{F_{24}\}_2 + [R_{BO}]_{2/2}\{F_{25}\}_2$$
$$+ [R_{CO}]_{2/2}\{W_3\}_2 = \{\dot{H}_{3O}\}_{1/2}$$

Equating applied to effective forces

$$\{W_3\}_2 + \{F_{24}\}_2 + \{F_{25}\}_2 = m\{A_4\}_{1/2}$$

$$bF_5 - aF_2 = mgc$$
$$aF_1 - bF_4 = J\omega\dot{\gamma}$$
$$I_o\ddot{\gamma} + k\gamma = L \qquad L = (k - I_o p^2)A\sin pt$$
$$F_1 + F_4 = -mc\ddot{\gamma}$$
$$F_2 + F_5 = -mc\dot{\gamma}^2$$
$$F_3 - mg = 0$$

$$(a + b)F_1 = J\omega Ap\cos pt + mbcAp^2\sin pt$$
$$(a + b)F_4 = -J\omega Ap\cos pt + macAp^2\sin pt$$
$$(a + b)F_2 = -mgc + mabAp^2\cos^2 pt$$
$$(a + b)F_5 + mgc - macAp^2\cos^2 pt$$

4.77.

$$\{\omega_4\}_{1/3} = \begin{bmatrix} \dot{\alpha} \\ \dot{\gamma}\sin\alpha + \omega \\ \dot{\gamma}\cos\alpha \end{bmatrix} = \begin{bmatrix} \omega_x \\ \omega_y + \omega \\ \omega_z \end{bmatrix}$$

$$\{H_4\}_{1/3} = \begin{bmatrix} I\omega_x \\ J(\omega_y + \omega) \\ I\omega_z \end{bmatrix} + \begin{bmatrix} h_x \\ h_y \\ h_z \end{bmatrix}$$

$$\{\dot{H}_4\}_{1/3} = \begin{bmatrix} \dot{h}_x + h_z\omega_y - h_y\omega_z \\ \dot{h}_y \\ \dot{h}_z + h_y\omega_x - h_x\omega_y \end{bmatrix}$$

$$\{L_{32}\}_3 = \begin{bmatrix} -c\dot{\alpha} - k\alpha \\ L_2 \\ L_3 \end{bmatrix} = \{\dot{H}_4\}_{1/3}$$

$$\{L_2\}_2 + \{L_{23}\}_2 = \{\dot{H}_2\}_{1/2}$$

(ctd.)

$$\{L_2\}_2 = \begin{bmatrix} 0 \\ 0 \\ -k_o\gamma + \ell \end{bmatrix} = \{\dot{H}_2\}_{1/2} + [\ell_3]_2\{\dot{H}_4\}_{1/3}$$

4.78.

$$\{\omega_5\}_{1/4} = [\ell_2]_4\{\omega_2\}_{1/2} + [\ell_3]_4\{\omega_3\}_{2/3} + \{\omega_4\}_{3/4} + \{\omega_5\}_{4/4}$$

$$= \begin{bmatrix} \dot{\alpha}\cos\beta - \Omega\sin\beta\cos\alpha \\ \dot{\beta} + \Omega\sin\alpha \\ \dot{\alpha}\sin\beta + \dot{\beta} + \Omega\cos\beta\cos\alpha + \omega \end{bmatrix} = \begin{bmatrix} \omega_x \\ \omega_y \\ \omega_z + \omega \end{bmatrix}$$

$$\{\dot{H}_5\}_{1/4} \simeq \begin{bmatrix} h\omega_z + I\dot{\omega}_x \\ -h\omega_x + I\dot{\omega}_y \\ J\dot{\omega}_z \end{bmatrix} = \{L_{54}\}_4 = \begin{bmatrix} L_1 \\ 0 \\ L_2 \end{bmatrix}$$

4.79.

$$\{\omega_4\}_{1/3} = \{\omega_2\}_{1/3} + \{\omega_4\}_{3/3}$$

$$= \omega\begin{bmatrix} 0 \\ \sin\alpha \\ \cos\alpha \end{bmatrix} + \omega\begin{bmatrix} 0 \\ -\{(a/r) + \sin\alpha\} \\ 0 \end{bmatrix} = \omega\begin{bmatrix} 0 \\ -a/r \\ \cos\alpha \end{bmatrix}$$

$$\{V_4\}_{1/3} = \begin{bmatrix} -\omega a \\ 0 \\ 0 \end{bmatrix} = \begin{bmatrix} -v \\ 0 \\ 0 \end{bmatrix} \quad \{A_4\}_{1/3} = v^2/a\begin{bmatrix} 0 \\ -\cos\alpha \\ \sin\alpha \end{bmatrix}$$

$$\{\dot{\omega}_4\}_{1/3} = (v^2\cos\alpha)/a^2\begin{bmatrix} (a/r) + \sin\alpha \\ 0 \\ 0 \end{bmatrix}$$

Moments about A

$$[R_{CA}]_{3/3}\{W_4\}_3 = m[R_{CA}]_{3/3}\{A_4\}_{1/3} + \{\dot{H}_4\}_{1/3}$$

$$gr\sin\alpha = \frac{v^2 r\sin\alpha}{a} + \frac{v^2 r\cos\alpha}{4a} + \frac{v^2 r\cos\alpha}{4a} + \frac{v^2 r^2 \sin\alpha\cos\alpha}{4a^2}$$

External forces equal to mass acceleration

$$\{W_4\}_2 + \{F_{41}\}_2 = m\{A_4\}_{1/2}$$

4.80. Moments about C

$$[R_{BC}]_{1/1}\{F_{23}\}_1 + [R_{AC}]_{1/1}\{F_{21}\}_1 = \{\dot{H}_2\}_{1/1}$$

(ctd.)

Answers

$\{F_{23}\}$ is perpendicular to the plane containing BC and BD. Hence

$$\{F_{23}\}_1 = F[R_{BC}]_{1/1}\{R_{BD}\}_{1/1}/|[R_{BC}]_{1/1}\{R_{BD}\}_{1/1}|$$

$$\{\dot{H}_2\}_{1/1} = [\ell_2]_1[I_2]_{2/2}[\ell_2]_1^T\{\dot{\omega}_2\}_{1/1}$$

since $\{\omega_2\}_1$ is a null vector. Also, since the external forces on body 2 have no moment about the x_2 axis, $\{\dot{\omega}_2\}_1$ must be perpendicular to BC and

$$\{R_{BC}\}_{1/1}^T\{\dot{\omega}_2\}_{1/1} = 0$$

External forces equal mass acceleration

$$\{W_2\}_1 + \{F_{21}\}_1 + \{F_{23}\}_1 = m\{A_2\}_{1/1} = m[R_{BC}]_{1/1}\{\dot{\omega}_2\}_{1/1}$$

$$\begin{bmatrix} [a_1] & [a_2] & [a_3] & [b_1] \\ [a_4] & [a_5] & [a_6] & [b_2] \\ [a_7] & [a_8] & 0 & [b_3] \end{bmatrix} = \begin{bmatrix} 0 \\ 0 \\ 0 \\ c \\ 0 \end{bmatrix}$$

$[a_1] = [\ell_2]_1[I_2]_{2/2}[\ell_2]_1^T \qquad [a_2] = [R_{CA}]_{1/1}$

$[a_3] = [R_{BC}]_{1/1}\{R_{DB}\}_{1/1}/|[R_{BC}]_{1/1}\{R_{DB}\}_{1/1}|$

$[a_4] = m[R_{CA}]_{1/1} \qquad [a_5] = \begin{bmatrix} -1 & 0 & 0 \\ 0 & -1 & 0 \\ 0 & 0 & -1 \end{bmatrix}$

$[a_6] = [R_{BC}]_{1/1}\{R_{DB}\}_{1/1}/|[R_{BC}]_{1/1}\{R_{DB}\}_{1/1}|$

$[a_7] = \{R_{BD}\}_{1/1}^T \qquad [a_8] = [0 \quad 0 \quad 0]$

$[b_1] = \{\dot{\omega}_2\}_{1/1} \qquad [b_2] = \{F_{21}\}_1$

$b_3 = F \qquad c = \{W_2\}_1$

4.81.
$$\{\dot{H}_2\}_{1/2} = \{0\} \quad \{\omega_2\}_{1/2} = \begin{bmatrix} \omega_x \\ \omega_y \\ \omega_z \end{bmatrix} \quad \{\dot{\omega}_2\}_{1/2} = \begin{bmatrix} \dot{\omega}_x \\ \dot{\omega}_y \\ \dot{\omega}_z \end{bmatrix}$$

$$B\dot{\omega}_x - (A - B)\omega_y\omega_z = 0 \qquad 1$$

$$A\dot{\omega}_y = 0 \qquad 2$$

$$B\dot{\omega}_z + (A - B)\omega_x\omega_z = 0 \qquad 3$$

$$\dot{\omega}_y = 0 \quad \text{and} \quad \omega_y = \text{constant} = \Omega$$

Differentiate Eq. 3 with respect to time

$$B\ddot{\omega}_z + \Omega(A - B)\dot{\omega}_x = 0 \qquad 4$$

Substitute Eq. 4 into Eq. 1

$$\ddot{\omega}_z + \frac{\Omega^2(A - B)^2}{B^2}\omega_z = 0$$

The equation in ω_x can be similarly obtained.

4.82.
$$\{\omega_4\}_{1/3} = \{\omega_2\}_{1/3} + \{\omega_3\}_{2/3} + \{\omega_4\}_{3/3}$$

$$= \begin{bmatrix} 0 \\ -\dot{\phi}\sin\beta \\ \dot{\phi}\cos\beta \end{bmatrix} + \begin{bmatrix} -\dot{\beta} \\ 0 \\ 0 \end{bmatrix} + \begin{bmatrix} 0 \\ \dot{\psi} \\ 0 \end{bmatrix} = \begin{bmatrix} -\dot{\beta} \\ \dot{\psi} - \dot{\phi}\sin\beta \\ \dot{\phi}\cos\beta \end{bmatrix}$$

$$\{H_{4O}\}_{1/3} = [I_4]_{4/3}\{\omega_4\}_{1/3} - m[R_{AO}]_{3/3}\{\omega_3\}_{1/3}$$

$$= ma^2/4 \begin{bmatrix} -5\dot{\beta} \\ 2(\dot{\psi} - \dot{\phi}\sin\beta) \\ 5\dot{\phi}\cos\beta \end{bmatrix}$$

$$\{H_{4O}\}_{1/3}\Big|_{t=0} = ma^2/2 \begin{bmatrix} 0 \\ \omega \\ 0 \end{bmatrix}$$

The component of angular momentum about y_3 is constant and therefore

$$\omega = \dot{\psi} - \dot{\phi}\sin\beta$$

or

$$\dot{\psi} = \omega + \dot{\phi}\sin\beta$$

$$\{H_{4O}\}_{1/2} = [\ell_3]_2\{H_{4O}\}_{1/3}$$

The component of angular momentum about z_2 is constant and therefore

Answers

$$0 = -2(\dot{\psi} - \dot{\phi}\sin\beta)\sin\beta + 5\dot{\phi}\cos^2\beta$$

giving

$$\dot{\phi} = 2\omega\sin\beta/5\cos\beta$$

and

$$\dot{\psi} = \omega(1 + 2\tan^2\beta/5)$$

$$2T_4 = \{\omega_4\}_{1/3}^T \{H_4 O\}_{1/3}$$

$$= ma^2/4 \begin{bmatrix} -\dot{\beta} & \omega & \dot{\phi}\cos\beta \end{bmatrix} \begin{bmatrix} -5\dot{\beta} \\ 2\omega \\ 5\dot{\phi}\cos\beta \end{bmatrix}$$

Conservation of energy

$$(2T_4 + 2V_4)\Big|_{\beta=\dot{\beta}=\dot{\phi}=0} = (2T_4 + 2V_4)\Big|_{\dot{\beta}=0}$$

Hence

$$\omega^2 = 10g\cos^2\beta/a\sin\beta$$

For $\beta = 30°$, $\omega = 12/\sqrt{a}$ rad/s.

4.83.
$$[I]\{\dot{\omega}\} + [\omega][I]\{\omega\} = \{0\}$$

$$\{\omega\}^T[I]\{\dot{\omega}\} + \{\omega\}^T[\omega][I]\{\omega\} = 0$$

The second term on the left hand side of this equation is zero and the first term is the rate of change of kinetic energy, which is zero. The kinetic energy is thus a constant.

$$\{\omega\}^T[I][I]\{\dot{\omega}\} + \{\omega\}^T[I][\omega][I]\{\omega\} = 0$$

The second term on the left hand side of this equation is zero and the first term integrates to

$$\{\omega\}^T[I][I]\{\omega\} = \text{constant.}$$

Since
$$\{H\} = [I]\{\omega\} \quad \text{and} \quad \{H\}^T = \{\omega\}^T[I]^T = \{\omega\}^T[I]$$

$$\{H\}^T\{H\} = \text{constant.}$$

4.84. Suffixes 1/1 are omitted throughout.

$$\{R_{AO}\} = \begin{bmatrix} 6.928 \\ 4 \\ 0 \end{bmatrix} \text{cm} \quad \{R_{BA}\} = \begin{bmatrix} -6.928 \\ -27.644 \\ 9.373 \end{bmatrix} \text{cm}$$

(ctd.)

$$\{R_{DB}\} = \begin{bmatrix} 0 \\ -3.485 \\ -9.373 \end{bmatrix} \text{ cm} \qquad \{V_{AO}\} = \begin{bmatrix} -400 \\ 692.82 \\ 0 \end{bmatrix} \text{ cm/s}$$

$$\{V_{BA}\} = \begin{bmatrix} 400 \\ -166.586 \\ -195.65 \end{bmatrix} \text{ cm/s} \qquad \{V_{BD}\} = \begin{bmatrix} 0 \\ 526.2 \\ 195.65 \end{bmatrix} \text{ cm/s}$$

$$\{V_3\} = \{V_A\} + 2/3\{V_{BA}\} = \begin{bmatrix} -133.3 \\ 581.72 \\ -130.43 \end{bmatrix} \text{ cm/s}$$

$$\{\omega_3^n\} = \begin{bmatrix} 7.744 \\ 2.66 \\ 13.568 \end{bmatrix} \text{ rad/s} \qquad \{\omega_4\} = \begin{bmatrix} -56.14 \\ 0 \\ 0 \end{bmatrix} \text{ rad/s}$$

$$\{A_{AO}^P\} = \begin{bmatrix} -692.8 \\ -400 \\ 0 \end{bmatrix} \text{ m/s}^2 \qquad \{A_{AO}^n\} = \begin{bmatrix} 400 \\ -692.8 \\ 0 \end{bmatrix} \text{ m/s}^2$$

$$\{A_{BA}^P\} = \begin{bmatrix} 17.4 \\ 69.42 \\ -23.54 \end{bmatrix} \text{ m/s}^2 \qquad \{A_{BD}^P\} = \begin{bmatrix} 0 \\ -109.83 \\ -295.4 \end{bmatrix} \text{ m/s}^2$$

$$\{A_{BA}^n\} = \begin{bmatrix} 275.4 \\ -40.85 \\ 80 \end{bmatrix} \text{ m/s}^2 \qquad \{A_{BD}^n\} = \begin{bmatrix} 0 \\ -954.4 \\ 354.8 \end{bmatrix} \text{ m/s}^2$$

$$\{\dot{\omega}_3^n\} = \begin{bmatrix} -212.4 \\ 350.7 \\ 877.3 \end{bmatrix} \text{ rad/s}^2 \qquad \{\dot{\omega}_4\} = \begin{bmatrix} 10182 \\ 0 \\ 0 \end{bmatrix} \text{ rad/s}^2$$

$$\{A_3\} = \{A_A\} + 2/3\{A_{BA}^P\} + 2/3\{A_{BA}^n\}$$

$$= \begin{bmatrix} -97.6 \\ -1074 \\ 39.6 \end{bmatrix} \text{ m/s}^2$$

Equations for the motion of body 3

$$[R_{BA}]\{F_{34}\} = m[R_{CA}]\{A_3\} + I\{\dot{\omega}_3^n\}$$

(ctd.)

Answers

$$\{F_{34}\} + \{F_{32}\} = m\{A_3\}$$

$$\{F_{34}\}^T \{V_{BD}\} = 0$$

$$\{F_{34}\} = \begin{bmatrix} 74 \\ -41.6 \\ -111.8 \end{bmatrix} N \quad \{F_{32}\} = \begin{bmatrix} -132.6 \\ -602.7 \\ 135.6 \end{bmatrix} N$$

The rate at which energy is supplied to body 2 can be equated to the rate of change of kinetic energy of body 3.

$$\{\omega_2\}^T \{L_2\} = I\{\omega_3\}^T \{\dot{\omega}_3\} + m\{V_3\}^T \{A_3\}$$

$$L_z = -36.45 \text{ N m}$$

Moments about F for body 2

$$\{L_2\} + [R_{EF}]\{F_{25}\} + [R_{AF}]\{F_{23}\} = \{0\}$$

$$\{F_{25}\} = \begin{bmatrix} -1185 \\ -534 \\ 0 \end{bmatrix} N \quad \{L_2\} = \begin{bmatrix} 0 \\ 0 \\ -36.4 \end{bmatrix} \text{ N m}$$

Equating external forces on body 2 to zero

$$\{F_{25}\} + \{F_{23}\} + \{F_{26}\} = \{0\}$$

$$\{F_{26}\} = \begin{bmatrix} -120.7 \\ 68.7 \\ 135.6 \end{bmatrix} N$$

Moments about H for body 4

$$[R_{BH}]\{F_{43}\} + [R_{EH}]\{F_{47}\} = \{0\}$$

$$\{F_{47}\} = \begin{bmatrix} 0 \\ -417.7 \\ -155.3 \end{bmatrix} N$$

Equating external forces on body 4 to zero

$$\{F_{43}\} + \{F_{47}\} + \{F_{48}\} = \{0\}$$

$$\{F_{48}\} = \begin{bmatrix} 74 \\ 376 \\ 43.5 \end{bmatrix} N$$

4.85. Terms in $[I_2]_{3/3}$

$$A_3 = m(b^2 + 3a^2)/6 = 84m/6 \qquad B_3 = m(b^2 + 2c^2)/6 = 17m/6$$
$$C_3 = m(3a^2 + 2c^2)/6 = 83m/6 \qquad F_3 = -mab/4 = -22.5m/6$$

Terms in $[I_2]_{4/4}$

$$A_2 = m(b^2 + a^2)/18 = 34m/18 \qquad B_2 = m(6c^2 + b^2)/18 = 33m/18$$
$$C_2 = m(6c^2 + a^2)/18 = 49m/18 \qquad F_2 = -mab/36 = -7.5m/18$$

$$[\ell_4]_3 \quad 3 \overset{\frown}{\alpha \text{ about } x_3} 4 \qquad \alpha = 17°$$

Terms in $[I_2]_{3/4}$

$$A_4 = 84m/6 \qquad B_4 = 10m/6 \qquad C_4 = 88.9m/6$$

$$[\ell_2]_5 \quad 2 \overset{\frown}{\alpha \text{ about } x_2} 5 \qquad \alpha = 21.58°$$

Terms in $[I_2]_{2/5}$

$$A_5 = 34m/18 \qquad B_5 = 30m/18 \qquad C_5 = 52m/18$$

4.86.

$$\{R_{BC}\}_{3/3} = \begin{bmatrix} 3.347 \\ 12.273 \\ 4.463 \end{bmatrix} \text{cm} \qquad \{R_{CA}\}_{2/2} = \begin{bmatrix} 11.653 \\ 42.727 \\ 15.537 \end{bmatrix} \text{cm}$$

$$[I_4]_{4/4} = [I_2]_{2/2} - m_2 [R_{AC}]^2_{2/2} + [I_3]_{3/3} - m_3 [R_{BA}]^2_{3/3}$$

$$[I_4]_{4/4} = \begin{bmatrix} 117 & -12.56 & -1.04 \\ -12.56 & 52.6 & -16.75 \\ -1.04 & -16.75 & 100 \end{bmatrix} \text{kg m}^2$$

$$[I_4]_{4/5} = \begin{bmatrix} 119.4 & 0 & 0 \\ 0 & 45.17 & 0 \\ 0 & 0 & 104.79 \end{bmatrix} \text{kg m}^2$$

$$[\ell_4]_5 = \begin{bmatrix} 0.9679 & 0.1688 & 0.1861 \\ -0.2142 & 0.9417 & 0.2594 \\ 0.1314 & 0.2909 & -0.9477 \end{bmatrix}$$

Bibliography

1. Brand, T. and Sherlock, A. (1970) *Matrices: Pure and Applied*, Arnold, London.
2. Gere, J. M. and Weaver, W. Jr. (1967) *Matrix Algebra for Engineers*, Van Nostrand, New York.
3. Barnett, S. (1979) *Matrix Methods for Engineers and Scientists*, McGraw-Hill, New York.
4. Derusso, P.M., Roy, R.J., and Close, C. M. (1965) *State Variables for Engineers*, Wiley, New York.
5. Morrison, J. L. M., and Crossland, B. (1964) *An Introduction to the Mechanics of Machines*, Longmans, London.
6. Green, W. G. (1955) *Theory of Machines*, Blackie, London.
7. Inglis, W. (1951) *Applied Mechanics for Engineers*, Cambridge University Press, Cambridge.
8. Kane, T. R. (1961) *Analytical Elements of Mechanics, Volume 1: Statics, Volume 2: Dynamics*, Academic Press, New York.
9. Smith, C. E. (1976) *Applied Mechanics: More Dynamics*, Wiley New York.
10. Pestel, E. C., and Thomson, W. T. (1968) *Dynamics*, McGraw-Hill, New York.
11. Thomson, W. T. (1961) *Introduction to Space Dynamics*, Wiley, New York.
12. Meriam, J. L. (1975) *Dynamics*, Wiley, New York.
13. Beer, F. P., and Johnston, E. R. (1972) *Dynamics*, McGraw-Hill, New York.
14. Greenwood, D. T. (1965) *Principles of Dynamics*, Prentice-Hall, New Jersey.
15. Wells, D. A. (1967) *Lagrangian Dynamics*, McGraw-Hill, New York.
16. Speigel, M. R. (1967) *Theoretical Mechanics*, Schaum, New York.
17. Meirovitch, L. (1970) *Methods of Analytical Dynamics*, McGraw-Hill, New York.
19. Wittenberg, J. (1977) *Dynamics of Systems of Rigid Bodies*, Teubner, Stuttgart.
20. Wrigley, W. et. al. (1969) *Gyroscopic Theory, Design and Instrumentation*, M. I. T. Press, Cambridge, Massachusetts.
21. Pars, L. A. (1965) *A Treatise on Analytical Dynamics*, Heineman, London.